RHEOLOGY

Edited by **Juan De Vicente**

Rheology
Edited by Juan De Vicente

Published by InTech
Janeza Trdine 9, 51000 Rijeka, Croatia

Publishing Process Manager Daria Nahtigal

Technical Editor Teodora Smiljanic

Cover Designer InTech Design Team

Additional hard copies can be obtained from orders@intechopen.com

Rheology, Edited by Juan De Vicente
p. cm.
ISBN 978-953-51-0187-1

For EU product safety concerns:
IN TECH d.o.o., Prolaz Marije Krucifikse Kozulić 3, 51000 Rijeka, Croatia,
info@intechopen.com or visit our website at intechopen.com.

Meet the editor

Juan de Vicente received Ph.D. degrees in Physics from the University of Granada (Spain) and the University of Nice-Sophia Antipolis (France) with European Mention and Extraordinary Doctorate Awards. He has carried out research stays at the LPMC - University of Nice, RRC - University of Wisconsin-Madison, and Vakgroep Reologie - Universiteit Twente under a FPU Predoctoral Fellowship. He has been Marie Curie Postdoc and Marie Curie ERG Fellow at Unilever Corporate Research and Imperial College London. He is recipient of the "Young Investigator Award" from the Social Council and the "Physics Research Award" from the Academy of Sciences. He is presently Associate Professor of Applied Physics at the University of Granada. His research interests include magnetorheology and soft-elasto-ferrohydrodynamic lubrication.

Contents

Preface

Rheology involves the study of the deformation and flow of matter. The goal is to establish relationships between stress and deformation for (non-Newtonian) materials where neither Newton's law nor Hooke's law suffice to explain their mechanical behaviour. Many materials exhibit a non-Newtonian behaviour and the area is relevant in many fields of study from industrial to technological applications such as concrete technology, geology, polymers and composites, plastics processing, paint flow, hemorheology, cosmetics, adhesives, etc ...

In this Intech E-book, 13 chapters on various rheology related aspects are compiled. A number of current research projects are outlined as the book is intended to give the readers a wide picture of current rheology research balancing between fundamentals and applied knowledge. For this purpose, the chapters are written by experts from the Industry and Academia.

The first part of the book is dedicated to polymers and deals with various advanced researches in this field. The first chapter, by Riedel et al. is devoted to the use of broadband dielectric spectroscopy to test macroscopic models for polymer dynamics. Chapters 2 and 3, by Tanaka and Grillet et al., revisit the fundamentals of gel formation and discuss the methods of characterizing polymer gels using rheological techniques to probe their adhesion and mechanical response. Chapter 4, by Salazar and Rodríguez focuses on the fracture behavior of controlled-rheology polypropylenes. Chapter 5, by Nitta and Yamana attempts to clarify the relationship between tensile mechanical nonlinearity and geometrical nonlinearity resulting from the transient change in Poisson's ratio in crystalline polymers. This first section finishes with Chapter 6, by Tai, and analyzes the rheological properties of poly(lactic acid) and poly(lactic acid-co-glycolic acid)/CO2 mixtures.

The second part of the book covers two important flow fields of outstanding interest in petroleum industry. On the one hand, chapter 7, by Fester et al. deals explicitly with pipe flows. In particular, they review the loss coefficient data for laminar flow on non-Newtonian fluids in pipe fittings. Chapter 8, by Sun et al. focuses on enhanced oil recovery methods and explores the flow behavior of polymer solutions through a porous media.

The third part of the book is devoted to drilling fluids. Chapter 9, by George et al. concerns the rheology of fiber-containing fluids specific to the oil and gas industry. They focus on drilling sweeps that are used to improve borehole cleaning.

The fourth part of the book covers some relevant aspects of food rheology. The gelation mechanism of ⊚-Carrageenan and its interaction with starch and non-starch polysaccharides is reviewed in chapter 10 by Tecante and Nuñez-Santiago. Later, in chapter 11, by Komlenic et al., it is demonstrated that rheological properties of dough strongly depend on acidification.

The last part of the book deals with liquids crystals and ceramic microelectronic devices. Chapter 12, by Mendoza et al. reviews recent theoretical results on the rheology of systems consisting of a flow-aligning nematic contained in cells and capillaries under a variety of different flow conditions and under the action of applied electric fields. Finally, chapter 13, by Alias and Shapee, stresses the impact of silver paste rheology in the fabrication of ceramic microelectronic devices (low temperature co-fired ceramic LTCC devices).

The format of an E-book is chosen to enable fast dissemination of new research, and to give easy access to readers. The chapters can be read individually.

I would like to express my gratitude to all the contributing authors that have made this book become a reality. I wish to thank also Intech staff and their team members for the opportunity to publish this work, in particular, Ana Pantar, Tajana Jevtic and Daria Nahtigal for their support which has made my job as editor an easy and satisfying one.

Finally, I gratefully acknowledge financial support by the Ministerio de Ciencia e Innovación (MICINN MAT 2010-15101 project, Spain), by the European Regional Development Fund (ERDF), and by the projects P10-RNM-6630 and P11-FQM-7074 from Junta de Andalucía (Spain).

Juan de Vicente
University of Granada
Spain

Part 1

Polymers

1

Polymer Rheology by Dielectric Spectroscopy

Clement Riedel[1], Angel Alegria[1], Juan Colmenero[1] and Phillipe Tordjeman[2]
[1]*Universidad del Pais Vasco / Euskal Herriko Unibertsitatea*
[2]*Université de Toulouse, INPT*
[1]*Spain*
[2]*France*

1. Introduction

The aim of this chapter is to discuss the main models describing the polymer dynamics at macroscopic scale and present some of the experimental techniques that permit to test these models. We will notably show how Broadband Dielectric Spectroscopy (BDS) permits to obtain rheological information about polymers, i.e. permits to understand how the matter flows and moves. We will focus our study the whole chain motion of cis-polyisoprene 1,4 (PI).

Due to dipolar components both parallel and perpendicular to the chain backbone, PI exhibits a whole chain dielectric relaxation (normal mode) in addition to that associated with segmental motion. The Rouse model (Rouse, 1953), developed in the fifties, is well known to describe rather correctly the whole chain dynamics of unentangled polymers. In the past, the validity of the Rouse model has ever been instigated by means of different experimental techniques (BDS, rheology, neutron scattering) and also by molecular dynamics simulations. However this study is still challenging because in unentangled polymer the segmental dynamics contributions overlap significantly with the whole chain dynamics. In this chapter, we will demonstrate how we have been able to decorelate the effect of the α-relaxation on the normal mode in order to test how the Rouse model can quantitatively describes the normal mode measured by BDS and rheology. The introduction of polydispersity is a key point of this study.

The reptational tube theory, first introduced by de Gennes de Gennes (1979) and developed by Doi and Edwads Doi & Edwards (1988) describes the dynamics of entangled polymers where the Rouse model can not be applied. Many corrections (as Contour Length Fluctuation or Constraints Release) have been added to the pure reptation in order to reach a totally predictive theory McLeish (2002); Viovy et al. (2002). In the last part of this chapter, we will show how the relaxation time of the large chain dynamics depends on molecular weight and discuss the effects of entanglement predicted by de Gennes theory on dielectric spectra. PI is a canonical polymer to study the large chain dynamics; however, due to its very weak relaxation, the measurement of this dynamics at nanoscale is still challenging.

2. Material and methods

2.1 Polyisoprene samples

Polymers are characterized by a distribution of sizes known as polydispersity. Given a distribution of molecular sizes (N_i molecules of mass M_i), the number-averaged molecular weight M_n is defined as the first moment of distribution (Eq. 1) and the mass-averaged molecular weight M_w as the ratio between the second and first moments of the distribution (Eq. 2).

$$M_n = \sum_i \frac{N_i\, M_i}{N_i} \tag{1}$$

$$M_w = \sum_i \frac{N_i\, M_i^2}{N_i\, M_i} \tag{2}$$

The polydispersity I_p is the ratio Mw/Mn.

Polymers can have dipoles in the monomeric unit that can be decomposed in two different components: parallel or perpendicular to the chain backbone. The dipole moment parallel to the chain backbone giving rise to an "end-to-end" net polarization vector will induce the so-called dielectric normal mode dielectric relaxation that can be studied using theoretical models. The dipole moment perpendicular to the chain backbone will lead to the segmental α-relaxation that can only be described using empirical models, since no definitive theoritical framework exists for this universal process.

We have chosen to study the relaxations of two different samples. The first is 1,4-cis-polyisoprene (PI). This isomer is a A-type polymer in the Stockmayer classification Stockmayer (1967): it carries both local dipole moments parallel and perpendicular to the chain backbone. The chapter on the large scale dynamics will be focused on the study of the dynamics of the normal mode of PI.

The polyisoprene (PI) samples were obtained from anionic polymerization of isoprene (Fig. 1). According to the supplier, Polymer Source, the sample is linear (no ramification) and its micro-composition is 80% cis, 15% trans and 5% other 1. We are working with the isomer *cis* that have a net end-to-end polarization vector. Before the experiments samples were dried in a vacuum oven at 70 žC for 24 hours to remove any trace of solvent.

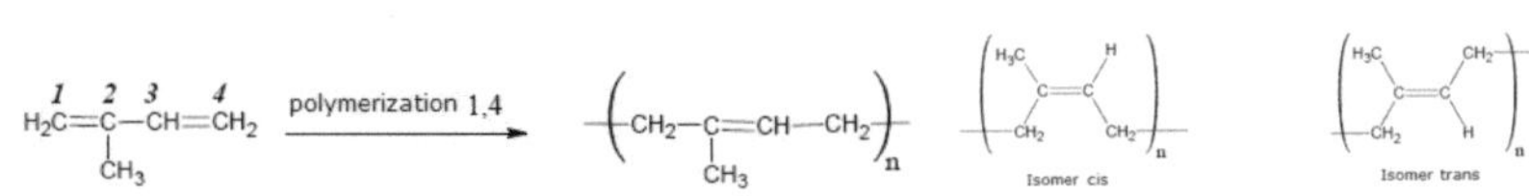

(a) Polymerization of isoprene (b) Isomer cis and trans

Fig. 1. Polymerization and isomer of isoprene

For our study, seven samples have been chosen to cover a large band of molecular weight with low polydispersity (Table 1). The polydispersity (determined from size-exclusion chromatography experiments) is given by supplier while the glass transition temperature (T_g) has been measured by differential scanning calorimetry. To avoid oxidation, PI samples were stored at -25 žC.

Sample	M_n [kg/mol]	M_w [kg/mol]	I_p	T_g [K]
PI-1	1.1	1.2	1.11	194
PI-3	2.7	2.9	1.06	203
PI-10	10.1	10.5	1.04	209
PI-33	33.5	34.5	1.04	210
PI-82	76.5	82	1.07	210
PI-145	138	145	1.07	210
PI-320	281	320	1.14	210

Table 1. Molecular weight, polydispersity and T_g of PI

2.2 Rheology

Elasticity is the ability of a material to store deformational energy, and can be viewed as the capacity of a material to regain its original shape after being deformed. Viscosity is a measure of the ability of a material to resist flow, and reflects dissipation of deformational energy through flow. Material will respond to an applied force by exhibiting either elastic or viscous behavior, or more commonly, a combination of both mechanisms. The combined behavior is termed viscoelasticity. In rheological measurements, the deformational force is expressed as the stress, or force per unit area. The degree of deformation applied to a material is called the strain. Strain may also be expressed as sample displacement (after deformation) relative to pre-deformation sample dimensions.

Dynamic mechanical testing involves the application of an oscillatory strain $\gamma(t) = \gamma_0 \, cos(\omega t)$ to a sample. The resulting sinusoidal stress $\sigma(t) = \sigma_0 \, cos(\omega t + \delta)$ is measured and correlated against the input strain, and the viscous and elastic properties of the sample are simultaneously measured.

If the sample behaves as an ideal elastic solid, then the resulting stress is proportional to the strain amplitude (Hooke's Law), and the stress and strain signals are in phase. The coefficient of proportionality is called the shear modulus G. $\sigma(t) = G \, \gamma_0 \, cos(\omega t)$ If the sample behaves as an ideal fluid, then the stress is proportional to the strain rate, or the first derivative of the strain (Newton's Law). In this case, the stress signal is out of phase with the strain, leading it by 90ř. The coefficient of proportionality is the viscosity η. $\sigma(t) = \eta \, \omega \, \gamma_0 \, cos(\omega t + \pi/2)$

For viscoelastic materials, the phase angle shift (δ) between stress and strain occurs somewhere between the elastic and viscous extremes. The stress signal generated by a viscoelastic material can be separated into two components: an elastic stress (σ') that is in phase with strain, and a viscous stress (σ'') that is in phase with the strain rate ($d\gamma/dt$) but 90ř out of phase with strain. The elastic and viscous stresses are sometimes referred to as the in-phase and out-of-phase stresses, respectively. The elastic stress is a measure of the degree to which the material behaves as an elastic solid. The viscous stress is a measure of the degree to which the material behaves as an ideal fluid. By separating the stress into these components, both strain amplitude and strain rate dependence of a material can be simultaneously measured. We can resume this paragraph by a set of equation:

$$\sigma(t) = \left(\frac{\sigma_0 \, cos(\delta)}{\gamma_0} \right) \gamma_0 \, cos(\omega \, t) + \left(\frac{\sigma_0 \, sin(\delta)}{\gamma_0} \right) \gamma_0 \, sin(\omega \, t) \quad (3)$$

$$G' = \frac{\sigma_0 \cos(\delta)}{\gamma_0} \quad G'' = \frac{\sigma_0 \cos(\delta)}{\gamma_0} \tag{4}$$

$$\underline{G}(\omega) = G' + iG'' \tag{5}$$

2.3 Broadband dielectric spectroscopy

The set-up of a broadband dielectric spectroscopy (BDS) experiment is displayed in Fig. 2. The sample is placed between two electrodes of a capacitor and polarized by a sinusoidal voltage $\underline{U}(\omega)$. The result of this phenomena is the orientation of dipoles which will create a capacitive system. The current $\underline{I}(\omega)$ due to the polarization is then measured between the electrode. The complex capacity $\underline{C}(\omega)$ of this system is describe by the complex dielectric function $\underline{\epsilon}(\omega)$ as:

$$\underline{\epsilon}(\omega) = \epsilon'(\omega) - i\,\epsilon''(\omega) = \frac{\underline{C}(\omega)}{C_0} = \frac{\underline{J}(\omega)}{i\,\omega\,\epsilon_0 \underline{E}(\omega)} = \frac{\underline{I}(\omega)}{i\,\omega\,\underline{U}(\omega)\,C_0} \tag{6}$$

where: C_0 is the vacuum capacitance of the arrangement

$\underline{E}(\omega)$ is the sinusoidal electric field applied (within the linear response)
$\underline{J}(\omega)$ is the complex current density $\epsilon_0 = 8.85\,10^{-12} A\,s\,V^{-1}\,m^{-1}$is the permittivity of vacuum

ϵ' and ϵ'' are proportional to the energy stored and lost in the sample, respectively.

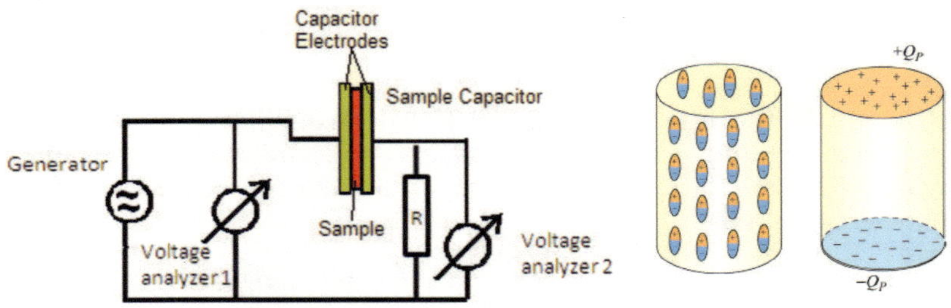

(a) Schema of the principle of BDS

(b) Polarization of matter and resulting capacitive system

Fig. 2. Schema of the principle of BDS and polaryzation of matter

In our experiments samples were placed between parallel gold-plated electrodes of 20 mm diameter and the value of the gap (between the electrodes) was fixed to 0.1mm (by a narrow PTFE cross shape piece).

Polarization of matter ($\overrightarrow{P}$) can be describe as damped harmonic oscillator. When the electromagnetic field is sinusoidal the dipole oscillates around its position of equilibrium. This response is characterized by $\underline{\epsilon}(\omega)$:

$$\overrightarrow{P} = (\underline{\epsilon}(\omega) - 1)\epsilon_0\,\overrightarrow{E}(\omega) \tag{7}$$

The complex dielectric function is the one-sided Fourier or pure imaginary Laplace transform of the correlation function of the polarization fluctuations $\phi(t) = [\overrightarrow{P}(t) \cdot \overrightarrow{P}(0)]$ (phenomological theory of dielectric relaxation Kremer & Schonals (2003))

$$\frac{\underline{\epsilon}(\omega) - \epsilon_\infty}{\epsilon_s - \epsilon_\infty} = \int_0^\infty \exp(-i\,\omega\,t)\left(-\frac{d\phi(t)}{dt}\right) dt \tag{8}$$

where ϵ_s and ϵ_∞ are are the unrelaxed and relaxed values of the dielectric constant.

3. Resolving the normal mode from the α-relaxation

For low molecular weight, the normal mode of PI overlaps at high frequency with the α-relaxation (Fig. 3). This overlapping is an intrinsic problem in checking the Rouse

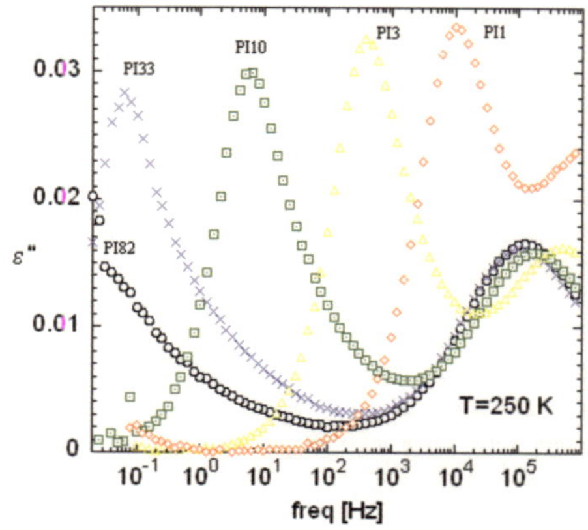

Fig. 3. Dielectric relaxation curves collected at 250 K on PI with different molecular weights.

model (that describes the normal mode dynamics, see next section), since its applicability is limited to chains with moderate molecular weight (below the molecular weight between entanglements). In fact, even by microscopic techniques with spatial resolution as neutron scattering, it is rather difficult to distinguish the border between chain and segmental relaxations Richter et al. (2005).Because the time scale separation between the two dynamical processes is not complete, a detailed analysis of the validity of the Rouse model predictions at high frequencies requires accounting accurately for the α-relaxation contribution.

The imaginary part of the dielectric α-relaxation can be analyzed by using the phenomenological Havriliak-Negami function:

$$\underline{\epsilon}(\omega) = \epsilon_\infty + \frac{\epsilon_s - \epsilon_\infty}{(1 + (i\,\omega\,\tau_{HN})^\alpha)^\gamma}. \tag{9}$$

In addition, a power law contribution ($\propto \omega^{-1}$) was used to account for the normal mode contribution at low frequencies, which is the frequency dependence expected from the Rouse model for frequencies larger than the characteristic one of the shortest mode contribution. Thus, we assumed that the high frequency tail of the normal mode follows a C/ω law and superimposes on the low frequency part of the alpha relaxation losses, being C a free fitting parameter at this stage. The α-relaxation time corresponding to the loss peak maximum was obtained from the parameters of the HN function as follows Kremer & Schonals (2003):

$$\tau_\alpha = \tau_{HN} \frac{\left[sin\left(\frac{\alpha\,\gamma\pi}{2+2\gamma}\right)\right]^{1/\alpha}}{\left[sin\left(\frac{\alpha\,\pi}{2+2\gamma}\right)\right]^{1/\alpha}} \tag{10}$$

The results obtained for samples with different molecular weight at a common temperature of 230 K are shown in Fig. 4. It is evident that in the low molecular weight range the time scale of the segmental dynamics is considerably faster than that corresponding to the high molecular weight limit. This behavior is mainly associated with the plasticizer-like effect of the enchain groups, which are more and more relevant as chain molecular weight decreases. The line in figure 4 describing the experimental behavior for the PI samples investigated is given by:

$$\tau_\alpha(M) = \frac{\tau_\alpha(\infty)}{(1+10000/M)^2} \tag{11}$$

As can be seen in Fig. 3 and 4, the α-relaxation contribution from long chains is nearly

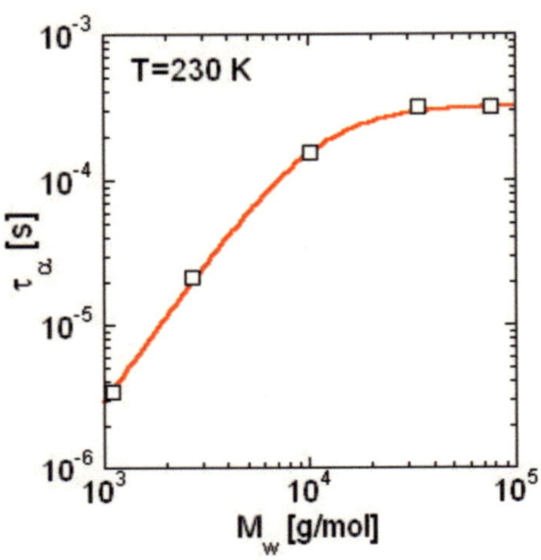

Fig. 4. α-relaxation times of polyisoprene at 230 K as a function of molecular weight. The line corresponds to Eq. 11

independent on molecular mass, but for low molecular masses (c.a. below 20 000 g/mol) it is shifted to higher frequencies and slightly broader. Because the rather small length scale involved in the segmental dynamics (around a nanometer) Cangialosi et al. (2007); Inoue et al. (2002a) it is expected that for a relatively low molecular weight sample, as PI-3, there would be contributions to the α-relaxation with different time scales originated because the presence of chains of different lengths. This could explain the fact that the α-relaxation peak of the PI-3 sample is slightly broader than that of a high molecular weight one, the PI-82 for instance (see Figure 3). Thus, in order to take this small effect into account we decided to describe the α-relaxation of the PI-3 sample as a superposition of different contributions. The contribution to the α-relaxation from a single chain of molecular weight M is assumed to be of the HN type (Eq. 9), with intensity proportional to the number of units (segments) in the chain, i.e. proportional to M. Under this assumption, the whole α-relaxation would be given as follows:

$$\epsilon_\alpha(\omega) = \int \frac{\Delta\epsilon(M)}{(1+(i\omega\tau_{HN})^\beta)^\gamma} g(M)dM \tag{12}$$

$$= \frac{\Delta\epsilon_\alpha}{M_N} \int \frac{M}{(1+(i\omega\tau_{HN})^\beta)^\gamma} g(M)dM \tag{13}$$

where g(M) is a Gaussian-like distribution introduced to take into account the effects of the actual molecular weight distribution of the sample:

$$g(M) = \frac{1}{\sqrt{2}\,\pi\sigma} \exp\left(-\frac{(M - M_n)^2}{2\sigma^2}\right) \tag{14}$$

$$\sigma = M_n \sqrt{\frac{M_w}{M_n} - 1} \tag{15}$$

g(M)dM is the number density of chains with molecular weight M, $\Delta\epsilon_\alpha$ is the total dielectric strength associated to the α-relaxation, and $1/M_n$ is just a normalization factor. The parameters β and γ in Eq. 13 were taken form the fitting of the α-relaxation losses of a high molecular weight sample, PI-82, (β = 0.71, γ = 0.50), i.e. the shape of each component has been assumed to be that obtained from the experiment in a sample with a high molecular weight and a narrow distribution. For such a sample no differences between the contributions of distinct chains are expected (see Figure 4). Furthermore, note that for this sample the α-relaxation is very well resolved from the normal mode and, therefore, its shape can be accurately characterized. Moreover, according to Eq. 11, the following expression for $\tau_{HN}(M)$ was used, $\tau_{HN}(M)/\tau_{HN}(\infty) = [1 + (10000/M)^2]^{-1}$, where a value of $\tau_{HN}(\infty) = \tau_{HN}(82000)$ is a good approximation. In order to avoid the unphysical asymptotic behavior ($\epsilon'' \propto \omega^\beta$) given by the HN equation at very low frequencies, the physical asymptotic behavior ($\epsilon'' \propto \omega$) was imposed for frequencies two decades lower than that of the peak loss frequency. This cut-off frequency was chosen because is the highest cut-off frequency allowing a good description of the α-relaxation data from the high molecular weight PI samples. The resulting

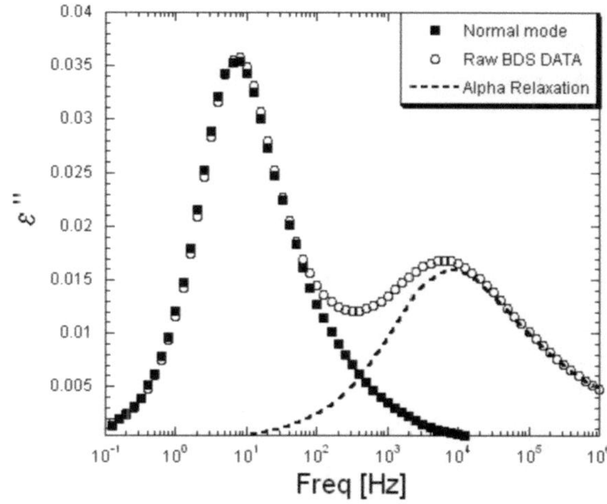

Fig. 5. Resolved normal of PI-3, the dotted line represents the modelled contribution of the α-relaxation. Reprinted with permsision from Riedel et al, Macromolecules 42, 8492-8499 Copyright 2009 American Chemical Society

curve is depicted in Fig. 5 as a dotted line. The value of $\Delta\epsilon_\alpha$ (single adjustable parameter) has been selected to fit the experimental data above f=5e4 Hz, where no appreciable contributions from the normal mode would be expected. After calculating the α-relaxation contribution, the normal mode contribution can be completely resolved by subtracting it from the experimental data. The so obtained results are depicted in Figure 5 for PI-3.

4. Rouse model: experimental test

The large chain-dynamics of linear polymers is one of the basic and classical problems of polymer physics, and thereby, it has been the subject of intensive investigation, both experimentally and theoretically, over many years Adachi & Kotak (1993); de Gennes (1979); Doi & Edwards (1988); Doxastakis et al. (2003); Hiroshi (2001); McLeish (2002); Watanabe et al. (2002a); Yasuo et al. (1988). Despite of the broad range of models and theoretical approaches existing in the literature there are many aspects of the problem that remain to be understood (see refs.Doi & Edwards (1988); Doxastakis et al. (2003); McLeish (2002); Rouse (1953) and references therein). Most of the current investigations are devoted to the problem of the dynamics of highly entangled polymer melts with different architectures and topologies, and to the rheology of polymer systems of industrial relevance Doi & Edwards (1988); Likhtman & McLeish (2002); Watanabe et al. (n.d.). Concerning the chain-dynamics of unentangled polymers, it is generally assumed that the well-known Rouse model Rouse (1953) provides a suitable theoretical description. The Rouse model represents a linear chain as a series of beds and springs subjected to entropic forces in a medium with a constant friction. Although this simple approach obviously fails in describing the melt dynamics of long chains at longer times, the Rouse model is also used for describing the fastest part of the response of these long chains and thereby it is a common ingredient of all available model and theories. In the past, the validity of the Rouse model has ever been instigated by means of different experimental techniques and also by molecular dynamics simulations. But as already mentioned in the previous section, a full and detailed test of the Rouse model is challenging because in unentangled polymer melts the segmental dynamics (α-relaxation) contributions overlap significantly with the high-frequency components of the chain dynamics. This fact, among others, restricts the use of rheology experiments to test accurately the Rouse model on unentangled polymer chains. It is very hard to obtain rheology data of unentanged polymers in the melt. This is due to the rapid relaxation times of the material and the broad spread of the effect of more local molecular mechanisms that affect the stress relaxation modulus at higher frequencies. We will first detail how we have been able to resolve the normal mode from the α-relaxation in BDS data. Then, we will start our study of the Rouse model by introducing the expression of both complex dielectric permittivity and shear modulus before presenting the full detail of the test of the Rouse model using both rheology and BDS.

4.1 Theory

The random walk is an extremely simple model that accounts quantitatively for many properties of chains and provides the starting point for much of the physics of polymer. In this model, we consider an ideal freely joined chain, made up of N links, each define by a vector $\overrightarrow{r}_i$ and b the average length between the links (Fig. 6). The different links have independent orientations. Thus the path of the polymer in space is a random walk.

The end-to-end vector $\overrightarrow{R}_N(t)$ is the sum of the N jump vectors $\overrightarrow{r}_{i+1} - \overrightarrow{r}_i$ which represent the direction and size of each link in the chain:

$$\overrightarrow{R}_N(t) = \sum_{i=1}^{N-1} \overrightarrow{r}_{i+1} - \overrightarrow{r}_i \tag{16}$$

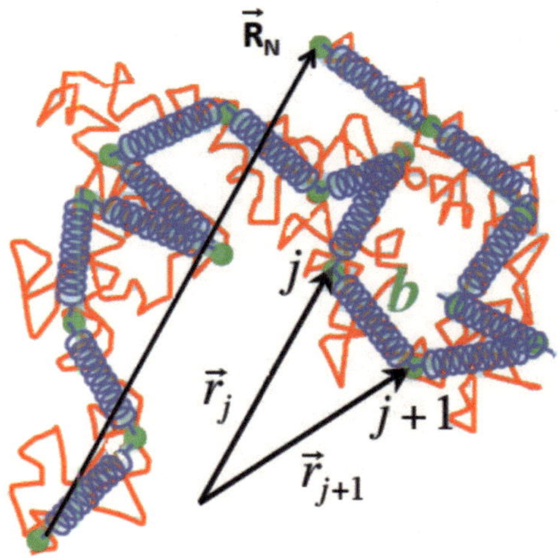

Fig. 6. Schematic representation of the spring model for polymer chain

The mean end-to-end distance is

$$\langle \vec{R}_N(t)^2 \rangle = N b^2 \tag{17}$$

The overall size of a random walk is proportional to the square root of the number of steps.

The correlation function of $\vec{R}_N(t)$ can be expressed as the sum of the vibration of the so-called Rouse modes:

$$\langle \vec{R}_N(t) \vec{R}_N(0) \rangle = \frac{2b^2}{N} \sum_{p:odd}^{N-1} cot^2 \left(\frac{p\pi}{2N} \right) \exp \left(\frac{-t}{\tau_p} \right) \tag{18}$$

with

$$\tau_p = \frac{\xi b^2}{12 k_b T sin^2 \left(\frac{p\pi}{2N} \right)} \tag{19}$$

In the limit of the long chains, for $p\pi<2N$, we obtain the well known expression of the end-to-end vector:

$$\langle \vec{R}_N(t) \vec{R}_N(0) \rangle \cong \frac{8 b^2 N}{\pi^2} \sum_{p:odd} \frac{1}{p^2} \exp \left(-\frac{t}{\tau_p} \right) \tag{20}$$

Due to the factor $1/p^2$ the correlation function of the end-to-end vector is dominated by the slowest mode and as $\tau_p = \frac{\tau_1}{p^2}$ the higher modes are shifted to higher frequencies.

For small deformation, the chain can be approximatively described using a gaussian configuration and the expression of the *xy* component of the shear stress is given by Doi & Edwards (1988):

$$\sigma_{xy} = \frac{3 \nu k_b T}{N b^2} \sum_{i=1}^{N} \langle (\vec{r}_{i+1} - \vec{r}_i)_x (\vec{r}_{i+1} - \vec{r}_i)_y \rangle \tag{21}$$

where ν is the number of chain per volume unit.

4.1.1 Expression of $\underline{\epsilon}(\omega)$ in the frame of the Rouse model

BDS measure the time fluctuation of the polarization. Polymers can have dipole moments parallel to the chain backbone, leading to a net "end-to-end" vector. The part of the polarization related with these dipole moments is proportional to the time fluctuation of the end-to-end vector: $\phi(t) = [\overrightarrow{P}(t) \cdot \overrightarrow{P}(0)] \propto [\overrightarrow{R_N}(t) \cdot \overrightarrow{R_N}(0)]$. Using the expression of the end-to-end vector (Eq. 18) in the phenomenological theory of the dielectric relaxation (Eq. 8) we obtain an expression of the frequency dependance of the dielectric permittivity in the frame of the Rouse model:

$$\epsilon'(\omega) \propto \frac{2b^2}{N} \sum_{p:odd}^{N-1} cot^2\left(\frac{p\pi}{2N}\right) \frac{1}{1+\omega^2\tau_p^2} \tag{22}$$

$$\epsilon''(\omega) \propto \frac{2b^2}{N} \sum_{p:odd}^{N-1} cot^2\left(\frac{p\pi}{2N}\right) \frac{\omega\tau_p}{1+\omega^2\tau_p^2} \tag{23}$$

The factor $cot^2(p)$ (which can be developed in $1/p^2$ when p<N) strongly suppress the contribution of high p modes. Therefore, the dielectric permittivity is sensitive to slow (low p modes) and facilitates resolving normal mode and segmental relaxation contributions.

4.1.2 Expression of $\underline{G}(\omega)$ in the frame of the Rouse model

We can calculate the expression of the shear modulus in the time domain from Eq.21, and using a Fourier transform, we obtain:

$$G'(\omega) \propto \sum_{p=1}^{N-1} \frac{\omega^2\tau_p^2/4}{1+\omega^2\tau_p^2/4} \tag{24}$$

$$G''(\omega) \propto \sum_{p=1}^{N-1} \frac{\omega\tau_p/2}{1+\omega^2\tau_p^2/4} \tag{25}$$

All the modes are contributing in the expression of the shear modulus. Therefore, in the case of small chains, rheology is not a convenient experimental technique to resolve the chain modes from the segmental dynamics.

4.2 Test of the Rouse model

4.2.1 Choice of the sample: Working below M_E

The Rouse model describe the dynamics of unentangled polymers only. Therefore it is important to characterize its domains of application. As we will see below, the use of the time temperature superposition (TTS, see next paragraph) to create master curves in rheology is questionable. Figure 7 represents the master plot obtained for PI-82 from different temperatures T=[40, 20, -15, -30, -50]žC (each represented by a different color) shifted to the reference temperature of -50žC. Master curves permits to obtain the full rheology of the polymer. At low frequencies we can see the Maxwell zone ($G' \propto \omega, G'' \propto \omega^2$). Then, the elastic plateau, where G'>G", is well defined and the polymer is therefore entangled. After the plateau, the Rouse zone is characterized by $G' \propto G'' \propto \omega^{1/2}$. Rheology allows measuring a value of the molecular weight of entanglement M_E by measuring G_n^0, value of G' at the

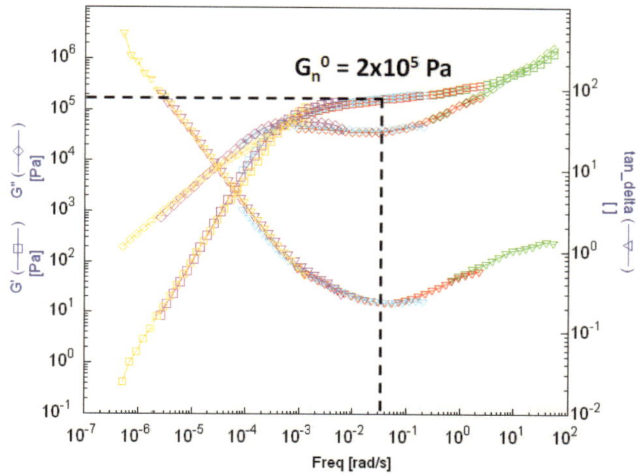

Fig. 7. Master curve of PI-82 at the reference temperature of -50žC

minimum of tan(δ) (around the middle of the elastic plateau). This measurement is made at one temperature, and therefore no master curve is needed to get the value of G_n^0. According to reference Doi & Edwards (1988):

$$M_E = \frac{\rho RT}{G_n^0} \tag{26}$$

where ρ=0.9 g/cm^3 is the density, R = 8.32 J/K/mol is the ideal gas constant and T is the temperature of measurement. Using the value obtained from Fig. 7: $G_n^0 = 2.10^5$ Pa, we obtain a value of M_E=9 kg/mol.

The dielectric normal mode reflects the fluctuations of the end-to-end vector and is dominated by the slowest chain normal mode. As higher modes are scaled and shifted to higher frequencies, the timescale of the normal mode peak, $\tau_N = \frac{2\pi}{f_N}$ (where f_N is the frequency of the maximum of the peak), measured by BDS provides a rather direct access to the Rouse time τ_1 when Rouse theory is fulfilled. Using these values of τ_N at 230 K we have checked whether below the molecular weight between entanglements the Rouse model predictions concerning the molecular weight dependence of the slowest relaxation times ($\tau_1 \propto M^2$) is verified. In Fig. 8 we present the values of the ratio τ_N/τ_α as a function of the molecular mass. The ratio between the longest relaxation time and the value of τ_α obtained at the same temperature in the same experiment is the way used trying to remove the possible variation in τ_N arising from the monomeric friction coefficient, which can be assumed as straightforwardly related with the changes in the glass transition temperature, and hence, with the noticeable effect of end chain groups in the segmental dynamics. Fig. 8 shows that below a molecular weight of around 7 000 g/mol the data scales approximately with M^2. This is just the Rouse model prediction, i.e., what is deduced from Eq. **??** for low-p values where the following approximation holds. This value is intermediate between that of the molecular mass between entanglements (M_e=9 kg/mol) obtained by rheology and that measured from neutron scattering experiments M_e=5 kg/mol Fetters et al. (1994).

Once we have confirmed the range of molecular masses where the molecular weight dependence of the longest relaxation time verifies Eq. **??**, we will test if the whole dielectric

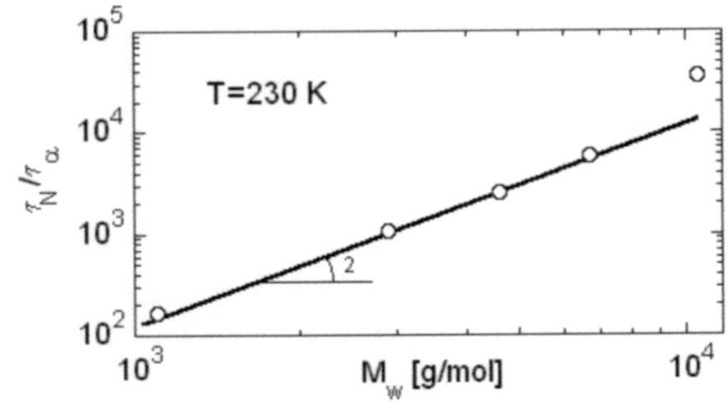

Fig. 8. Ratio τ_N/τ_α , as a function of the molecular weight at 230 K. Solid line represents the behavior predicted by the Rouse model where $\tau_N = \tau_1 \propto M^2$. Reprinted with permsision from Riedel et al, Macromolecules 42, 8492-8499 Copyright 2009 American Chemical Society

normal mode conforms the Rouse model predictions. We have selected the sample PI-3 for this test because, on the one hand, it has a molecular weight sufficiently below of the molecular weight between entanglements (so all the molecular masses of the distribution are below M_E and, on the other hand, it shows a normal mode that is rather well resolved from the segmental α-relaxation contributions to the dielectric losses. Using higher molecular weight samples yield the possibility that the high molecular weight tail of the distribution was above the entanglement molecular weight. On the contrary, for lower molecular weight samples the stronger superposition of the normal mode and the α-relaxation will make the comparison less conclusive (see Fig. 3). For the test, we have taken the data recorded at a temperature of 230 K where the normal mode contribution is completely included within the experimental frequency window, being at the same time the α-relaxation contribution also well captured and then subtracted (see previous section and Figure 5).

4.2.2 Time temperature superposition (TTS)

Figure 9 a shows the high level of accuracy obtained by means of the present BDS experiments when measuring the rather weak dielectric relaxation of a PI-3 sample, for both the normal mode and the α-relaxation. From simple inspection of the data it is apparent that the normal mode peak shifts by changing temperature without any significant change in shape (time temperature superposition is verified for this process), which is one of the predictions of the Rouse model (see eq. 18). However, it is also evident that the shift of the normal mode peak is distinct than that of the α-relaxation one, i.e. the TTS fails for the complete response. **?** This fact is illustrated in Figure 9 b where data at two temperatures where both processes are clearly visible in the frequency window are compared. For this comparison the axes were scaled (by a multiplying factor) in such a way that the normal peaks superimpose. Whereas the superposition in the normal mode range is excellent, the same shift makes the alpha relaxation peak positions to be about half a decade different. The temperature dependance of the normal mode peak can be described by Williams-Landel-Ferry (WLF) Williams et al. (1995) equation:

$$log(a_T) = \frac{C1(T - T_{ref})}{T - T_{ref} + C2} \tag{27}$$

When considering the temperature dependence of the shift factor we found that, whereas for high molecular masses (above that between entanglements) the WLF parameters are the same

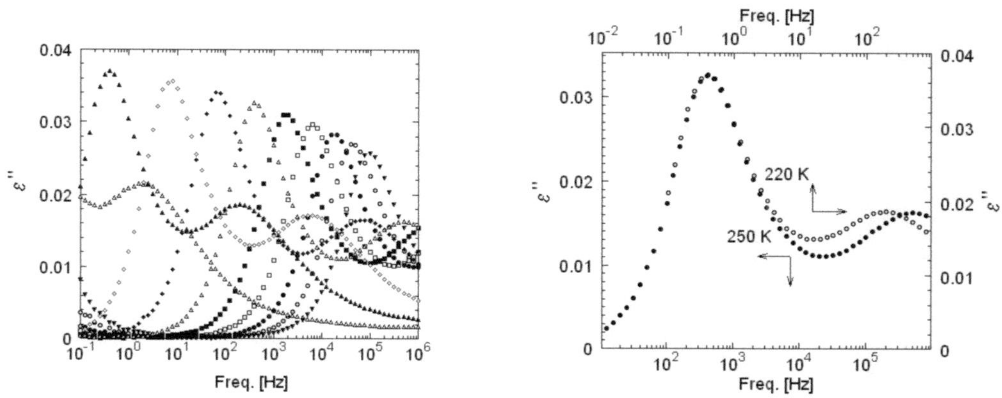

(a) Dielectric spectra of PI3 from 220K (left) to 300K (right) ΔT = 10K.

(b) Comparison of normal mode and α-relaxation frequency shifts for PI3 when temperature is changed from 220 K to 250 K.

Fig. 9. Shift of the dielectric response of PI-3 with temperature. Reprinted with permsision from Riedel et al, Macromolecules 42, 8492-8499 Copyright 2009 American Chemical Society

within uncertainties (WLF parameters with T_g as the reference temperature $C1 = 30.20.7$ and $C2 = 57.00.2K$), for lower molecular weight samples the value of C1 remains the same but C2 becomes noticeably smaller, being 49.2 K for PI-3. This is likely related with the significant variation of Tg in the low molecular range (see Table 1).

The data resulting from the rheology experiments performed on a PI-3 sample of 1.3 mm thickness using two parallel plates of 8 mm diameter are shown in Figure 10. To produce this plot a master curve at a reference temperature T_{ref} = 230 K was built imposing the horizontal shift factor a_T determined from BDS (see above) and a vertical shift factor b_T=T_r/T. Curves have been measured at 215, 220, 225, 230 and 240K, each color represent a temperature. It is apparent that the superposition so obtained is good in the terminal relaxation range, although at the highest frequencies, where the contributions of the segmental dynamics are prominent, the data superposition fails clearly. The use of master curves in rheology experiments is a standard practice because of the rather limited frequency range and the general applicability of the TTS principle to the terminal relaxation range. However, the presence of the segmental dynamics contribution at high frequencies when approaching T_g could make the application of TTS questionable because the different temperature shifts of global and the segmental dynamics (see above). Thus, using rheology data alone the high frequency side of the terminal region could be highly distorted.

4.2.3 Determination of the bead size

To use Eq. 23 we need the previous determination of N, which is not known a priori, as it requires the estimation of the bead size. Adachi and co-workers Adachi et al. (1990) estimated the bead size of PI on the basis of an analysis of the segmental relaxation in terms of a distribution of Debye relaxation times. These authors suggested that a PI bead would contain about 7 monomers. However, a generally accepted approach is to consider that the bead size would be of the order of the Kuhn length, b_k. Inoue et al. (2002b) The literature value of b_k for

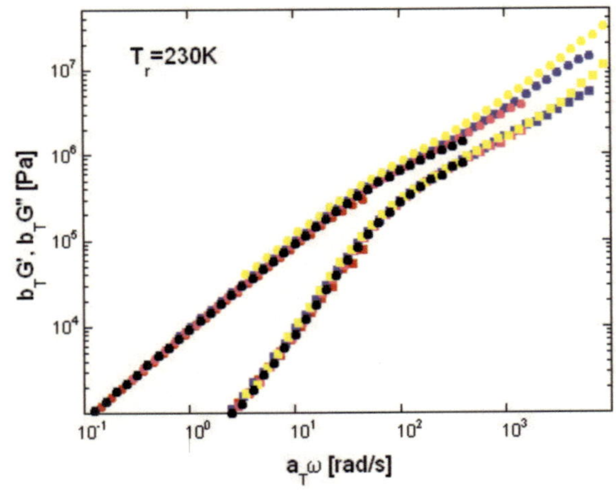

Fig. 10. Rheological master curve of PI-3(reference temperature T_r =230 K) of the real: ■ and imaginary: • parts of the shear modulus.

PI is 0.84 nm,Rubinstein & Colby (2003) which corresponds to a molecular mass of the Kuhn segment of 120 g/mol. This means that a Kuhn segment would contain about 1.5 monomers, about a factor of about 5 less than Adachi et al. estimated. On the other hand, recent results has evidenced that the α-relaxation in the glass transition range probes the polymer segmental motions in a volume comparable to b_k^3. Cangialosi et al. (2007); Lodge & McLeish (2000) Thus, we will introduce a "segmental" Rouse time $\tau_S = \tau_{p=N}$, which apparently have not a clear physical meaning (the fastest Rouse time would be p=N-1) but it can be related with the so-called characteristic Rouse frequency ($W = 3k_bT/b^2$) as $\tau_S^{-1} = 4W$. Nevertheless, τ_S is here used as a convenient parameter for the further analysis. We can identify τ_S at T_g with the α-relaxation time determined at this same temperature, i.e. $\tau_S(Tg)$=5 s. In this way, by using the WLF equation describing the temperature dependence of the normal mode peak, the obtained value of τ_S at 230 K would be 1.1e-4 s. Using this value and that obtained for τ_1 from the normal peak maximum in Eq. **??** a value of N=24 results. Note that according with this value the bead mass would be around 121 g/mol in very good agreement with literature results for the Kuhn segment mass,Rubinstein & Colby (2003) and consequently the corresponding bead size will be nearly identical to b_k.

4.2.4 Monodisperse system

Thus, in order to test the validity of the Rouse model in describing the dielectric normal mode relaxation data we assumed N=24 in Eq. 23, which would be completely predictive, except in an amplitude factor, once the slowest relaxation time τ_1 was determined from the loss peak maximum. The so calculated curve is shown as a dashed line in Figure **??**. It is very clear that the calculated curve is significantly narrower than the experimental data, not only at high frequencies where some overlapping contribution from the α-relaxation could exists, but more importantly also in the low frequency flank of the loss peak. This comparison evidence clearly that the experimental peak is distinctly broader than the calculation of the Rouse model based in Eq. 7, confirming what was already envisaged in Figure 7a of ref. Doxastakis et al. (2003). Nevertheless, an obvious reason for this discrepancy could be the fact that the actual sample has some (small) polydispersity. Despite of the fact that PI samples with a low polydispersity

(1) were chosen, even samples obtained from a very controlled chemistry contains a narrow distribution of the molecular weights, which was not considered in the previous calculation.

4.2.5 Introduction of polydispersity

Now, we are in a situation where it becomes possible to perform a detailed comparison between the experimental normal mode relaxation and the calculated Rouse model expectation. In order to incorporate the small sample polydispersity, the response expected from the Rouse model has been calculated as a weighted superposition of the responses corresponding to chains with different molecular weights. Since the molecular masses of the chains in the PI-3 sample are all below the molecular weight between entanglements, the molecular weight dependence of τ_p will be that given by Eq. 19 for all of the different chains. For calculating $\tau_p(M)$ we have used in Eq. 19 a common value τ_s irrespective of the molecular weight of the particular chain. Furthermore, we calculated $\tau_1(Mw)$ as the reciprocal of the peak angular frequency of the experimental normal mode of this sample.

In this way the Rouse model remains completely predictive and the corresponding dielectric response to be compared with normal mode contribution from the actual sample will be given by:

$$\epsilon''_N(\omega) = \frac{\Delta\epsilon_N}{M_n} \int M \frac{2b^2}{N} \sum_{p:odd}^{N-1} cot^2 \left(\frac{p\pi}{2N}\right) g(M)dM \tag{28}$$

where the contribution to the normal mode from a given chain is proportional to the chain dipole moment (to the end-to-end vector), and thus again proportional to M.

As can be seen in Figure 11 (solid line), the sum over all the modes and all the molecular weight provides a satisfactory description of the experimental dielectric losses, namely at frequencies around and above the peak. The excellent agreement evidences that taking into account the (small) polydispersity of the sample under investigation is necessary to provide a good description of the dielectric normal mode contribution by means of the Rouse model, without any adjustable parameter other than the Rouse time, which (for the average molecular mass) is essentially determined from the reciprocal of the maximum loss angular frequency.

4.2.6 BDS and rheology in the same experiment

Once we have found that the Rouse model can account accurately for the chain dynamics as observed by dielectric spectroscopy, the question that arise is if using the very same approach it would be possible to account also for other independent experiments, namely rheology. A key point to perform such a test is to be sure that all the environmental sample conditions remains the same. To be sure about this, we performed simultaneous dielectric and rheology experiments at 230 K. The sample thickness was smaller to balance the data quality of both dielectric and rheological results. We found that a thickness of 0.9 mm using two parallel plates of 25 mm diameter provides a rather good compromise. The output of these experiments at 230 K is shown in Figure 12. Despite of the very different geometry used, the rheological results are in close agreement with those obtained before. In this simultaneous experiment, the dielectric normal mode is clearly resolved so the peak position defining the time scale can be determined with low uncertainty, although again the accuracy of the dielectric relaxation data is not so good as that obtained in the data presented before.

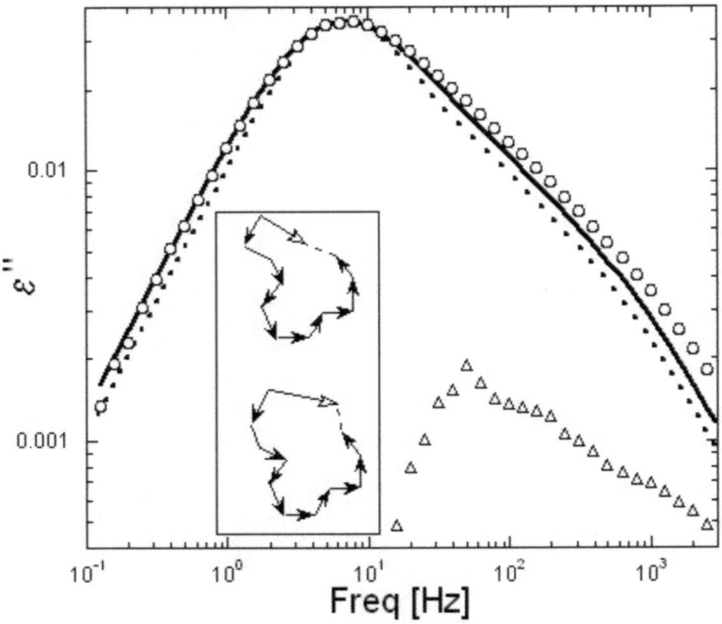

Fig. 11. Resolved normal mode relaxation of PI-3 sample. The lines represent the behaviour predicted by the Rouse model including for the actual sample polydispersity (solid line) and without it (dotted line). Triangles correspond to the difference between the experimental losses and the Rouse model predictions. The inset shows schematically how the presence of configuration defects allows fluctuations of the whole chain dipole moment without variation of the end-to-end vector. Reprinted with permsision from Riedel et al, Macromolecules 42, 8492-8499 Copyright 2009 American Chemical Society

Nevertheless, this experiment will be essential in testing the ability of the Rouse model in accounting simultaneously for both the dielectric and rheology signatures of the whole chain dynamics. The approach used was to determine τ_1 from the dielectric losses, according with the description used above that is able to accurately account for the complete dielectric relaxation spectrum, and to use a similar approach to generate the corresponding rheology behavior. Thus, the only unknown parameter needed to perform the comparison with the experimental G'(ω) and G''(ω) data will be G_∞ (the high frequency limit of the modulus in terminal zone), which in fact it is not needed for calculating tan (δ)= G''(ω)/G'(ω). The equations used were obtained from Eq. RouseRheo1 and RouseRheo2 following the same procedure that the one from BDS:

$$G'(\omega) = \int \frac{G_\infty}{M_n} \sum_{p=1}^{N-1} \frac{\omega^2\tau_p^2/4}{1+\omega^2\tau_p^2/4} g(M)dM \tag{29}$$

$$G''(\omega) = \int \frac{G_\infty}{M_n} \sum_{p=1}^{N-1} \frac{\omega\tau_p/2}{1+\omega^2\tau_p^2/4} g(M)dM \tag{30}$$

The results of the comparison between the calculated responses and the experimental data are shown in Fig. 12. A satisfactory description is obtained for G'(ω) and G''(ω) by means

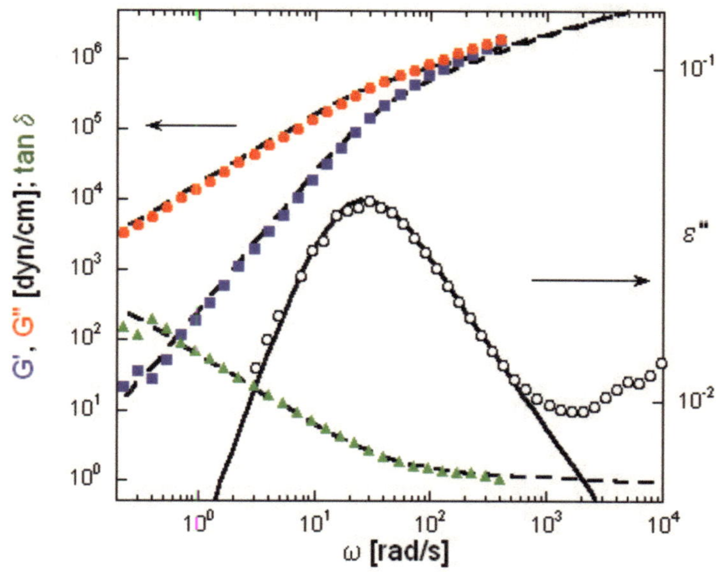

Fig. 12. Simultaneous rheology (G′: ■, G″: •, tan(δ) : ▲) and BDS experiments (ϵ'': ○) on PI-3 at 230 K. Lines correspond to the description obtained using the Rouse model, with a single set of parameters, for all the data sets.

of the same Rouse model parameters used in describing the dielectric normal mode. More interestingly, the description of tan(δ) is also rather good, for which the previous calculation is compared with the experimental data without any arbitrary scaling. Despite the good agreement obtained, it is worthy of remark that the ability of rheology experiments for checking the Rouse model in full detail is much more limited than the dielectric one because both the narrower frequency range accessible and the stronger overlapping of the segmental dynamics contributions.

4.3 Discussion

Despite the overall good agreement, Figure 11 evidences that the experimental losses are slightly larger (maximum difference about 10%) than the Rouse model prediction in the high frequency side. Although one could consider that this is simply due to the contribution of the overlapping α-relaxation that has not been properly subtracted, the fact that the maximum of these extra losses intensity occurs at frequencies two decades above the segmental relaxation peak (see triangle in Figure 11) seems to point out to other origin. In agreement with this idea, that the frequency distance above the NM peak where the Rouse model description starts underestimating the experimental losses does not depend much on temperature. Thus, extra contributions from some chain-modes are detected in the dielectric normal mode. In fact the peak intensity of the extra losses occurs at the position of the p=3 mode (see triangles in Fig. 11), i.e. the second mode contributing to the dielectric normal mode. All this might evidence the fact that the Rouse model approach ignores several aspects of the actual chain properties as that of the chain stiffness Brodeck et al. (2009) or the lower friction expected to occur at the chain ends Lund et al. (2009). In this context it is noticeable that the contribution from the α-relaxation extended considerably towards the frequency range where the normal mode is more prominent. In fact the cut-off frequency used for describing the α-relaxation in Fig. 5 was close to 100 Hz, i.e, where the differences between the normal mode response and the Rouse model are more pronounced. This result indicates that for most of the high-p chain modes the segmental relaxation is not completed. This is in contrast to the assumptions of the Rouse model where it is considered that all the internal motions in the chain segment are so fast that their effect can be included in the effective friction coefficient. It is noteworthy that a higher frequency cut-off would be not compatible with the experimental data of the high molecular weight PI samples, and would produce a more prominent underestimation of the dielectric normal mode losses. On the contrary, a lower frequency cut-off would improve slightly the agreement between the normal mode data and the Rouse model prediction but would imply and higher coupling between the segmental dynamics and the whole chain motion. Small deviations from the Rouse model predictions have also been reported from numerical simulations, molecular dynamics calculations and detected by neutron scattering experiments, Brodeck et al. (2009); Doxastakis et al. (2003); Logotheti & Theodorou (2007); Richter et al. (1999) although these deviations become evident only for relatively high p-values. Note, that dielectric experiments being mainly sensitive to the low-p modes can hardly detect such deviations. On the other hand, experiments in solution have also evidenced differences between the experimental data and the predictions of the Rouse model,Watanabe et al. (1995) which were tentatively attributed to chain overlapping effects since the deviations occur above a given concentration. Nonetheless, there are also possible experimental sources for the small extra high frequency contributions to the dielectric losses as it would be the presence in the actual polymer of a fraction of monomeric units others than the 1,4-cis ones (up to 20%). The motion of such units, having a much smaller component of the dipole moment parallel to the chain contour, would produce small amplitude fluctuations of the whole dipole moment uncorrelated with the fluctuations of the end-to-end vector (see inset in Figure 11). The chain motions around these 'configuration defects' would generate a relatively weak and fast contribution to the dielectric normal mode that could explain the experimental data. Unfortunately, the actual sample microstructure prevents to definitively address whether the deviations from the Rouse model predictions are actually indicative of its limitations.

5. Dynamics regimes as a function of the molecular weight and effects of entanglement

As shown in the previous section, the polymer dynamics follows different regimes as a function of the molecular weight and molecular weight distribution. Abdel-Goad et al Abdel-Goad et al. (2004), using rheology measurements coupled with an empirical Winter-relaxation BSW-model obtained three different exponents (1, 3.4 and 3) in the molecular weight dependence of the zero shear viscosity. Previous BDS studies on the molecular weight dependence of the normal mode relaxation time showed a crossover from the unentangled dynamics to the entanglement regime Adachi & Kotak (1993); Boese & Kremer (1990); Hiroshi (2001;?); Yasuo et al. (1988). The reptational tube model was first introduced by de Gennes de Gennes (1979) and developed by Doi and Edwads Doi & Edwards (1988) to described this phenomena of entanglement. Many corrections (as Contour Length Fluctuation or Constraints Release) have been added to the pure reptation in order to reach a totally predictive theory McLeish (2002); Viovy et al. (2002). However, none of BDS experiment has been able to access the crossover to exponent 3 expected by the pure reptation theory which, as aforementioned, has been detected for the viscosity. In this section, we will detail how careful BDS experiments allow detecting the two different crossovers from the Rouse up to the pure reptation regime. Entanglement effects in BDS spectra will be also analyzed. Finally, the possible influence of the narrow molecular weight distribution of the samples on the dielectric loss shape will be discussed.

5.1 Low molecular weight

As can be seen in Fig. 3 the fast variation of the normal mode relaxation peak prevents detecting it in the frequency window at 250K for molecular weight higher than 80 kg/mol. Thus, we will first focus the analysis on the samples with lower molecular weight. As already mentioned, for low molecular masses the changes in the local friction coefficient (arising because the significant changes in the end chain groups accompanying the changes in molecular weight) influences the relaxation time. Therefore, just as in Fig. 8, the ratio of the normal mode time scale to that of the α-relaxation was evaluated (Fig. 13). To increase the plot

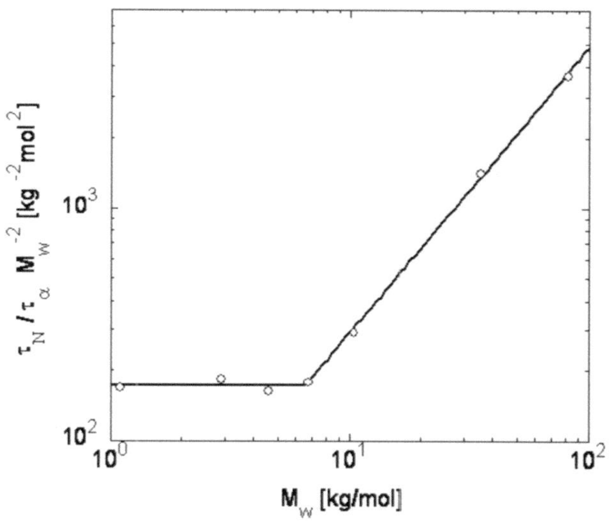

Fig. 13. Longest and segmental relaxation time ratio as a function of the molecular weight at 250 K. The vertical axis is scaled by M^2 to emphasize the transition from the Rouse to the intermediate regime. The solid line corresponds to the description of the data with a sharp crossover between two power law regimes with different exponents. Reprinted with permsision from Riedel et al, Rheologica Acta 49, 507-512 Copyright 2010 Springer

sensitivity to changes between the different regimes a factor of M_w^{-2} has been applied to the ratio between the characteristic times (reciprocal of the peak angular frequency) of the two relaxation processes. As already mentioned, the M_w^2 dependence expected for unentangled polymers on the basis of the Rouse theory is fulfilled for samples with molecular weight below 7 kg/mol. The exponent describing the higher molecular mass range considered in this plot was 3.20.1 which is distinctly lower, but close, to the 3.4 usually found in rheological experiments Ferry (1995). It is noteworthy that imposing this exponent to fit our data will result in a higher value of the crossover molecular weight, thus increasing the discrepancy with the reported/admitted value of M_e=5kg/mol measured by neutron scattering Fetters et al. (1994). It is noteworthy that, as it is well know, the ratio τ_N / τ_α will change with temperature Ding & Sokolov (2006). Nevertheless, the previous results will not change significantly using data at other temperatures because the changes in the value τ_N/τ_α will be very similar for all the samples having different molecular weights and, consequently, the resulting molecular weight dependence would be unaltered.

5.2 High molecular weight

After analyzing the molecular weight dependence of the end-to-end fluctuations in the low and moderate molecular weight range, now we will focus the attention in the highest accessible molecular weights. In the high molecular weight range, the comparison among the different samples has to be performed at a significantly higher temperature due to the dramatic slowing down of the chain dynamics. The more suitable temperatures are those where the normal mode loss peak of the sample with the highest molecular weight occurs in the low frequency range of the experimental window. An additional factor that have to be taken into account is the fact that by increasing temperature the conductivity contribution to the dielectric losses becomes more prominent. The conductivity contribution appears as a ω^{-1} increasing of the dielectric losses. This is an important issue even for high quality samples when the experiments require accessing to the low frequencies at temperatures far above T_g. This situation is illustrated in Fig. 14 for the raw data of the PI sample having the highest investigated molecular weight. It is apparent that at 340 K the normal mode

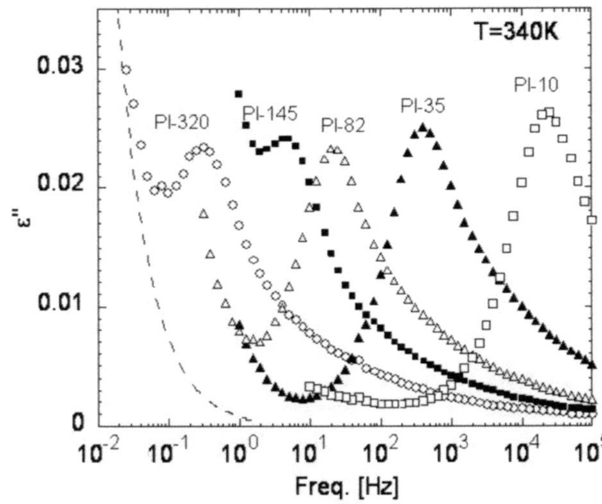

Fig. 14. Normal mode of the high molecular weight PI samples. Dashed line represents the calculated conductivity contribution to the dielectric losses for the PI-320 sample. Reprinted with permsision from Riedel et al, Rheologica Acta 49, 507-512 Copyright 2010 Springer

peak can be well resolved from conductivity for this sample but it would be hard to resolve the normal mode relaxation at this temperature for a sample with a significantly higher molecular weight. Furthermore, increasing temperature would not improve the situation since the overlapping of the conductivity contribution with the normal mode relaxation will also increase. This is a serious limitation of the dielectric methods for investigating the slowest chain dynamics in highly entangled systems. Nevertheless, as shown in Fig. 14, resolving the normal mode peak was possible for all the samples investigated although the contribution from conductivity will increase the uncertainty in the peak position for the samples with very high molecular weight. Figure 15 shows the molecular weight dependence of the slowest relaxation time for the high molecular weight regime obtained from the data presented in Fig. 14. Trying to increase the sensitivity of the plot to possible changes in behavior the data

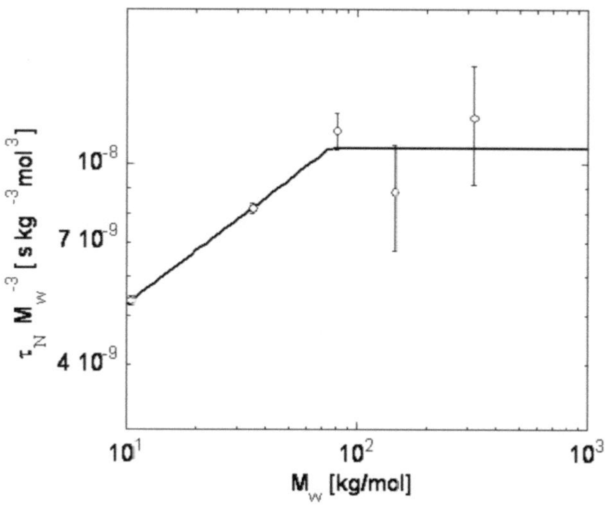

Fig. 15. Longest relaxation time from the higher molecular weight PI samples. The graph is scaled to M^3 to emphasize the crossover from the intermediate to the reptation regime. The solid line corresponds to the description of the data with a sharp crossover between to power law regimes with different exponents. Reprinted with permsision from Riedel et al, Rheologica Acta 49, 507-512 Copyright 2010 Springer

have been multiplied by $M_{\bar{w}}^{-3}$, which would produce a molecular weight independent result for a pure reptation regime. Despite of the uncertainties involved, our results evidence that for the highest molecular weight samples the molecular mass dependence approach the pure reptation regime expectation. The line in Fig. 4 corresponds to a crossover from and exponent 3.35 to a pure reptation-like regime. The small difference between this exponent and that obtained above from Figure 2 is more likely due to the fact that the sample with the lower molecular weight considered in Fig. 4 have a significantly lower glass transition temperature, an effect not been considered in Fig. 4. The crossover molecular weight obtained from Fig. 4 is 7510 kg/mol, i.e. it corresponds to about 15 times M_e. This value is slightly lower than that determined from viscosity data Abdel-Goad et al. (2004).

5.3 Discussion

The results previously described showed three different regimes for the molecular weight dependence of the chain longest relaxation time in PI, one below 7 kg/mol following the Rouse model prediction as expected for a non-entangled polymer melts, other above 75 kg/mol where the reptation theory provides a good description and an intermediate one, where the polymer is entangled but other mechanisms (like contour length fluctuations or constraints release) in addition to reptation would control the whole chain dynamics. In the rheological experiments above referred Abdel-Goad et al. (2004) it was shown that the viscosity of high molecular weight PI samples conforms well the reptation theory predictions.

Thus, we decided to test up to what extent the pure reptation theory is able to describe the normal mode relaxation spectrum of the highly entangled PI samples. This test can evidence the ability of the reptation theory to capture the main features of the slowest chain dynamics, despite the well documented failure of the reptation theory in accounting for the whole chain dynamics, even in the high molecular weight range. This is clearly evidenced by the reported mismatching of the normalized dielectric and rheological spectra Watanabe et al. (2002b). To this end, we compared our experimental data on the high molecular mass samples with the corresponding reptation theory predictions for the dielectric permittivity. This relation is similar to the one obtained in the frame of the Rouse model (Eq. 23. As the chain are long, N>p and the cot^2 can be developed in $1/p^2$. The value of $\tau_p = \tau_1/p^2$ is replaced by τ_d/p^2 where τ_d is the disentanglement time (reptation) time, which would correspond in good approximation to τ_N.

$$\epsilon''(\omega) \propto \sum_{p:odd}^{N-1} \frac{1}{p^2} \frac{\omega\,\tau_d/p^2}{1+(\omega^2\,\tau_d/p^2)^2} \tag{31}$$

Figure 16 shows the direct comparison between the experimental data for some of the samples investigated (symbols) having all of them the lowest available polydispersity index (1.05) and the pure reptation theory prediction (solid line). Both vertical and horizontal scaling factors have been applied to obtain a good matching of the peaks. It should be noted that the possible conductivity contributions to the normal mode relaxation were subtracted. The inset shows separately the data of the highest molecular weight sample because it has a markedly broader molecular weight distribution (polydispersity index 1.14). From Fig. 16 it becomes apparent

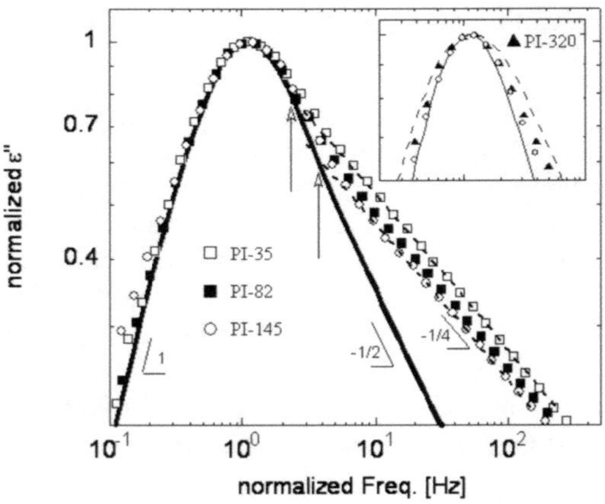

Fig. 16. Comparison of the BDS data of high molecular weight (symbols) with pure reptation theory (solid line). Dashed straight line showing the ij power law behavior of the samples with PI-33 and PI-145 are also shown. The vertical arrows indicate the crossover frequency between both regimes. The inset presents for comparison the highest molecular weight data (PI-320) with a high polydispersity (1.14) with that of smaller polydispersity (1.04). The dashed line represents what would be the reptation theory expectation when a very crude approximation is used to account the effect of the molecular weight. Reprinted with permsision from Riedel et al, Rheologica Acta 49, 507-512 Copyright 2010 Springer

that in the high frequency side of the loss peak deviations from the reptation predictions on the end-to-end vector fluctuations persist even for the highest molecular weight investigated. Whereas the high frequency behavior expected from the reptation theory is a power law with exponent -1/2, the experimental data present and exponent -1/4 (see Figure 16), which would be a signature of the relevance of chain contour length fluctuations at least in this high frequency side of the normal mode relaxation. Nevertheless, it is also clear that the range of these deviations reduce when increasing molecular weight. The vertical arrows in Fig. 5 shows that a factor of 5 increasing in molecular weight makes the crossover frequency to increase in a factor of about 2. By inspection of the rheological data reported by Abdel-Goad et al, Abdel-Goad et al. (2004) it is also apparent that in the very high molecular weight range where the viscosity scales as predicted by reptation theory the terminal relaxation is far from being properly described by this theory. Eventually the normal mode description by the pure reptation theory could be obtained only at extremely higher molecular weights, for which, as aforementioned, the dielectric experiments will not be suitable for investigating the extremely slow chain dynamics. Concerning this, it has been shown McLeish (2002) that for polyethylene de frequency dependence of the loss shear modulus verifies the reptation prediction only for a molecular weight as high as of 800 kg/mol, which for this polymer corresponds to about 400 times M_e, i.e., it would correspond to about 3000 kg/mol for PI. Taking the above-calculated shift of the crossover frequency into account, for this limiting molecular weight the crossover frequency would occur at around 20 Hz and the failure of the repetition theory description would be hardly detectable by using the same scale as in Figure 16. Figure 16 also shows that both the maximum and the low frequency side of the loss peak is well accounted by the reptation theory without any evident deviation, except for the sample having a broader distribution of molecular weight, which shows a distinctly broader normal mode peak (see inset of Fig. 16). This comparison evidences that the molecular weight distribution have a noticeable effect on the normal mode spectrum shape. We remind that the effect of the molecular weight distribution on the normal mode was properly accounted for in an unentangled PI sample by assuming that the contributions from chains in the sample with distinct molecular weight simply superimpose (see previous section on the Rouse model). When we tried the same approach with the higher molecular weight samples (dashed line in the inset of Fig. 16) it becomes evident that the situation for well-entangled polymers is different. Even by using the smallest polydispersity (1.04) the calculated response overestimates by far the broadening of the peak for the sample with highest polydispersity (1.14). Thus, for highly entangled polymers the effect of the molecular weight distribution on the normal mode is less evident than that observed in the unentangled polymer case. In fact, the complete disentanglement of a chain involves also the motions of the chains around, which would have a different molecular weight, being therefore the resulting time scale some kind of average of those corresponding to the ideally monodisperse melts. As a result the longest relaxation time in highly entangled melts should not dependent greatly on the molecular weight distribution provided it is not very broad.

6. Conclusion

In this chapter, we have used broadband dielectric spectroscopy (BDS) and rheology to study properties of linear polymers. We have focused our study on the large chain dynamics (normal mode), described by the reptational tube theory and the Rouse model when the polymer is entangled or not, respectively.

By means of broad-band dielectric spectroscopy we have resolved the normal mode relaxation of a non-entangled 1,4-cis-poly(isoprene) (PI) and therefore accessed to the end-to-end vector dynamics over a broad frequency/time range. A remarkably good comparison of the data with the Rouse model predictions is found if the effect of the actual narrow distribution of molecular masses of the sample investigated is accounted. The very same approach was found to provide a good description of a simultaneous dielectric/rheology experiment. The small excess contributions found in the high frequency side of the experimental dielectric normal mode losses could be associated, at least partially, to the sample microstructure details. Therefore, we conclude that the Rouse model accounts within experimental uncertainties for the end-to-end dynamics of unentangled PI, once the molecular weight distribution effects are considered. A more sensitive test would require an unentangled nearly monodisperse full 1,4-cis-polyisoprene sample, which can hardly be available.

Further experiments have permitted to detect two crossovers in the molecular weight dependence of the end-to-end relaxation time. The first corresponds to the crossover from the range where the Rouse theory is applicable to the entangled limited range, being the crossover molecular weight 6.5 $\pm$ 0.5 kg/mol, i.e. slightly above the molecular weight between entanglements. The crossover from the intermediate range to the behavior predicted by the pure reptation theory is found at around 75 $\pm$ 10 kg/mol, which corresponds to 15 times the molecular weight between entanglements. Despite of the fact that the reptation theory is able to describe the molecular weight dependence of the slowest relaxation time for these high molecular weight samples, the shape of the normal mode spectrum is still markedly different from that expected by this theory. Eventually, only at a much higher molecular weight (hundred times the molecular weight between entanglements) the reptation theory could completely describe the normal mode relaxation associate to the chain dynamics. Unfortunately, dielectric experiments in this range are not feasible.

To conclude, in this chapter we have shown that dielectric spectroscopy and rheology permitted to obtain physical information about polymer. Even if these two techniques measure different properties, the origin of these properties, is the same: the dynamics. As they permit to extend our knowledge on polymer dynamics, these techniques will allow a better understanding and control of polymer properties.

7. References

Abdel-Goad, M., Pyckhout-Hintzen, W., Kahle, S., Allgaier, J., Richter, D. & Fetters, L. J. (2004). Rheological properties of 1,4-polyisoprene over a large molecular weight range, *Macromolecules* 37(21): 8135–8144.

Adachi, K. & Kotak (1993). Dielectric normal mode relaxation, *Progress in polymer science* 18(3): 585.

Adachi, K., Yoshida, H., Fukui, F. & Kotaka, T. (1990). Comparison of dielectric and viscoelastic relaxation spectra of polyisoprene, *Macromolecules* 23(12): 3138–3144. doi: 10.1021/ma00214a018.

Boese, D. & Kremer, F. (1990). Molecular dynamics in bulk cis-polyisoprene as studied by dielectric spectroscopy, *Macromolecules* 23(3): 829–835.

Brodeck, M., Alvarez, F., Arbe, A., Juranyi, F., Unruh, T., Holderer, O., Colmenero, J. & Richter, D. (2009). Study of the dynamics of poly(ethylene oxide) by combining molecular dynamic simulations and neutron scattering experiments, *The Journal of Chemical Physics* 130(9): 094908.

Cangialosi, D., Alegria, A. & Colmenero, J. (2007). Route to calculate the length scale for the glass transition in polymers, *Physical Review E* 76(1): 011514. Copyright (C) 2010 The American Physical Society Please report any problems to prola@aps.org PRE.

de Gennes, P. G. (1979). *Scaling Concepts in Polymer Physics*, Cornell University Press, Ithaca and London.

Ding, Y. & Sokolov, A. P. (2006). Breakdown of time temperature superposition principle and universality of chain dynamics in polymers, *Macromolecules* 39(9): 3322–3326.

Doi, M. & Edwards, S. F. (1988). *The theory of polymer dynamics*, Clarendon, Oxford.

Doxastakis, M., Theodorou, D. N., Fytas, G., Kremer, F., Faller, R., Muller-Plathe, F. & Hadjichristidis, N. (2003). Chain and local dynamics of polyisoprene as probed by experiments and computer simulations, *The Journal of Chemical Physics* 119(13): 6883–6894.

Ferry, J. D. (1995). *Viscoelastic Properties of Polymers*, Oxford University Press, New York.

Fetters, L. J., Lohse, D. J., Richter, D., Witten, T. A. & Zirkel, A. (1994). Connection between polymer molecular weight, density, chain dimensions, and melt viscoelastic properties, *Macromolecules* 27(17): 4639–4647. doi: 10.1021/ma00095a001.

Hiroshi, W. (2001). Dielectric relaxation of type-a polymers in melts and solutions, *Macromolecular Rapid Communications* 22(3): 127–175. 10.1002/1521-3927(200102) 22:3 <127::AID-MARC127>3.0.CO;2-S.

Inoue, T., Uematsu, T. & Osaki, K. (2002a). The significance of the rouse segment: Its concentration dependence, *Macromolecules* 35(3): 820–826. doi: 10.1021/ma011037m.

Inoue, T., Uematsu, T. & Osaki, K. (2002b). The significance of the rouse segment: Its concentration dependence, *Macromolecules* 35(3): 820–826. doi: 10.1021/ma011037m.

Kremer, F. & Schonals, A. (2003). *Broadband Dielectric Spectroscopy*, Springer, Berlin.

Likhtman, A. E. & McLeish, T. C. B. (2002). Quantitative theory for linear dynamics of linear entangled polymers, *Macromolecules* 35(16): 6332–6343. doi: 10.1021/ma0200219.

Lodge, T. P. & McLeish, T. C. B. (2000). Self-concentrations and effective glass transition temperatures in polymer blends, *Macromolecules* 33(14): 5278–5284. doi: 10.1021/ma9921706.

Logotheti, G. E. & Theodorou, D. N. (2007). Segmental and chain dynamics of isotactic polypropylene melts, *Macromolecules* 40(6): 2235–2245. doi: 10.1021/ma062234u.

Lund, R., Plaza-Garcia, S., Alegria, A., Colmenero, J., Janoski, J., Chowdhury, S. R. & Quirk, R. P. (2009). Polymer dynamics of well-defined, chain-end-functionalized polystyrenes by dielectric spectroscopy, *Macromolecules* 42(22): 8875–8881. doi: 10.1021/ma901617u.

McLeish, T. C. B. (2002). Tube theory of entangled polymer dynamics, *Advances in Physics* 51(6): 1379.

Richter, D., M.Monkenbusch, Arbe, A. & Colmenero, J. (2005). *Neutron Spin Echo in Polymer Systems*, Springer, Berlin.

Richter, D., Monkenbusch, M., Allgeier, J., Arbe, A., Colmenero, J., Farago, B., Bae, Y. C. & Faust, R. (1999). From rouse dynamics to local relaxation: A neutron spin echo study on polyisobutylene melts, *The Journal of Chemical Physics* 111(13): 6107–6120.

Rouse, P. E. (1953). A theory for the linear elasticity properties of dilute solutions of coiling polymers, *The Journal of Chemical Physics* 21(7).

Rubinstein, M. & Colby, R. H. (2003). *Polymer Physics*, Oxford University, New York.

Stockmayer, W. (1967). Dielectric disperson in solutions of flexible polymers, *Pure Applied Chemesitry* p. 539.

Viovy, J. L., Rubinstein, M. & Colby, R. H. (2002). Constraint release in polymer melts: tube reorganization versus tube dilation, *Macromolecules* 24(12): 3587–3596. doi: 10.1021/ma00012a020.

Watanabe, H., Matsumiya, Y. & Inoue, T. (2002a). Dielectric and viscoelastic relaxation of highly entangled star polyisoprene: Quantitative test of tube dilation model, *Macromolecules* 35(6): 2339–2357. doi: 10.1021/ma011782z.

Watanabe, H., Matsumiya, Y. & Inoue, T. (2002b). Dielectric and viscoelastic relaxation of highly entangled star polyisoprene: Quantitative test of tube dilation model, *Macromolecules* 35(6): 2339–2357. doi: 10.1021/ma011782z.

Watanabe, H., Sawada, T. & Matsumiya, Y. (n.d.). Constraint release in star/star blends and partial tube dilation in monodisperse star systems, *Macromolecules* 39(7): 2553–2561. doi: 10.1021/ma0600198.

Watanabe, H., Yamada, H. & Urakawa, O. (1995). Dielectric relaxation of dipole-inverted cis-polyisoprene solutions, *Macromolecules* 28(19): 6443–6453. doi: 10.1021/ma00 123a009.

Williams, M. L., Landel, R. F. & Ferry, J. D. (1995). Mechanical properties of substances of high molecular weight in amorphous polymers and other glass-forming liquids, *Journal of American Chemical Society* 77(19): 3701–3707.

Yasuo, I., Keiichiro, A. & Tadao, K. (1988). Further investigation of the dielectric normal mode process in undiluted cis-polyisoprene with narrow distribution of molecular weight, *The Journal of Chemical Physics* 89(12): 7585–7592.

2

Viscoelastic Properties for Sol-Gel Transition

Yutaka Tanaka
University of Fukui, Dept. of Engineering, Japan

1. Introduction

Gel formation is inseparable from viscoelastic properties. Research works of gels usually accompany viscoelastic methodology, because the gel formation closely related to the disappearance of fluidity in colloidal system, which frequently occurs in a polymer solution.(Almdal et al., 1993) Change in viscoelastic properties is significant during the gelation process of a polymer solution, where three dimensional networks coming from cross-links between chains have the key role.

The system is liquid-like and often called as sol before cross-linking starts, and remains a liquid until the viscosity becomes infinite. At that moment, called as gel point, there is at least one molecule with an infinite molecular weight. Beyond the gel point, evolution of elastic modulus follows with the progress of cross-links. As for networks, various formations can be observed, those are thermoreversible, often seen in natural polymer, irreversible, physically cross-linked, chemically cross-linked, super molecular - developed recently in advanced materials - and so on. The dramatic change in viscoelasticity is of interest from both practical and physicochemical points of view, and many research works have been carried out. Also, several reviews have been published for gels and sol - gel transition. Ross-Murphy and Clark gave a review on physical gels. (Clark & Ross-Murphy, 1987) Extensive monographs on thermoreversible gelation were produced by Guenet and Nijenhuis.(Guenet, 1992; Nijenhuis, 1997) As for the structure and properties of polymer networks, Stepto compiled research works.(Stepto, 1998)

With a consideration on these existing research results, this chapter approaches to the viscoelasticity of gelation with several topics; those are theoretical and experimental works on power laws observed for various properties in sol-gel transition region, delay of gel point and the ring formation for the gelation of RA_{fa} + $R'B_{fb}$ type polymerisation, network formation by the end-linking of star polymer and the viscoelastic behaviour.

2. Evolution of viscoelasticity near sol-gel transition

2.1 Sol – gel transition and power law

The gelation has attracted many research workers, especially in the field of polymer science, which probably is because gels are complex systems. In fact, in the respective gels their own complexities are presented. To overcome the complexities of network structure and describe successfully the formation and properties of gels, various theoretical models have so far been proposed. In Section 2.1.1, Bethe lattice was demonstrated as an example of plain models to express the gel point prediction. In Section 3.1, a network formation model of rather specific structure is described to account for the delay of gel point.

As for the description of gelation, the framework of scaling is frequently used to account for both experimental and theoretical research results. As is well known, power law appears for many physical parameters to capture polymer properties in the scaling law. Also in the model of Bethe lattice, quite a few power law dependences can be seen for gel fraction, degree of polymerisation etc. In Section 2.1.2, those relationships are shown to discuss the analogues between properties of gel and other physical phenomena.

Intensive research works have been carried out by Winter et al., over many years to elucidate the power law observed for the sol-gel transition, which is shown in Section 2.1.3. (Winter, 1997) The concepts of mechanical self-similarity, and further topological self-similarity as its extension, have been introduced by their studies, where the same properties are found at different length scales within the observed network. Afterwards, their research results have been accepted broadly. Many works follow to find the experimental results of various polymer gels to obey power law relaxation. With these research results, theoretical works in relation to sol-gel transition is discussed in the following three sections.

2.1.1 Bethe lattice, an example of percolation model to account for gelation

It is Flory who first gave theoretical framework on the gelation. The idea starts with the generating of branching during polymerisation. The branchings take place repeatedly which eventually lead to the formation of the infinite network.(Flory, 1941) Afterwards, modified ideas have been proposed by many researchers.(Gordon & Ross-Murphy,1975; Miller&Macosko, 1978)

Another trend towards the model of a satisfactory interpretation of the experimental data comes from physics of phase transition which regards the gelation as an example of critical phenomenon. This trend is developing with the viewpoint of scaling approaches for the static and dynamic properties of gels. The scaling concepts introduced into the theory of polymer solutions were subsequently extended to the description of swollen networks.

It is possible to consider the transition from sol to gel with the percolation model, which illustrates the process of gel formation by the linkage of two dots where many dots are spaced regularly. An example of the experimental system is the monomer sample including no solvent, where the functionality of the monomer is larger than three.

In the model, the monomers lie on the lattice points. The linkage, referred to as bond, randomly occurs only between neighbouring monomers. Let the probability of the bond formation be p, and the definition of s-cluster is given with s, the number of monomers included in the cluster. Then, s-clusters of the finite size appear at $p < p_c$ for a given value of p_c, which corresponds with sol. For $p > p_c$, at least one infinite cluster and s-clusters of various sizes appear, which corresponds with gel. And consequently, sol-gel transition is the change in phase from sol comprised of s-clusters of finite size to gel having at least one infinite cluster. The probability of the bond formation is the predominant parameter whether or not the system in the lattice turns to gel. This process is called as the bond percolation. p_c of a threshold corresponds with gel point. The examples of real experimental parameters of p are temperature, pH, concentration, the extent of reaction of the monomer, etc.

The physical quantities relating to gelation can be derived using a lattice model of Bethe lattice which is an example of percolation models. Figure 1 shows Bethe lattice where the coordination number, z, is $z=3$, meaning that the dot corresponds with the monomer whose functionality is equal to three; the dot is called as the site in the model. Bonds of z emanate from the site of the centre, also sites are at the opposite end of the bonds. Then, another z bonds emanate from each of these sites again; one of z bonds is tied with the site of the centre, the other $(z - 1)$ bonds extend outwards. Accordingly, the branching takes place repeatedly in Bethe lattice.

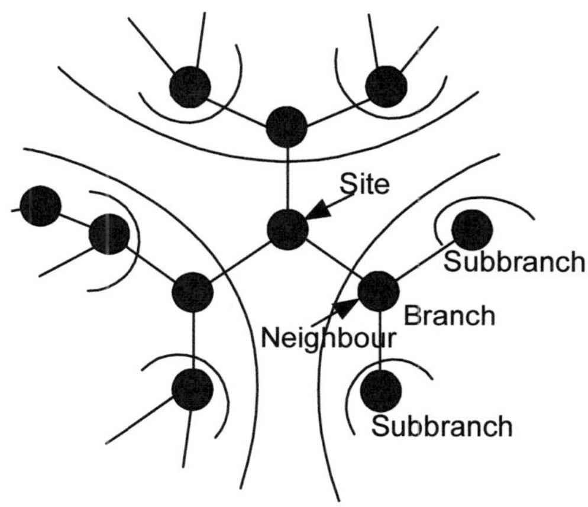

Fig. 1. The surroundings of the origin of a large Bethe lattice.

It is a remarkable feature that Bethe lattice has no closed loops. If a site was chosen randomly and moved to the neighbouring site, then new sites of $(z - 1)$ come in. It is always new sites as long as moving forwards. The definitions of branch, neighbour, subbranch are shown in Figure 1 as they will be used later.

To define the threshold of p_c statistically, the probability of path continuation is introduced. It is the probability that a randomly chosen site has at least one path emanating from it that continues to infinity. This probability increases as the increase in p, and becomes equal to unity at p_c . Now, we will consider the path continuation from the randomly chosen site to its neighbour. There are $(z - 1)$ sites available around the neighbour except for the first site chosen randomly; the neighbour can form a bond to each of these sites with the probability of p. Therefore, the randomly chosen site can form bonds of $(z - 1)\times p$ with its neighbours on average. At the critical percolation point, we have $(z - 1)\times p_c = 1$ and the gel point can be given as,

$$p_c = 1/(z - 1) \tag{1}$$

Next, the percolation probability, denoted as P is derived; that is, the probability that a randomly chosen site is in the cluster of infinite size. P corresponds with the gel fraction in the experimental system. Let the probability that the particular branch emanating from a randomly chosen site does not continue to infinity, be H. H is the sum of two cases. One is that the bond does not form between the randomly chosen site and the branch, which probability is $(1-p)$. The other one is that the bond forms, but the branch does not continue to infinity, which probability is pH^2 . We have, consequently,

$$H = pH^2 + (1-p) \tag{2}$$

Furthermore, the probability that the neighbour of the randomly chosen site continue to infinity is given as $(1- H^3)$, because H^3 means there is no path to infinity from the neighbour through each of three branches. The probability of bond formation between the randomly chosen site and neighbour is p, and thus $P = (1- H^3)\times p$ is obtained. From the solution of eq.(2), as eq.(2) is a quadratic equation of H, P can be expressed as a function of p.

$$P = 0 \text{ for } p < p_c,\ P = \frac{(p-1)^3 - p^3}{p^2} \text{ for } p > p_c \tag{3}$$

Eq.(3) is equivalent with the result of the weight fraction of gel Flory derived using the branching model.(Flory, 1941)

Next, the averaged cluster size, S is derived; S corresponds with the weight average molecular weight in the experimental system. The result of the expressions of S is described here,

$$S = \frac{p(1+p)}{(1-2p)} \tag{4}$$

Using eqs. (3) and (4) the critical behaviours for P and S can be examined near the threshold of p_c . If the probability of bond formation is larger than p_c and close to p_c , then only one infinite cluster exists and P becomes small. In fact, provided that z=3, i.e., p_c=(1/2), P=0 at p=p_c (see the solution of eq.(2)), then we have,

$$P \propto (p - p_c).$$

As for the averaged cluster size, if p is smaller than p_c and approaches to p_c , then S increases rapidly. That is, S diverges at p=p_c (see eq.(4)), which situation corresponds to the divergence of the weight-average molecular weight at gel point for the experimental system. The equation of S near the gelation threshold is given as,

$$S \propto (p_c - p)^{-1}.$$

From the arguments regarding Bethe lattice described above, several expressions of critical behaviour can be obtained for physical parameters of sol-gel transition. Likewise, the consideration on 3-dimentional lattice will give the expressions of P and S in the form of power law. Further, more generalised formulas can be obtained for d-dimentional lattice of d>3 as,

$$P \propto (p - p_c)^{\beta} \tag{5a}$$

$$S \propto (p - p_c)^{\gamma} \tag{5b}$$

Other physical parameters like number of s-cluster, radials of cluster, correlation length etc will be described in the next section.

For the gel point prediction which is applicable to the experimental data, it is necessary to introduce solvent molecules to the lattice model shown above. The introduction of solvent is the process to lay both monomer and solvent molecules randomly to lattice dots (that is, sites), which is known as the site percolation. In the lattice model, the site percolation assumes that, whenever a monomer is laid in the neighbour, the mutual bond is necessarily produced between neighbours. The probability of the site occupation by the monomer is the dominant factor of the gelation. In the experimental system, the probability of occupation corresponds with the monomer concentration. The cluster of infinite size appears above the threshold of the probability of occupation.

The site-bond percolation can be given as an extension of the site percolation, which accounts for the sol-gel transition of the systems comprised of monomer and solvent. The result of the site-bond percolation over a lattice model is shown in Figure 2, where the probability of site occupation, ϕ, and that of the bond formation, p were varied independently. The prediction of sol-gel transition like Figure 2 has been compared with the experimental results for the solution of polystyrene and carbon disulfide. (Tan et al., 1983)

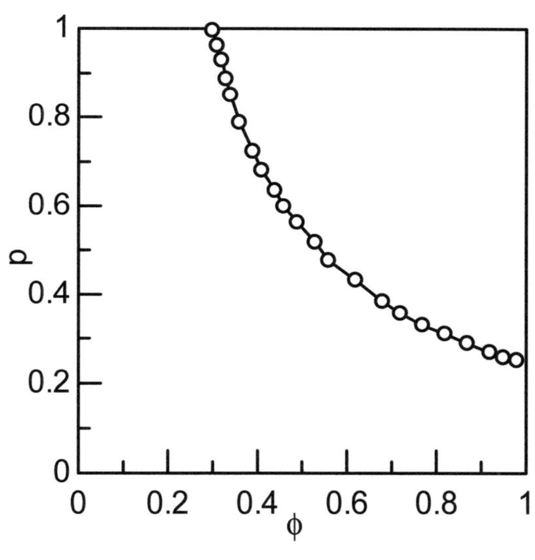

Fig. 2. Phase diagram of random site- bond percolation.

2.1.2 Power law near sol-gel transition

The gelation process is analogous to the percolation with some simplicities. The analogues are studied by Stauffer et al., for physical parameters appeared in the different transitions of percolation, sol-gel transition, liquid-gas transition, the ferromagnet. (Stauffer, 1981) In particular, the power law dependences of several physical parameters near the transition point were taken as the research object. Those relationships are usually in the form of power function of the relative distance from the critical point as seen below. Table 1 shows the analogies in the critical behaviour. The power laws of the gel fraction and the degree of polymerisation are already seen in eqs.(5a and 5b). Likewise, the critical exponents of physical quantities regarding sol-gel transition can be given as follows.

$$G \propto |(p - p_c)/p_c|^{\beta} \quad (6)$$

$$DP_w \propto |(p - p_c)/p_c|^{-\gamma} \quad (7)$$

$$\xi \propto |(p - p_c)/p_c|^{-\nu} \quad (8)$$

$$E \propto |(p - p_c)/p_c|^{t} \quad (9)$$

$$\eta \propto |(p - p_c)/p_c|^{k} \quad (10)$$

η is the viscosity of the solution and given in the pre-gel stage. E is the elasticity and given in the post-gel stage.

It is not until the comparison with the experimental data are made that the significance of the power law dependence, and hence the significance of the physical parameters of eqs.(6-10) are clarified. Several experimental results for the exponents of power law have been compiled by Brauner.(Brauner, 1979) The gelling substances taken up in his work were, polystyrene cross-linked with divinylbenzene, poly(vinyl chloride), poly(diallyl phthalate), deoxy hemoglobin, amylose etc. It turns out from the data compilation that values of β, γ, k scatter well depending on experimental systems. One of the causes of the data scattering is that the exponent is sensitive to the choice of p_c and the experimental determination of the exponent is unstable. In this regard, Gordon et al. also pointed out to be cautious on the estimation of the transition point; they examined the soluble fraction as a function of heating time for the coagulation of milk.(Gordon&Torkington, 1984) In addition, the range of the variable of p will influence on the estimation of the exponent; the exponent might be determined using p and other values in the outside of the critical behaviour. As a result, it is still a controversial issue whether the experimental values are equivalent with those predicted by percolation.

Gelation	Percolation	liquid-gas	ferromagnet
G : gel fraction	P_∞ : percolation probability	$\Delta\rho$: density difference	M_0 : spontaneous magnetisation
DP_w : degree of polymerisation	S : mean size of finite cluster	κ : compressibility	χ : susceptibility
ξ : correlation length	ξ : correlation length	ξ : correlation length	ξ : correlation length
p_c : gel point	p_c : percolation threshold	T_c : critical temperature	T_c : curie temperature
p : conversion	p : probability	T : temperature	T : temperature

Table 1. Analogies in the critical behaviour of different phase transitions.

An example of experimental determination of t value is shown in Figure 3.(Nakamura, 1993) The sample is gellan, polysaccharide, one of the gelling agents. Its aqueous solution shows thermoreversible sol-gel transition, which depends on concentration and temperature. The aqueous solution is sol at high temperature and gel at low temperature. Taking the correspondence between the solution and the percolation, the probability of the site occupation is proportional to the concentration of gellan, the probability of the bond formation is roughly proportional to temperature. Therefore, the measurement of temperature dependence of the elasticity corresponds to the examination of bond percolation under the constant probability of the site occupation. The following equation was applied to the result of the measurement, then data were plotted in Figure 3.

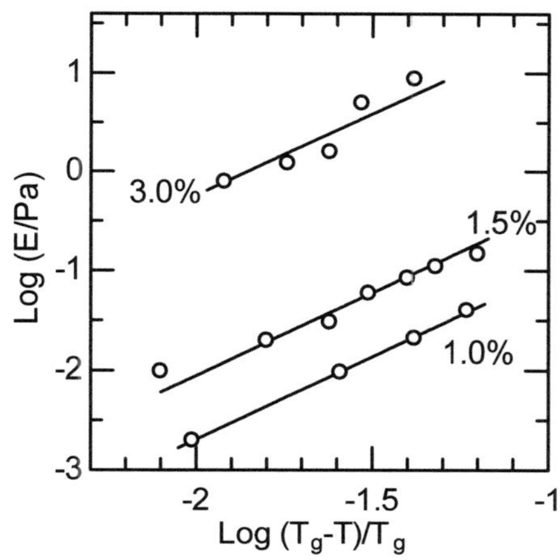

Fig. 3. Logarithmic plot of elasticity as a function of the reduced temperature for aqueous gellan solutions of 1.0%, 1.5%, 3.0%.

$$E = C \cdot \left| \frac{T_g - T}{T_g} \right|^t \tag{11}$$

C is the proportional constant, T is temperature of the aqueous solution, T_g the temperature of gelation. The result shows t=1.5 being independent of the concentration, which is close to the prediction of the percolation.

Some unsolved problems remain in the critical behaviour of sol-gel transition. Those are, the relations between different exponents, the exponent and molecular mechanism, and so on. Nevertherless, the applicability of power law to several physical properties, the analogy between sol-gel, liquid-gas, ferromagnet transitions, the relevance to the volume phase transition of gel(Tanaka, 2011), these matters contain research objects of much interests.

2.1.3 Mechanical self similarity and scale of observation

Several power functions were found between various physical quantities in relation to polymer gel. Of these power laws, the discovery of the relations for frequency and dynamic viscoelasticity of $G'(\omega)$ and $G''(\omega)$ has attracted much interest, and many research workers followed the frequency power law of the gelation. The application of the frequency power law to the gel point was first proposed by Winter et al.(Winter and Mour, 1997) Their experimental results suggested, at first, new methods for localising the gel point by the detection of a loss tangent independent of the frequency. Moreover, they discussed that the topological self similarity in the polymer network can be related to the straight line in $G'(\omega)$

and $G''(\omega)$ over wide frequencies in the double logarithmic plot, described as, (Winter, 1987; Vilgis&Winter, 1988)

$$G'(\omega) \propto \omega^n,\ G''(\omega) \propto \omega^n\ . \tag{12}$$

In their studies, dynamic mechanical experiments in small amplitude oscillatory were carried out for the end-linking polydimethylsiloxane and polyurethane. The extent of reaction, p, of the end-linking is the variable to dominate sol-gel transition. The gel point of the polydimethylsiloxane network appears at $p=p_c$. A simple power law was found to govern the viscoelastic behaviour of the critical gel, described by a complex modulus

$$G^*(\omega, p_c) = \Gamma(1-n)\times S_g(i\omega)^n\ . \tag{13}$$

The "gel strength" S_g depends on the mobility of the chain segments. Γ is the usual gamma function. The relaxation exponent of n adopts values between 0 and 1. i is the imaginary unit. The power function of frequency implies the property of mechanical selfsimilarity where $G'(\omega)$ and $G''(\omega)$ values are scaled with frequency.

The evolution of rheology can be observed continuously during the entire gelation process. For the state of the sample, the gelation process can be divided into three; those are pre-gel state, critical gel state and post-gel state. Then, it becomes understandable the evolution of rheology by taking the critical gel as the reference state. The three states are shown in Figure 4 schematically using the dynamic viscosity of η^* as the representative rheological property.

$$\eta^* = \frac{\sqrt{G'^2 + G''^2}}{\omega} \tag{14}$$

These three states are evidenced in the experimental data of polydimethylsiloxane. The progress of viscoelastic data with gelation can be accounted for using the plot of $|\eta^*|$ and ω. At high frequency, the power law behaviour of $|\eta^*| \propto \omega^{-1}$ is exhibited. The sample appears to be at the gel point in this frequency window. However, at the pre-gel states, deviations from the power law dependence are seen as the frequency is lowered and the finite size of the largest cluster is recognisable. The low frequency behaviour is still that of a typical viscoelastic liquid which is described as,

$$G'(\omega) \propto \omega^2,\ G''(\omega) \propto \omega\ , \text{ in the limit } \omega \to 0. \tag{15}$$

Therefore the zero frequency viscosities, η_0, are determined for respective p values as shown in Figure 4. The characteristic frequency of ω^* can be defined by the intersection of η_0 with the power law of the critical gel. ω^* divides between the gel and the liquid behaviours. The value of ω^* decreases with approaching to the gel point.

In the post-gel region, $p>p_c$ the polymer exhibits a finite equilibrium modulus of g_∞. A characteristic frequency $\underline{\omega}^*$, can be defined again(see Figure 4) for respective p values. At high frequencies above some crossover $\underline{\omega}^*$, the sample shows the behaviour of critical gel state and at frequency below it, the sample behaves like a typical viscoelastic solid ($G' = g_\infty$, $G'' \propto \omega$ at $\omega \to 0$). $\underline{\omega}^*$ value increases with increasing the extent of cross-linking.

In the evolution of viscoelastic properties, the most interesting part of the rheological behaviour shifts to lower and lower frequencies as the transition becomes closer. The shift to lower frequencies is so pronounced that the actual transition through the gel point cannot be observed in an experiment, because ω* has shifted below the lower limiting frequency of the rheometer. Only far beyond the gel point in the post-gel region, the characteristic frequency becomes large again, so the solid behaviour becomes measurable within the frequency range of the instrument.

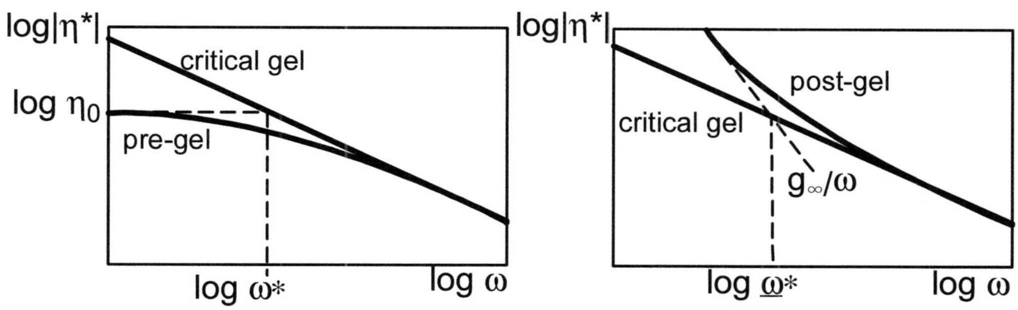

Fig. 4. Schematic representation for the relation of |η*| vs ω for three states of pre-gel, critical gel, post-gel, and the definition of the characteristic frequency, ω*, $\underline{\omega}$* of the pre-gel and the post-gel respectively.

Although the viscoelastic properties are sensitive to the frequency scales, the basic frequency-length relationship always can be applied, in the sense that high frequency probes smaller length scales, and low frequency probes large length scales. This is usually ruled by a dispersion relation between frequency, time and length.

$$\frac{1}{\omega} \propto t \propto L^{\alpha} \tag{16}$$

t is relaxation time and *L* is the scale of observation. α is an exponent, specific to the model considered. The mechanical selfsimilarity is the relation between viscoelasticity and frequency. Eq.(16) indicates the relation of frequency and $L^{-\alpha}$. From these fragments, the concept of selfsimilarity for network structure in critical gel are considered.(Cates, 1985; Muthukumar, 1985)

The power law behaviour of the critical gel correlates to a length scale range. Beyond the very wide frequency range, where power law appears at gel point, a lower frequency limit is expected to be given by the sample size and an upper frequency limit is given by the transition of single chain behaviour. If the frequency is large enough, length scales much smaller than self similar regions are probed, for example, finding the range of a single chain between two cross-links. Power law relaxation is observed when *L* has a value in the range between these lower and upper boundaries of lengths in the material.

The characteristic frequency ω* corresponds to a correlation length,

$$\omega^* \propto \xi^{-\alpha} \tag{17}$$

which is characteristic for the network structure. The structure is hypothesised to be selfsimilar at scales below ξ, as supported by mechanical selfsimilarity. Also a possible model has been given to support the hypothesis. The critical gel is treated as polymeric fractal, that is, a selfsimilar object. ξ has to be identified by the size of the selfsimilar regions, i.e. the critical correlation length. We have seen the relation between ξ and p_c in Section 2.1.2, as $\xi \propto |(p - p_c)/p_c|^{-\nu}$, which implies $\omega^* \rightarrow 0$ if $p \rightarrow p_c$ and agrees with the above arguments.

The evolution of ξ with the progress of end-linking reaction is a direct expression of the evolution of network structure. In the pre-gel state, the molecules are taken into clusters by the end-linking reaction, which are weak solids of small spatial dimension. With the increase in reaction time, one gets larger and larger clusters. Assuming there are large clusters in the melt of smaller clusters and uncrosslinked chains, the mechanical experiment is sensitive to the size of a typical cluster. At longer reaction times this size becomes larger and larger until it extends across the entire sample. The size of the typical cluster becomes infinite at the gel point where the critical gel is formed. The network structure of the critical gel is self similar and so only a very large length matters. Generally one calls such self similar objects "fractals". The straight lines in $G'(\omega)$, $G''(\omega)$ over all ω indicate that there is no dominant length scale in the polydimethylsiloxane sample, i. e. the correlation length is infinite and on each scale of observation a similar structure can be found.

2.2 Experimental approach for sol-gel transition

In the previous section, many power law dependences could be seen near gel point for various properties.In particular, the notion of mechanical selfsimilarity and structural selfsimilarity derived by the power law attracted many research workers. The experimental method that investigates dynamic viscoelasticity and frequency near gel point was adopted and many results have been reported for wide range of materials. Also the scaling concepts have been used frequently as an analysing method. The materials are; food, surfactant, liquid crystals, elastomer etc, which suggests that gelation is the universal phenomena and common to the compounds of large fields, and further the power law is applicable to the wide range. A part of examples are shown here; gelatin(Michon et al, 1993; Hsu & Jamieson, 1990), carrageenan (Hossain et al, 2001), ethyl(hydroxyethyl)cellulose and surfactant (Kjøniksen et al., 1998), soy protein (Caillard, et al., 2010), poly(vinyl alchol) (Kjøniksen & Nystroem, 1996), poly(urethane) network (Nicolai et al., 1997), liquid crystalline network (Valentva et al., 1999), pluronic triblock copolymer end-modified (Bromberg et al., 1999), poly(acrylonitrile) (Tan et al., 2009), asphalts (Vargas & Manero, 2011), hyperbranched poly(ε-caprolactone) (Kwak et al., 2005). The analytical approach which is common to these experimental examinations attempts, using a common scale, to estimate the gelation of different solvents, different chemical species, cross-links of different kinds. Among the research reports of conventional polymers, the experimental works for poly(vinyl chloride) (Li et al., 1997-1998) demonstrates obvious results on the relationships between viscoelasticity, concentration, temperature, molecular weight and so forth. The methods to determine scaling exponents experimentally are also shown for viscosities and elastic moduli. Those results are described in the following sections.

2.2.1 Concentration and gel strength

Aoki et al. prepared PVC gels from well-characterised poly(vinyl chloride) samples and the solvent of bis(2-ethylhexyl) phthalate (DOP). In the experiment, four PVC samples were used and the gelation of the solution was examined with PVC concentration as the variable to dominate sol-gel transition. The weight-average molecular weight (M_w) of the samples are 39.4×10^3, 87.4×10^3, 102×10^3, 173×10^3, they are denoted as PVC4, 9, 10, 17, respectively.

$G'(\omega)$ and $G''(\omega)$ data are shown in Figures 5 and 6 for the solutions of PVC4. The polymer concentration ranges from 9.8 to 154 g/L as indicated. At very low concentrations, the viscoelastic behaviour of PVC4 solutions follows that of a viscoelastic liquid as described by

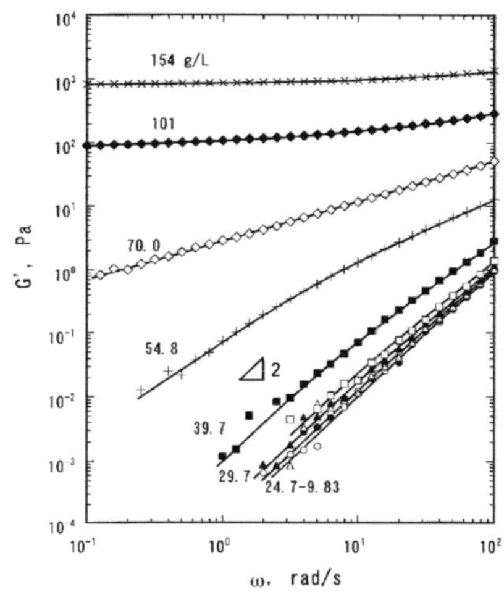

Fig. 5. Storage modulus G' of PVC/ DOP as a function of ω for various concentrations of PVC4 as indicated. Reproduced, with permission of the author.

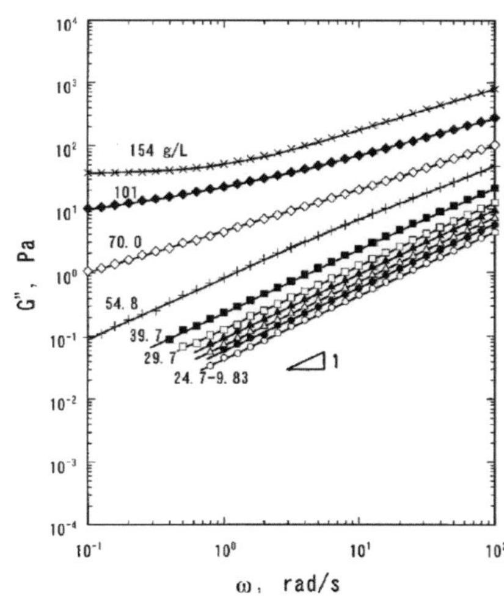

Fig. 6. Loss modulus G'' of PVC/DOP as a function of ω for various concentrations of PVC4 as indicated. Reproduced, with permission of the author.

eq.(15), while the deviation from the relation of eq.(15) becomes more pronounced with the increase in concentration. The result, $G'(\omega)$=const. at $\omega\rightarrow 0$ indicates the formation of a gel plateau. The slope of the $G'(\omega)$ vs ω curve (at $\omega\rightarrow 0$) in the double logarithmic plot changes from 2 to 0, whereas that of the $G''(\omega)$ vs ω curve at ($\omega \rightarrow 0$) varies from 1 to 0. Furthermore, it is obvious at the gel point that both slopes take a common value between 0 and 1. The viscoelastic behaviour for other systems, PVC9, -10 and -17 was virtually same. The range of polymer concentration where the sol–gel transition is observed was dependent on the molecular weight of PVC.

Determination of gel point for PVC gel was conducted on the basis of the frequency-independence of tan δ, which is known as the method using a multifrequency plot of tan δ vs gelation time, temperature, or concentration, depending on which variable governs the gelation process. The gel point has been well determined using this method when the gelation time or temperature was a controlling variable for gelation.(Koike, 1996; Peyrelasse, 1996) Figure 7 shows the multifrequency plot of tan δ vs PVC concentration, where all curves pass through the single point at the certain PVC concentration, which can be defined as the gel point, c_g.

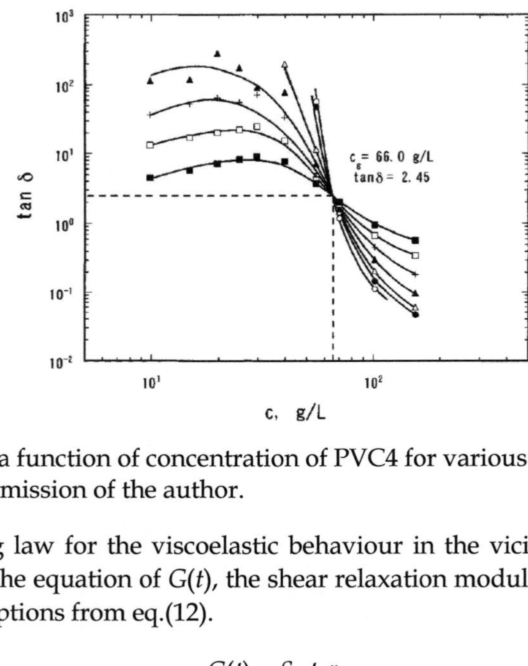

Fig. 7. plot of tan δ as a function of concentration of PVC4 for various frequencies. Reproduced, with permission of the author.

The well-used scaling law for the viscoelastic behaviour in the vicinity of the gel point is expressed in eq.(12) The equation of $G(t)$, the shear relaxation modulus, shown below is one of the deduced descriptions from eq.(12).

$$G(t) = S_g\, t^{-n} \tag{18}$$

S_g is called as the gel strength and has an unusual unit of Pa s^n. n is named the critical relaxation exponent because n determines the stress relaxation rate at the gel point. One may simply understand that S_g is the relaxation modulus at the gel point when the relaxation time t equals 1s. The expression of S_g as $S_g = G(t) \times t^n$ may be of help to understand the physical meaning of S_g. n=0 gives $S_g = G_0$, the elastic modulus, that describes the rigidity of the sample. Also S_g represents the viscosity for n=1. A similar expression can be applied at the gel point for $G'(\omega)$ and $G''(\omega)$.

$$G'(\omega) = \frac{G''(\omega)}{\tan(n\pi/2)} = S_g \omega^n \Gamma(1-n)\cos(n\pi/2) \tag{19}$$

The complex modulus seen in eq.(13) is rewritten to the storage and loss moduli in equation (19). Because the critical exponent n is determinable from tan δ value at the gel point shown in Figure 7, S_g can be calculated from $G'(\omega)$ and $G''(\omega)$ data at the gel point.(N.B. $\delta = n\pi/2$) .

Equation (19) suggests that there exists a crossover of $G'(\omega)$ and $G''(\omega)/\tan(n\pi/2)$ at the gel point. $G'(\omega)$ value at the gel point is obtained by plotting $G'(\omega)$ and $G''(\omega)/\tan(n\pi/2)$ against PVC concentration, and then S_g would be easily calculated. Figure 8 is the plot of this kind for the PVC solutions. Since the solution at the gel point obeys the frequency independence of tan δ, all crossover points appear well at the gel point being consistent with Figure 7.

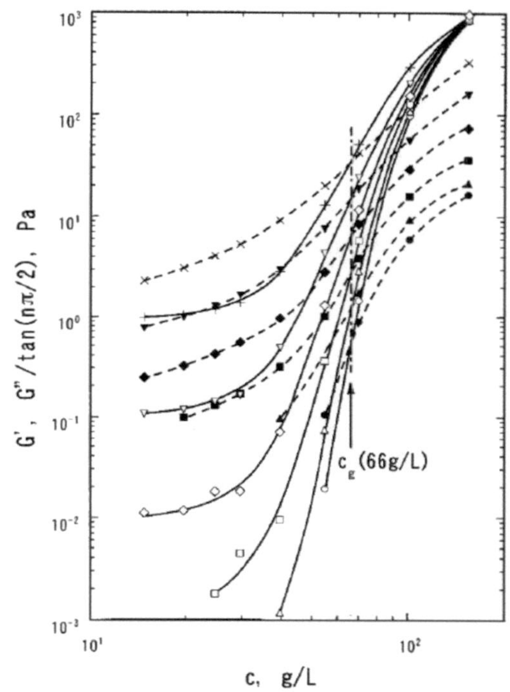

ω (rad/s)	0.316	1	3.16	10	31.6	100
G' (Pa)	○	△	□	◇	▽	+
G"/tan(nπ/2) (Pa)	●	▲	■	◆	▼	×

Fig. 8. Plots of G' and $G''/\tan(n\delta/2)$ against polymer concentration c for the PVC4 system. ω, the frequency was varied from 0.316 to 100 rad/s. n=0.75 was used to calculate $G''/\tan(n\delta/2)$. c_g, gel point is indicated by the arrow. Reproduced, with permission of the author.

2.2.2 Divergence of viscosity

In the previous section, detail of the frequency power law was described for PVC solution as an example of experimental gelation system. Moreover, rheological studies are seen in this section by focusing on how the zero shear viscosity, η_0, diverges when the gelling system goes close to the gel point from the solution of low concentration.

Concerning the divergence of η_0, the scaling law of eq.(10) has already mentioned as a critical behaviour near the gel point. Equation (10) was issued for $p<p_c$, and p can be, for example, the extent of reaction for crosslinking, gelation time, temperature; k is the critical exponent determining the critical characteristics near the gel point. Experimental determination for k is not difficult provided that p_c is known. If p_c is unknown, eq.(20) is useful for the simultaneous determination of p_c and k.

$$\frac{-\eta_0^{-1}}{\mathrm{d}\eta_0^{-1} / \mathrm{d}p} = \frac{p_c - p}{k} \tag{20}$$

Equation (20) is given by transforming eq.(10), then the values of p_c and k are acquired from the slope and intercept of the plot of $-\eta_0^{-1}/(\mathrm{d}\eta_0^{-1}/\mathrm{d}p)$ vs p. Aoki et al. demonstrated that this plot works efficiently to determine p_c and k from the data of $G'(\omega)$ and $G''(\omega)$ of PVC solution.(Li, 1997a)

In their work, three samples of PVC4, PVC9, PVC17 were used for dynamic viscoelasticity measurements of pre-gel samples. Figure 9 shows the plot of η^* vs ω for PVC4 solutions of various concentrations. η^* is derived using eq.(14). Solutions of c lower than 59.9g/L show η^* of the frequency-independence at $\omega \rightarrow 0$, which enable to determine η_0. The range of frequency-independence shifts to lower frequency side as the increase in c. Eventually, the frequency-independent behaviour shifts below the lower limiting of the rheometer, which has already mentioned in Figure 4. Note that c_g of PVC4 solution is 66g/L. In fact, PVC4 of c=62.9g/L, slightly lower than c_g, does not show the frequency independence. Also for solutions of other PVC systems of the concentrations slightly lower than c_g, similar behaviour is observed.

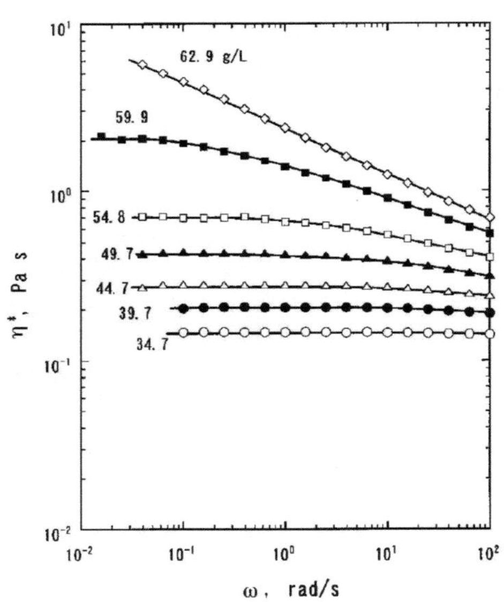

Fig. 9. Complex viscosity η^* as a function of ω for PVC4/DOP samples. The PVC concentration ranges from 34.7 to 62.9g/L as indicated. Reproduced, with permission of the author.

Before constructing the plot of eq.(20), η_0^{-1} was plotted against c as shown in Figure 10 to calculate $d\eta_0^{-1}/dc$. N.B. p corresponds to c in this experiment. Then $-\eta_0^{-1}(d\eta_0^{-1}/dc)^{-1}$ was calculated. Figure 11 shows that linear relations are obtained for $-\eta_0^{-1}(d\eta_0^{-1}/dc)^{-1}$ and c for PVC4, PVC9, PVC17 solutions. The slope of the straight line is equivalent with $-1/k$, the intercept of x-axis is equivalent with c_g. As compared to c_g value obtained by the method of frequency independence of loss tangent(see Fig. 7), c_g value of the method of Fig. 11 varied within a deviation of about ±7%. The variation for k values of these two methods were also within about ±7%, one is obtained from the plot of η_0 vs $(c_g - c)/c_g$ in double logarithmic scale with c_g of Fig. 7 method, the other one is from Fig. 11 method.

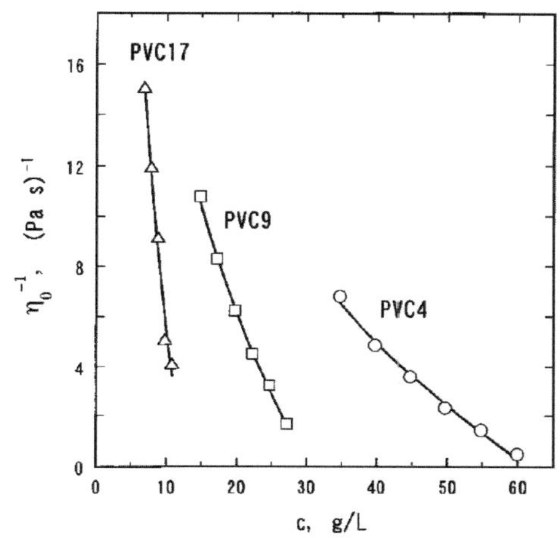

Fig. 10. η_0^{-1} plotted as a function of PVC concentration, c, for PVC4, PVC9 and PVC17 solutions. The solid lines are obtained by fitting to the linear function of log c.

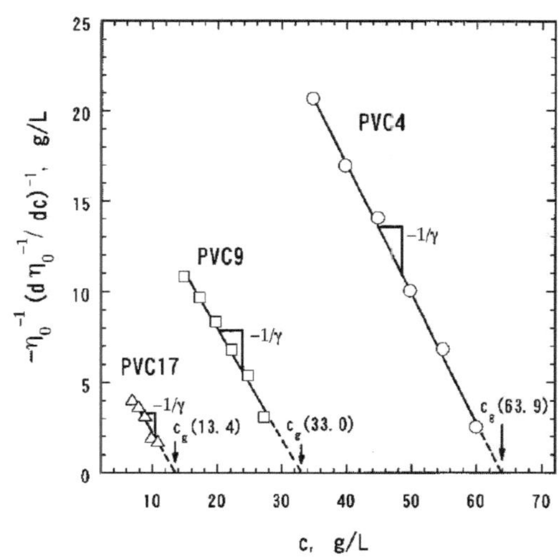

Fig. 11. Plots of $-\eta_0^{-1}(d\eta_0^{-1}/dc)^{-1}$ vs polymer concentration c for PVC4, PVC9, and PVC17 systems. For each gelling system, a linear fitting of the data gives simultaneously the value of exponent k and the critical concentration c_g .

The advantage of Fig. 11 method is the simultaneous determination of c_g and k. However, there is also disadvantage in Fig. 11 method, that is, the error level is high in the determination of $d\eta_0^{-1}/dc$ from Figure 10 in which the ranges of η_0^{-1}/dc and c are not wide enough.

3. Delay of gel point and ring formation

In relation to properties of gels, many research works have been carried out to predict gel point for the gelation of RA_{fa} + $R'B_{fb}$ type polymerisation. RA_{fa} and $R'B_{fb}$ are monomers with functionalities of f_a and f_b respectively. Although the system appears to be limited to the network formed by polymerising monomer, it can be generalised by selecting f_a and f_b appropriately or modifying the reactive species. Flory presented the most simple description of gel point as shown in eq.(21), often called as the classical theory.(Flory, 1941)

$$(f_a - 1)(f_b - 1)\times p_a p_b = 1 \tag{21}$$

A more simple expression has already described in eq.(1). p_a and p_b are the extent of reaction for reactive groups A and B, respectively. In deriving eq.(21), some ideal reactions are considered, in which following two assumptions are used; all like groups have equal reactivities, and there is no intramolecular reaction. Conversely, if unequal reactivity and the intramolecular reaction are included in the polymerisation, the extent of reaction at gel point shifts to higher value than that of eq.(21), and beyond the gel point the equilibrium modulus is lowered. The intramolecular reactions cause ring structures in network which are elastically inactive chains. Accordingly, it is possible to estimate the amount of intramolecular reaction and to quantify the unequal reactivity by establishing suitable model and measuring the gel point shift experimentally.

Recently, Tanaka et. al developed the gelation model for the polymerisation of epoxy and diamine including the effects of unequal reactivity and intramolecular reaction, where the ring-forming parameter was introduced to characterise the competition between intermolecular and intramolecular reactions.(Tanaka, et al., 2009) This section is concerned with the gelation model developed and its application to the polymerisation of polyoxypropylene (POP) diamine and the diglycidyl ether of bisphenol A (DGEBA), RA_4 + $R'B_2$ type polymerisation.

3.1 Gel point in the network formation by RA_4 + $R'B_2$ type polymerisation

A reaction scheme was defined for a POP diamine with four reactive hydrogen atoms, and DGEBA having two epoxy groups. There are two features required for defining the scheme; to follow the progress of the reaction by the extent of reaction, p, and to incorporate the ring-forming parameter as shown later.

In the course of network formation, the primary amine in POP diamine and an epoxy group in DGEBA react with a rate constant k_1, then the secondary amine formed and an epoxy group react with a rate constant k_2 to form the tertiary amine. Hydrogen atoms are categorised as being in states H1, H2 and HR according to the reactions. That is, the hydrogen atoms that belong to unreacted amino groups are defined as H1, the unreacted hydrogen atoms that belong to the semireacted amino groups are defined as H2, and reacted hydrogen atoms are defined as HR. In the reaction of rate constant k_1, two H1 atoms are lost, and one H2 atom and one HR atom are formed. In the reaction of k_2, one H2 atom is lost and one HR atom is formed.

The concentrations of H1, H2, HR atoms and epoxy groups, are written as c_{H1}, c_{H2}, c_{HR} and c_{EP} respectively. Let the concentration of H1 and epoxy groups before the reaction(i.e. t=0) be $c_{H1}{}^0$, $c_{EP}{}^0$, then p_{HR}, the extent of reaction of hydrogen atoms is,

$$p_{HR} = c_{HR} / c_{H1}{}^0 . \tag{22}$$

p_{HR} is the variable to dominate the gelation, and is directly measurable by FT-IR spectroscopy for the reaction of POP diamine and DGEBA.(Mijovic, 1995) Furthermore, p_{HR} can easily be incorporated into the equation of chain growth expressed in terms of the probability of paths shown below.

The rates of change of the concentrations of H1, H2 and HR are given by the following equations;

$$\mathrm{d}c_{H1}/\mathrm{d}t = -2\, k_1 \times c_{H1} \times c_{EP} \tag{23}$$

$$\mathrm{d}c_{H2}/\mathrm{d}t = k_1 \times c_{H1} \times c_{EP} - k_2 \times c_{H2} \times c_{EP} \tag{24}$$

$$\mathrm{d}c_{HR}/\mathrm{d}t = k_1 \times c_{H1} \times c_{EP} + k_2 \times c_{H2} \times c_{EP} \tag{25}$$

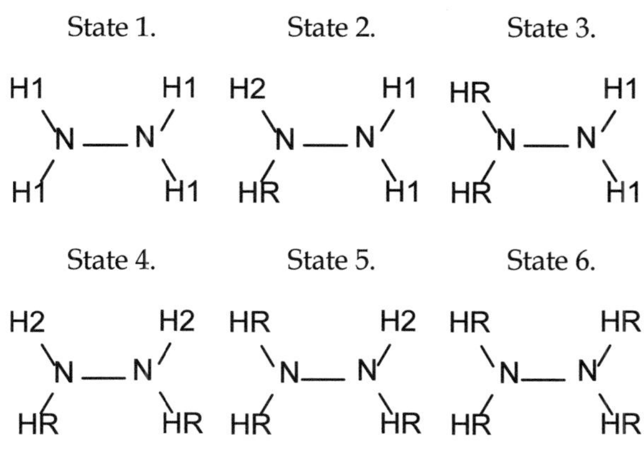

Fig. 12. Classification of the state of reaction for POP diamine unit in the progress of polymerisation. Numbers are placed as State i, i=1-6.

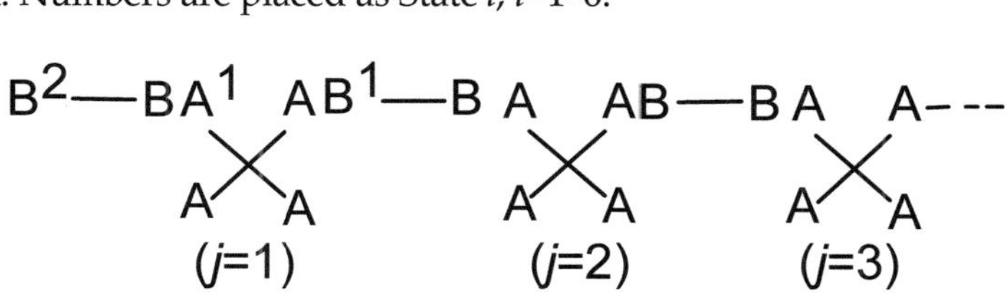

Fig. 13. Linear sequence of reactive units used to show chain growth and to define the size of the ring structure, j. The reactive group A corresponds to H atom of the amino group and the group B to epoxy group.

The progress of the reaction can be monitored by solving the rate equations of eqs.(23-25) with experimental conditions. That is, the relations of p_{H1} vs p_{HR} and p_{H2} vs p_{HR} are obtained with the parameter of ρ (=k_2/k_1), the rate constant ratio, where p_{H1}=(c_{H1}/c^0_{H1}), p_{H2}=(c_{H2}/c^0_{H1}), p_{EP} =(c_{EPR}/c^0_{EP}). p_{H1} decreases monotonously, while p_{H2} increases in the first half then decreases during the progress of polymerisation to completion. ρ is the index of the unequal reactivity; ρ=1 if H1 and H2 are in equal reactivity, the increase in ρ more than 1 means the reactivity of H2 is higher than that of H1.

Six states of the diamine unit can be defined as shown schematically in Figure 12. The classification of the diamine unit shown here is similar to that of Dušek et al.(Dušek, 1975). However, its use to define the gel point is different. The gel point in the study of Dušek et al. is defined as the point of divergence of the weight- average molar mass. Whereas the gel point is derived in terms of the probabilities of continuing of paths in this work so as to incorporate experimentally measurable value of p_{HR} into them.

If the mole fractions of the six states are written as X_i , i=1–6, they can be expressed as functions of p_{H1}, p_{H2} and p_{HR}. Hence, the X_i are functions of p_{HR}.

	State 2		State 3		State 4		State 5		State 6
H atom	H1, H2	HR	H1	HR	H2	HR	H2	HR	HR
Probability*	$\frac{3}{4}$	$\frac{1}{4}$	$\frac{1}{2}$	$\frac{1}{2}$	$\frac{1}{2}$	$\frac{1}{2}$	$\frac{1}{4}$	$\frac{3}{4}$	1
Number of paths	1	0	2	1	2	1	3	2	3

* Probability to chose the specified H atom.

Table 2. Probability of choosing and number of paths for each H atom of amino group in the chain growth.

To define the gel point, the probability of path continuation is introduced. The assumed linear sequence of structural units used to define gelation in the reaction of DGEBA and POP diamine is shown in Figure 13, which follows ARS theory.(Ahmad, 1980) The reactive group A corresponds to an H atom of the amino group and the group B to an epoxy group. The variable j gives the size of the ring. Suppose that the chain in Figure 13 grows from right to left, then γ_a is defined as the probability of a continuing path from a diamine unit attached to a randomly chosen H atom to the next diamine unit. Likewise, γ_b is defined as that of a continuing path from DGEBA unit attached to a randomly chosen epoxy group to the next DGEBA unit.

In order to obtain the expression of network formation, we need to consider the probability of continuing paths from a randomly chosen group to a statistically equivalent group; for example, a path from B^1 to B^2 in Figure 13. Let the fractions of A and B groups be X_a and X_b respectively, then the probability of continuing path, γ can be given as below;

$$\gamma = X_a\gamma_a + X_b\gamma_b \,. \tag{26}$$

In order to derive the expression of γ_a, the probability of chain growth from a diamine unit to next diepoxy unit is considered in terms of the states of diamine unit and the count of the number of continuing paths. If a diamine unit of State 2 is chosen, more specifically, an HR atom in State 2 is chosen, then it gives no paths out. If it is either an H1 or H2 atom, it gives one path out, which is HR in State 2. Because the probability that either an H1 or H2 atom is chosen in State 2 is 3/4 (see Figure 12), the contribution from State 2 to the chain growth is $3X_2/4$; it is given by the product of the number of paths and the probability. The probabilities that individual H atom is chosen and the number of paths out in each state of POP diamine unit are summarised in Table 2. The probability of chain growth from diepoxy to next diamine unit is p_{EP}. Hence, γ_a can be expressed by p_{EP} and the total of the contributions from the respective state of diamine unit.

$$\gamma_a = (\frac{3}{4} X_2 + \frac{3}{2} X_3 + \frac{3}{2} X_4 + \frac{9}{4} X_5 + 3X_6) \times p_{EP} \tag{27}$$

Similarly, γ_b can be expressed using X_i , i=1–6 and p_{EP} .

The calculation results of γ as a functions of p_{HR} are shown in Figure 14. ρ values used for the calculations are 1, 0.5, 0.1. The γ value increases from 0 to 3 as p_{HR} increases from 0 to 1. The result that γ=3 as p_{HR}=1 comes from the fact that functionalities of POP diamine and DGEBA are 4 and 2 respectively; when all the groups are reacted POP diamine units are all in State 6 which has 3 paths of continuation from a randomly chosen reacted amino group.

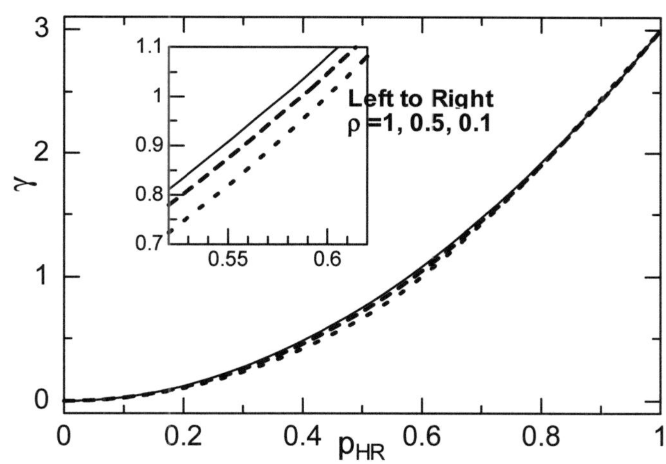

Fig. 14. The results of calculation for γ as functions of p_{HR} with different ρ values; solid thin line: ρ=1, dashed line: ρ=0.5, dotted line: ρ=0.1 The inset figure magnifies near γ=1.

Curves of γ vs p_{HR} depend on ρ, as can be seen in the inset figure, which is the effect of unequal reactivity on the path continuation. p_c, the value of p_{HR} at γ=1 is p_c=0.577 for ρ=1 which agrees with the value calculated from eq.(21) Note that we have p_a=p_b as the experimental condition. As ρ decreases p_c shifts to a higher value, which is confirmed to come from the dependence of the γ_b *vs.* p_{HR} curve on ρ. The shift of p_c is the delay of gel point. The decrease in ρ means less reactions of k_2; that is, formation of State 3, 5 and 6 is delayed. State 5 and 6 brings about bifurcation that leads to the gelation. In other words, the delay of gel point with the decrease in ρ is caused by the delay of formation of State 5 and 6.

3.2 Ring-forming parameter

For an RA_4+$R'B_2$ type polymerisation, competition always occurs between intermolecular and intramolecular reaction. In the competition between intermolecular and intramolecular reaction, let $c_{b,int}$ be the internal concentration of B groups from the same molecule around an A group being on the point of reacting, and let $c_{b,ext}$ be the concentration of B groups from other molecules. Then, the ring-forming parameter λ_b can be given concerning B group as;

$$\lambda_b = c_{b,int} / (c_{b,ext} + c_{b,int}), \tag{28}$$

$c_{a,int}$, and $c_{a,ext}$ can be similarly defined when a B group is about to react with an A group. The definition of λ_a is as follows.

$$\lambda_a = c_{a,int} / (c_{a,ext} + c_{a,int}) \tag{29}$$

The size of the ring structure can be given by j as shown in Figure 13. For example, the smallest ring structure is comprised of a pair of POP diamine and DGEBA units. This structure is regarded as that of j=1. The second smallest ring consists of two of each unit and is regarded as the structure of j=2. In consequence, $c_{b,int}$ is the total of the concentration of each size.

$$c_{b,\,int} = \sum_{j=1}^{\infty} c_{b,int,j} \tag{30}$$

$c_{b,int,j}$, j=1, 2, ---- is the concentration of the B groups to form each size of the ring structure around an A group. Likewise, $c_{a,int}$ can be written by the total of the concentration of the A groups to form each size of the ring structure around the B group.

$$c_{a,\,int} = \sum_{j=1}^{\infty} c_{a,int,j} \tag{31}$$

In counting the concentration $c_{b,int,\,j}$, it is necessary to consider a specific feature of the ring structure; that is, it can be thought of as a chain which end-to-end distance is equal to zero. It is assumed that the distribution of end-to-end distance can be written by independent Gaussian sub-chain statistics of eq.(32) for the polymer chain including POP diamine and DGEBA.

$$P(r) = \left(\frac{3}{2\pi \nu b^2}\right)^{3/2} \exp\left(-\frac{3r^2}{2\nu b^2}\right) \tag{32}$$

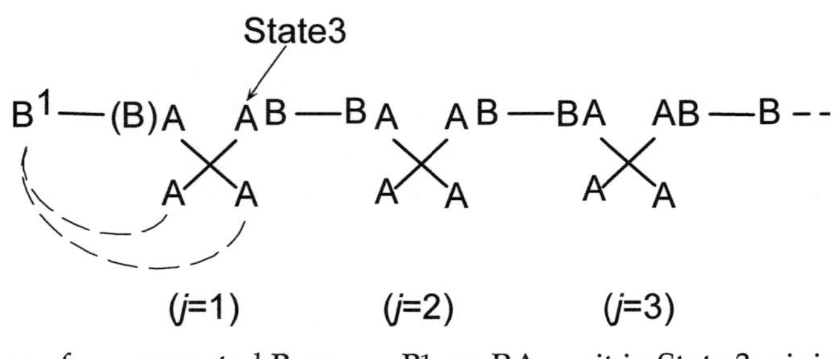

Fig. 15. Occurrence of an unreacted B group, B^1, on RA_4 unit in State 3, giving the possibility of A group reacting intramolecularly to form a ring structure of the smallest size.

$$<r^2>=\nu b^2 \tag{33}$$

r is end-to-end vector. b is the effective bond length of the chain of ν bonds. In particular, if a definition is given on P_{ab} with substituting r=0 and dividing by N_{av} , Avogadro's number, then it becomes a useful parameter in counting the concentration of $c_{a,int,j}$ and $c_{b,int,j}$.

$$P_{ab} = \frac{1}{N_{AV}}\left(\frac{3}{2\pi < r^2 >}\right)^{3/2} \tag{34}$$

Similar approach to describe intramolecular reaction has been reported for RA_3 + $R'B_2$ type polymerisation.(Jacobson & Stockmayer, 1950) P_{ab} means the concentration of intramolecular B groups of j=1 around an A group; it can also be said that it means the concentration of intramolecular A groups of j=1 around a B group.

Consequently, eq.(30) can be calculated as follows;

$$c_{b,int} = \sum_{j=1}^{\infty}\frac{1}{N_{AV}}\left(\frac{3}{2\pi < r^2 >}\right)^{3/2} = P_{ab}\sum_{j=1}^{\infty}\left(\frac{1}{j}\right)^{3/2} = 2.61P_{ab}. \tag{35}$$

In the above calculation Truesdell function was taken into account.(Truesdell, 1945) Similarly, $c_{a,int}$ =2.61P_{ab} can be given and an expression of λ_a is derived;

$$\lambda_a = P_{ab}/c_a^0 \tag{36}$$

where c_a^0 is the concentration of A group before the reaction and $c_a^0=c_{H1}^0$ in this work.

3.3 The gel point expression including ring formation

In order to obtain an expression for γ into which the ring-forming parameter was incorporated, path continuations both from diamine to diamine units and from diepoxy to diepoxy units, and the intramolecular reactions are considered simultaneously. Suppose that the randomly chosen group is H atom in the amino group, i.e. A group. Then, the contribution to the probability of continuing path, γ_a, is considered taking account of the internal concentration according to the states of diamine unit. Hence, we have λ_{ai}, with $c_{a,int,i}$, i=2-6 .

$$\lambda_{a\,i} = c_{a,int,\,i} / (c_{a,ext} + c_{a,int,\,i}) \tag{37}$$

For example, $c_{a,int\ 2}$ is the concentration of intramolecular A group when a B group continuing from POP diamine unit in State 2 is about to react. If there is no ring formation, the diamine unit of State 2 contributes $(3/4)X_2 \times p_{EP}$ to the path continuation. The formation of the ring reduces the contribution by a factor of λ_{a2}; that is, the contribution of the diamine unit of State 2 to γ_a can be expressed as $(3/4)X_2 \times p_{EP}(1-\lambda_{a2})$.

From the similar arguments for λ_{ai}, i=3-6 as well as λ_{bi}, the equations corresponding to eq.(27), and hence eq.(26) can be given to be solved under γ=1. Figure 15 illustrates the possible intramolecular reaction to form a ring structure of the smallest size when a randomly chosen unreacted A group is on an RA_4 unit in State 3 to calculate $c_{a,int,\,3}$.

As a consequence of the network model, the relation between p_c and P_{ab}/c_{H1}^0 (=λ_a) is obtained (see Figure 16.) which tells the internal concentration of the reactive group A shown by eq.(25)

at the moment that the solution of POP diamine and DGEBA comes to gel point of p_c . For ρ=1, Figure 16 shows p_c=0.577 at P_{ab}/c_{H1}^0=0, which corresponds with the gel point of eq.(21). p_c increases with (P_{ab}/c_{H1}^0), which implies that the more intramolecular reactions cause the delay of gel point, as P_{ab} determines how likely the growing polymer chain forms a ring structure.

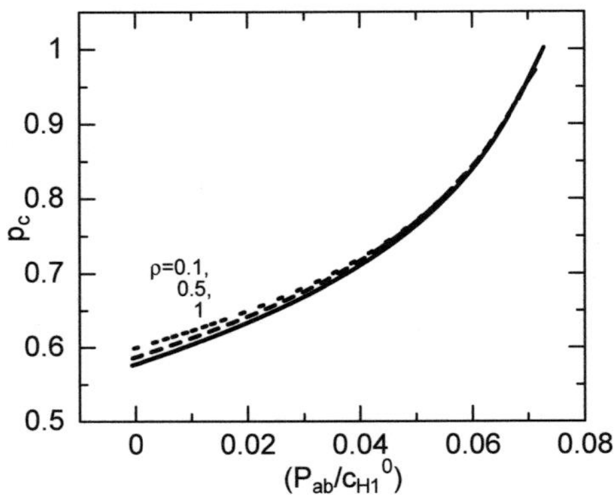

Fig. 16. The relation between p_c and (P_{ab}/c_{H1}^0) calculated for different ρ values. ρ=0.1, 0.5, 1 are used.

p_c of Figure 16 can be applied to the gel points of experimentally measured values to fined the ring-forming parameter of λ_a; note that c^0_{H1} is also an experimental value evaluated from the concentration of POP diamine in the solution. Figure 17 shows the results for POP diamine of various chain lengths; the plots lie on straight lines through the origin which corresponds with the gel point expressed by eq.(21). From the slope of the plot P_{ab} is determined. Because P_{ab} is a function of $<r^2>$, the chemical structure of monomer reactants determine the value of P_{ab}.

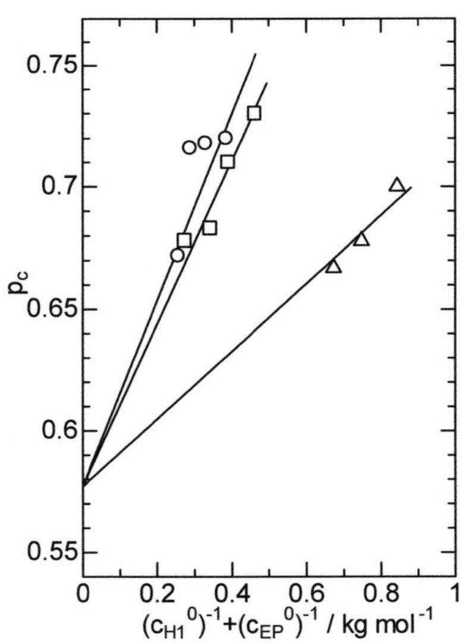

Fig. 17. p_c plotted against $(c_{H1}^0)^{-1}+(c_{EP}^0)^{-1}$ for polymerisation systems of various POP diamines Nominal molecular weight for the POP chain: M_n=230(Circle), 400(Square), 2000(Triangle) .

4. Network formation by end-linking of star polymer

For entangled linear flexible polymers, the viscoelastic relaxation in response to mechanical deformation is the long-standing problem and has been extensively studied. (Ferry, 1970) The response is solid-like at high frequencies for a given temperature, but often some relaxation occurs due to localised motion. At lower frequencies the system relaxes due to cooperative motion of the segments in the chain backbone. The lower the frequency is, the larger the size of the chain backbone that can relax its conformation in response to the deformation. The situation written above is of more interest for polymers having bifurcation. Furthermore, the extension of the idea to the connection of polymers having bifurcation is also possible.

The end-linking of star polymers is an example of the connection of polymers having bifurcation, and generates a branched polymer of complex architecture. Increasing the connectivity extent (p) leads to the formation of larger branched polymers until at a critical value (p_c) a gel is formed. Viscoelastic behaviour of the star polymer is also of importance in terms of polymer chain dynamics of the melt and concentrated solution. Some recent experimental studies clarifies the molecular origins of the position of plateau and terminal flow for star polymers having arm chains which molecular weight is higher than the entanglement molecular weight.

As for the end-linking, an experimental work has already been published for the viscoelastic relaxation of poly(propylene sulfide) star by Nicol et al., they discussed the relation of p and viscoelasticity in detail.(Nicol et al, 2001) It is a 3-arm star polymer and can be end-linked by the reaction with hexamethyl diisocyanate(HMDI) in the presence of a small amount of catalyst of dibutyl tin dilaurate. Also, it is regarded as a network formation of RA_3 + $R'B_2$ type reaction.(The values are functionalities.) The characteristic features to remark for poly(propylene sulfide) star, abbreviated as PPS star, is low T_g(≈ -37°C).

In relation to the experimental studies of the star polymer, an advanced calculation method for the viscoelasticity has been presented recently. The method based on the theory of tube model can be applied to the polymers of complex architecture. It is called as Branch-On-Branch Rheology (BOB) and has an advantage where branched polymers of different species can be mixed.(Das, 2006a; Das, 2006b) In addition, the polymer generate routine was incorporated in BOB to produce branched polymer by end-linking star polymers, which makes it possible to calculate $G'(\omega)$ and $G''(\omega)$ data for polymers of complex architecture.(Tanaka, 2008) This section is concerned with the comparison of $G'(\omega)$ and $G''(\omega)$ data between those calculated and experimental values which were supplied by the author of the original article. The data shown here are for the end-linked polymers with p lower than p_c, the gel point of connectivity. That is, those are viscoelastic behaviour of pre-gel state.

4.1 Poly(propylene) sulfide star and end-linking

Three samples of PPS star are taken as the objects of BOB calculation. The code names are T5, T6, T7. The weight-average molecular weights (M_w, in the nominal unit of kg/mol) of PPS star are 18.9, 35.5, and 71.2 for T5, T6, and T7 respectively. Numbers of repeating unit in the arm, n_a, are 72, 128, 263 for T5, T6, T7. The entanglement molecular weight, M_e of PPS is 4000. In the preparation of PPS star, it is inevitable that the dimer of the star polymer is produced by the coupling reaction of two -SH groups located at the chain ends. The coupling ratio ratios are 13, 12, 17% for T5, T6, T7.

Concerning the end-linking reaction, hydrogen atoms of -SH groups in the chain ends can be classified as H(1), H(2), H(3) and H(R), as shown in Figure 18. The classification enables us to describe the progress of reaction and the variation of the state of reaction of PPS star. Four states can be realised for PPS star according to the reaction; the mole fraction of each state was defined as x_i, i=1, 2, 3, 4. In the macromolecule generated through the reaction, PPS star of the respective state will be allocated as follows; State 2 is in the end of the macromolecules, State 3 is in the backbone, State 4 is in the branching.

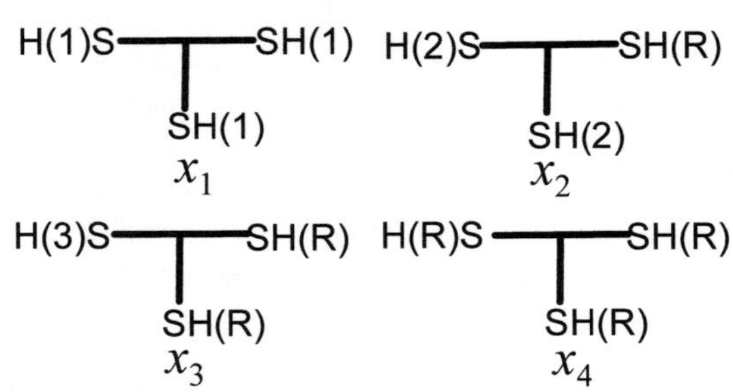

Fig. 18. Schematic representation for the state of reaction of PPS star unit. Reactive hydrogen atoms were classified as H(1), H(2), H(3) and H(R). x_1–x_4 are fractions of the respective state of PPS star unit.

As for the hydrogen atom, H(1) is an unreacted atom, H(R) is the reacted atom. H(2) and H(3) are also unreacted hydrogen atoms but included in PPS star having one and two H(R) respectively. The value of p can be given as follows with the concentration of the hydrogen atoms, c_{H1}, c_{H2}, c_{H3}, c_{HR}.

$$p = c_{HR} / (c_{H1} + c_{H2} + c_{H3} + c_{HR}) \tag{37}$$

The relationships between the number of the hydrogen atoms and the state of reaction can be written as; three H(1) are in State 1, two H(R) are in State 2 and one H(3) is in State 3. Also, H(R) are one in State 2, two in State 3 and three in State 4. As a practical meaning, p is equivalent with the experimentally determinable value of the connectivity extent. Further, rate equations were constructed for reactions from State 1 to State 2, from State 2 to State 3, from State 3 to State 4 with giving rate constants of k_1, k_2, k_3 for respective reaction.

By solving the rate equations, the relation of x_1 and p was obtained with choosing suitable value for the ratio of rate constants, ρ_0 and ρ_1, as a fitting parameter.

$$\rho_0 = k_2/k_1 \,, \rho_1 = k_3/k_2 \tag{38}$$

Figure 19 shows the result Of calculation for the relation of x_1 and p for PPS star unit of State 1. The experimental value of x_1 was also displayed in the figure; note that x_1 is the fraction of PPS star monomer and can be estimated with SEC experiment. It clearly shows that ρ_0=0.667 and ρ_1=0.5 reproduced well the experimental result of x_1. These values correspond with the ratio of the number of unreacted -SH group, which implies the reaction between -SH group and isocyanate group took place in proportional to the number of unreacted -SH group. Namely, the effect of induced unequal reactivity is negligible for the end-linking of PPS star. In the calculation of viscoelasticity shown below, the effect of unequal reactivity was not taken into account.

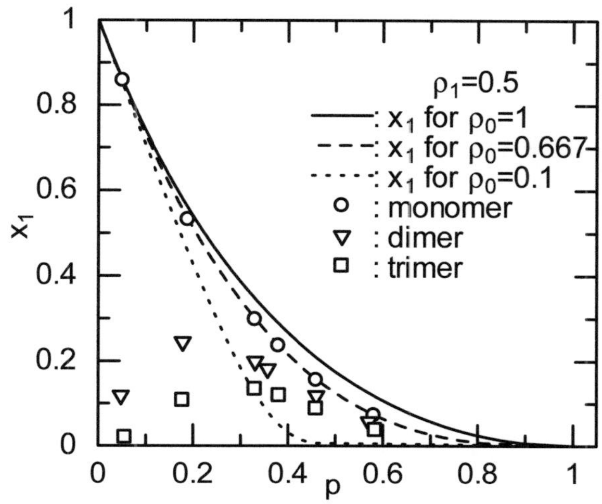

Fig. 19. Results of calculation on the mole fraction for PPS star unit of State 1 defined in Figure 18. Data obtained with ρ_1=0.5, ρ_0=0.1, 0.667 and 1 are displayed. The weight fractions of the first three oligomers in the end-link reaction determined by SEC measurement were added for comparison.

4.2 Evolution of viscoelasticity by end-linking

The experimental results were shown in Figure 20(a), (b), (c) as the master curves of $G'(\omega)$ and $G''(\omega)$ for PPS star of T5, T6, T7 before end-linking, respectively. From DSC measurement, glass transition temperature (T_g) of PPS star was determined around 236K. The maximum peaks for $G''(\omega)$ which correspond with T_g appear in the master curves of higher frequency regions for T6 and T7. As lowering the frequency, both $G'(\omega)$ and $G''(\omega)$ gradually decrease and come to the plateaus.

The plateau modulus, G_0, was calculated using the equation written below, then viscoelasticity curve was calculated by BOB rheology for three arms star polymer, which results were also shown in Figure 20.

$$G_0 = \frac{4}{5}\frac{dRT}{M_e} \tag{39}$$

d is density of PPS star, R is gas constant, M_e the entanglement molecular weight.

The effect of coupling reaction was considered in the viscoelasticity calculation of three arms star polymer. As mentioned above, the preparation of the star polymer accompanies the coupling reaction which mainly generates the dimer. It is obvious that the dimer is in the architecture of H-polymer, therefore the calculations were carried out with mixing three arms star and H-polymer. The mixing ratio was chosen with reference to the coupling ratio of PPS star. As for the specification of H-polymer, the molecular weights of the side arm and cross bar, M_{arm} and M_{bar} respectively, were taken as follows;

$$M_{arm} = M_{unit} \times n_a \text{ , } M_{bar} = M_{unit} \times 2n_a \tag{40}$$

where M_{unit} is the molecular weight of the repeating unit.

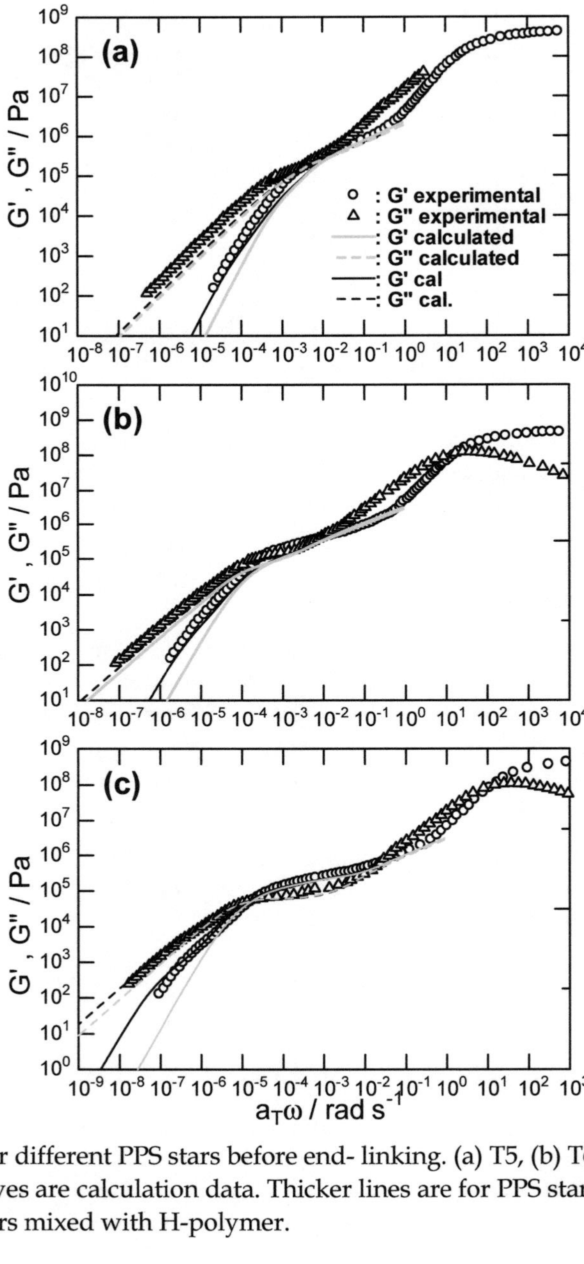

Fig. 20. G' and G'' for different PPS stars before end- linking. (a) T5, (b) T6, (c) T7. Plots are measured data, curves are calculation data. Thicker lines are for PPS stars alone, thinner lines are for PPS stars mixed with H-polymer.

As a result of BOB calculation before mixing H-polymer (the grey thicker lines in Figure 20 which are denoted as PPS stars alone), the experimental and calculation data agreed well around the plateaus for T5, T6, T7. However, the difference between the experiment and calculation became distinguishable as the increase in n_a in the frequency range lower than the plateau.

Contrary to the data of PPS star alone, for those of H-polymer mixing (the black thinner lines), the terminal flows shifted to lower frequency sides in comparison with those of PPS star alone; the shift increased with the increase in n_a. It is concluded that the reproducibility between the experimental and calculation data was improved by the mixture of H-polymer.

The experimental result of T7 after end-linking was shown in Figure 21 for the sample having the connectivity extent of 0.41. The calculation data of p=0.41 were added to the figure. The master curve of the experimental data was obtained for the reference temperature of T_{ref}=237K and for the frequency range ($a_T\omega$) of $10^{-11} \le a_T\omega \le 10^{-2}$ rad s^{-1} . Any clear plateau region was not shown for $G'(\omega)$. As for $G''(\omega)$, two ambiguous plateau regions appeared around $a_T\omega = 10^{-4}$ and 10^{-8} rad s^{-1} . Although they are ambiguous, the existence of two or more plateaus has supported for other polymers having complex architectures. The calculation data for end-linked star polymer reproduced the plateau region well. In addition to T7, good agreement for plateau regions were seen between the experimental and calculation data for other PPS stars.

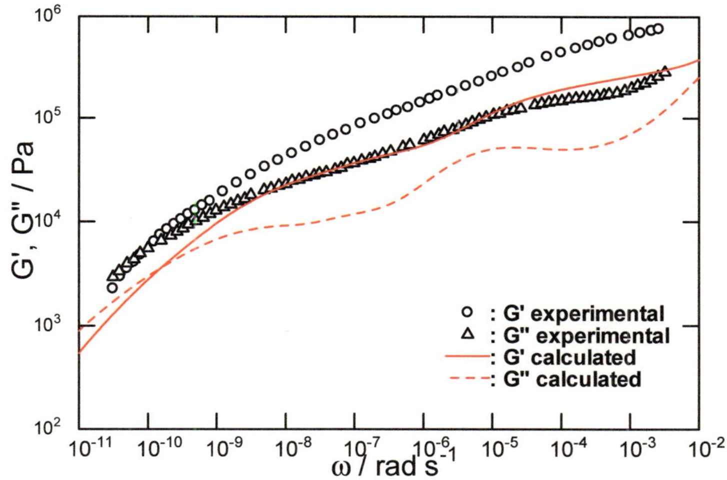

Fig. 21. G' and G'' for end-linked PPS star (T7) at p=0.41. (p: the connectivity extent)

The prediction power of BOB calculation was evidenced by the results shown above for viscoelasticity over wide frequency range. However, the viscoelasticity calculation is presently limited for pre-gel state. It has already found that the calculation values inevitably disagree with $G'(\omega)$ and $G''(\omega)$ data of PPS star when p approaches to the gel point determined by eq.(21). Furthermore, the architecture of the branched polymer becomes complex as the increase in functionalities for RA_{fa} + $R'B_{fb}$ type reaction, which BOB does not cover with. These are future works toward the comprehensive establishment of the calculation methods for the gelling system.

5. Concluding remarks

The rheological method is suitable very much to capture the properties of gel, taking account that rheology is mainly concerned with mechanical properties for substance under deformation. And therefore, research works of a huge number are seen for rheology of sol – gel transition. Of these papers, those related with power law dependences between physical parameters and the variables to control the transition are introduced in Section 2. The phenomenology of the relaxation pattern seems to be completed and is accepted widely. However, little is known yet about the molecular origin of the experimental data of viscoelasticity. Theoretical approach to the power law dependences is expected to help understanding the molecular origin. Such understanding would be desirable from a fundamental point of view.

Furthermore, the molecular scheme for the gel point prediction and viscoelasticity calculation in the course of the network formation were described in Section 3 and 4, respectively. Although some simpler models are in demand, the frameworks currently used are too complicated to use conventionally. However, the effect of unequal reactivity on the delay of gel point could be derived by drawing the detailed molecular scheme. Conversely, it is necessary to set the model up to details to meet with the realistic experimental data. Such molecular parameters allows us to prepare materials near the gel point with a wide range of properties for applications, like adhesives, absorbents, vibration dampers, sealants, membranes etc. With suitable design, it will be possible to control network structures, relaxation character, and then mechanical properties to the requirements.

6. References

Ahmad, Z.; Stepto, R.F.T. (1980) Approximate theories of gelation, Colloid Polym. Sci., Vol.258, 663-674.

Almdal, K.; Dyre, J.; Hvidt, S.; Kramer, O.(1993) Towards a Phenomenological Definition of the Term 'Gel', *Polym. Gels. Networks*, Vol.1, 5-17.

Brauner, U. (1979) Search for critical exponents at the gel-to-sol transition, *Makromol. Chem.*, Vol.180, 251-253.

Bromberg, Lev; Temchenko, Marina (1999) Self-Assembly in Aqueous Solutions of Poly (ethylene oxide)-b-poly(propylene oxide)-b-poly(ethylene oxide)-b-poly(vinyl alcohol), *Langmuir*, Vol.15(25), 8633-8639.

Caillard, R.; Remondetto, G. E.; Subirade, M. (2010) Rheological investigation of soy protein hydrogels induced by Maillard-type reaction, *Food Hydrocol.*, Vol.24(1), 81-87.

Cates, M. E. (1985) Brownian dynamics of self-similar macromolecules , *Journal de Physique* (Paris), 46(7), 1059-1077.

Clark, A. H., Ross-Murphy, S. B. (1987) Structural and mechanical properties of biopolymer gels, *Adv. Polym. Sci.*, Vol.83, pp.57-192

Das C.; Inkson N. J.; Read D. J.; Kelmanson, M. A.; McLeish, T. C. B. (2006) Computational linear rheology of general branch-on-branch polymers, *J. Rheol.* 50(2), 207-219.

Das C.; Read D. J.; Kelmanson, M. A.; McLeish, T. C. B. (2006) Dynamic scaling in entangled mean-field gelation polymers, *Phys. Rev. E.* 74, 011404-1 – 011404-10.

Dušek, K Ilavskŷ, M. Lunňák. S. (1975) Curing of epoxy resins. I. Statistics of curing of diepoxydes with diamines, *J. Polym. Sci. Polym. Symp.*, Vol.53, 29-44.

Ferry J. D. (1970) Viscoelastic properties of polymers, 2nd ed., Wiley, New York.

Flory, P. J. (1941) Molecular Size Distribution in Three Dimensional Polymers. I. Gelation., *J. Am. Chem. Soc.*, Vol.63, pp. 3083-3090.; Molecular Size Distribution in Three Dimensional Polymers. II. Trifunctional Branching Units., *ibid.* Vol.63, pp. 3091-3096.; Molecular Size Distribution in Three Dimensional Polymers. III. Tetrafunctional Branching Units., *ibid.* Vol.63, pp. 3096-3100.

Gordon, M., Ross-Murphy, S. B. (1975) The structure and properties of molecular trees and networks, *Pure Appl. Chem.*, Vol.43, pp.1-26.

Gordon, M.; Torkington, J. A.(1984) Scrutiny of the critical exponent paradigm as exemplified by gelation, *Pure. Appl. Chem.*, Vol.53, 1461-1478.

Guenet, J-M. (1992) *Thermoreversible gelation of polymers and biopolymers*, Academic Press, ISBN 0-12-305380-5, London

Hossain, K. S.; Miyanaga, K.; Maeda, H.; Nemoto, N. (2001) Sol-Gel Transition Behavior of Pure ι-Carrageenan in Both Salt-Free and Added Salt States, *Biomacromolecules*, Vol.2(2), 442-449.

Hsu, Shan Hui; Jamieson, Alexander M. (1990) Viscoelastic behavior at the thermal sol-gel transition of gelatin, *Polymer*, Vol.34(12), 2602-2608.

Jacobson, H.; Stockmayer, W. H. (1950) Intramolecular Reaction in Polycondensations, J. Chem. Phys., Vol.18, 1600-1606.

Kjøniksen, A. L.; Nyström, B. (1996) Effects of Polymer Concentration and Crosslinking Density on Rheology of Chemically Cross-Linked Poly(vinyl alcohol) near the Gelation Threshold, *Macromolecules* (1996), Vol.29(15), 5215-5222.

Kjøniksen, A.-L.; Nyström, B.; Lindman, B. (1998) Dynamic Viscoelasticity of Gelling and Nongelling Aqueous Mixtures of Ethyl(hydroxyethyl)cellulose and an Ionic Surfactant, *Macromolecules* , Vol.31(6), 1852-1858.

Koike, A.; Nemot, N.; Watanabe, Y.; Osaki, K. (1996) Dynamic Viscoelasticity and FT-IR Measurements of End-Crosslinking α, ω-Dihydroxyl Polybutadiene Solutions near the Gel Point in the Gelation Process, *Polym. J.*, Vol.28, 942-950.

Kwak,S. Y.; Choi, J.; Song, H. J. (2005) Viscoelastic Relaxation and Molecular Mobility of Hyperbranched Poly(E-caprolactone)s in Their Melt State, *Chem. Mater.* Vol.17, 1148 -1156.

Li, L.; Aoki, Y. (1997) Rheological Images of Poly(vinyl chloride) Gels. 1. The Dependence of Sol−Gel Transition on Concentration, *Macromolecules*, Vol.30, 7835-7841.

Li, L.; Uchida, H.; Aoki, Y.; Yao, M. L. (1997) Rheological Images of Poly(vinyl chloride) Gels. 2. Divergence of Viscosity and the Scaling Law before the Sol−Gel Transition, *Macromolecules*, Vol.30, 7842-48.

Muthukumar, M. (1985) Dynamics of polymeric fractals, *J. Chem. Phys.* 83, 3161-3168

Michon, C.; Cuvelier, G.; Launay, B. (1993) Concentration dependence of the critical viscoelastic properties of gelatin at the gel point, *Rheol. Acta*, Vol. 32(1), 94 - 103.

Mijovic, J.; Andjelic, S. (1995) A study of reaction kinetics by near-infrared spectroscopy. 1. Comprehensive analysis of a Model Epoxy/Amine System, Macromolecules, Vol.28, 2787-2796.

Miller, D. R., Macosko, C. W. (1978) Average Property Relations for Nonlinear Polymerization with Unequal Reactivity, *Macromolecules*, Vol.11, pp.656-662.

Nakamura, K.; Harada, K.; Tanaka, Y. (1993) Viscoelastic properties of aqueous gellan solutions: the effects of concentration on gelation, *Food Hydrocolloids*, Vol.7, 435-447.

Nicol, E.; Nicolai, T.; Durand, D. (2001) Effect of Random End-Linking on the Viscoelastic Relaxation of Entangled Star Polymers, *Macromolecules*, Vol.34, 5205-5214.

Nicolai, T.; Randrianantoandro, H.; Prochazka, F.; Durand, D. (1997) Viscoelastic relaxation of polyurethane at different stages of the gel formation. 2. Sol-gel transition dynamics, *Macromolecules*, 30(19), 5897-5904.

Nijenhuis, K. (1997) Thermoreversible Networks, *Adv. Polym. Sci.*, Vol.130, ISBN 3-540-61857-0.

Peyrelasse, J.; Lamarque, M.; Habas, J.P.; Bounia, N. E. (1996) Rheology of gelatin solutions at the sol-gel transition, Phys. Rev. E, Vol.53, 6126-6133.

Stauffer, D. (1981) Can percolation theory be applied to critical phenomena at gel points., *Pure. Appl. Chem.*, Vol.53, 1479-1487.

Stepto, R. F. T. (Ed.), (1998) *Polymer Networks – Principles of their Formation Structure and Properties*, Blackie Academic & Professional, ISBN 0-7514-0264-8, London.

Tan, H.; Moet, A.; Hiltner, A.; Baer, E. (1983) Thermoreversible gelation of atactic polystyrene solutions, *Macromolecules*, Vol.16, 28-34.

Tan, L.; Liu, S.; Pan, D. (2009) Viscoelastic Behavior of Polyacrylonitrile/Dimethyl Sulfoxide Concentrated Solution during Thermal-Induced Gelation, *J. Phys. Chem. B*, 113(3), 603-609.

Tanaka, Y. (2009) Viscoelastic Behaviour for End-Linking of Entangled Star Polymer: Application of the Calculation for Branched Polymer to End-Linked Polymer, *Nihon Reoroji Gakkaishi*, Vol.37, 89-95.

Tanaka, Y.; Xin Yue, J.; Goto, M. (2011) Thermoreversible Sol-Gel Transition of Aqueous solution of Sodium Polyacrylate cross-linked by Aluminium Ions: Derivation of Gel Strength for Aqueous Solution, *Proceedings of 12th IUMRS Int. Confer. Asia*, Taipei, Sep, 2011.

Truesdell, C.(1945) On a Function Which Occurs in the Theory of the Structure of Polymers, Annals of Mathematics, Vol.46, 144-157.

Valentova, H.; Bouchal, K.; Nedbal, J.; Ilavsky̌, M., (1999) Gelation and dynamic mechanical behavior of liquid crystalline networks, J. Macromol. Sci., Phys., B38(1 & 2), 51-66.

Vargas, M. A.; Manero, O. (2011) Rheological characterization of the gel point in polymer-modified asphalts, *J. Appl. Polym. Sci.*, Vol.119(4), 2422-2430.

Vilgis, T. A.; Winter, H. H. (1988) Mechanical selfsimilarity of polymers during chemical gelation, *Colloid Polym. Sci.*, 266, 494-500.

Winter, H. H. (1987) Evolution of rheology during chemical gelation, *Prog. Colloid Polym. Sci.*, 75, 104-110.

Winter, H. H.; Mours, M. (1997) Rheology of Polymers Near Liquid-Solid Transitions, *Adv. Polym. Sci.*, Vol.134, 165-234. ISBN 3-540-62713-8.

3

Polymer Gel Rheology and Adhesion

Anne M. Grillet, Nicholas B. Wyatt and Lindsey M. Gloe
Sandia National Laboratories,
USA

1. Introduction

Polymer gels are found in many applications ranging from foods (Ross-Murphy, 1995; Tunick, 2010) and drug delivery (Andrews & Jones, 2006) to adhesives (Creton, 2003) and consumer products (Solomon & Spicer, 2010). By manipulating the gel's microstructure, a wide variety of physical properties can be achieved ranging from hard rubbery plastics to soft hydrogels. Silicone-based polymer gels in particular have found wide utilization in consumer products ranging from medical implants to cooking utensils. Here we will discuss methods of characterizing polymer gels using rheological techniques to probe their adhesion and mechanical response. Further, we will link the observed adhesion and mechanical behavior to the gel microstructure.

Polymer gels are crosslinked networks of polymers which behave as viscoelastic solids. Because the polymer network is crosslinked, the gel network consists of one very large branched polymer which spans the entire gel. While gels can be soft and deformable, they also hold their shape like a solid. Depending on the physical structure of the polymer network, polymer gels can be classified as strong, weak or pseudo gels (Ross-Murphy, 1995). Chemically crosslinked polymer gels are considered strong gels. The crosslinks are permanent and cannot be reformed if broken. Weak gels contain crosslinks which can be broken and reformed such as colloidal gels and some biopolymer gels (Spicer & Solomon, 2010; Richter, 2007). Entangled polymer systems are sometimes referred to as pseudo gels because, over a range of time scales, physical entanglements between polymer chains mimic chemical crosslinks giving these materials gel-like properties (Kavanagh & Ross-Murphy, 1998). However, the equilibrium response of a pseudo gel to a constant applied stress is to flow like a fluid.

Polymer gel properties can be controlled by manipulating the microstructure of the polymer backbone and the surrounding liquid, if any. The strength of a gel, which is characterized by the equilibrium modulus, is generally proportional to the density of crosslinks with stiffer gels having a higher density of crosslinks (Gottleib et al., 1981). A gel can be made softer by increasing the spacing between crosslinks either by increasing molecular weight of the polymer chain connecting the crosslinks or diluting the gel with a liquid. Liquid in the gel which is not part of the crosslinked network is referred to as the sol and may consist of a solvent such as water, short chain polymers or long entangled polymers. The crosslinked polymer network is frequently referred to as the gel. Alternately, defects can be added to the network. For a given crosslink density, an ideal end-linked polymer gel where all polymer chains are connected at both ends to crosslinks and all crosslinks are connected fully to the polymer network will have the highest modulus (c.f. Figure 1a) (Patel et al., 1992). If there is an imbalance between the number of polymer chains and crosslinker, then defects are introduced into the network such as loops and dangling ends (c.f. Figure 1b), which results in a softer gel. Gels which are formed by random processes such as irradiation will form networks with many defects (c.f. Figure 1c). Thus, for a given application, there are many ways to adjust the properties of a polymer gel to optimize performance by controlling gel microstructure and processing conditions.

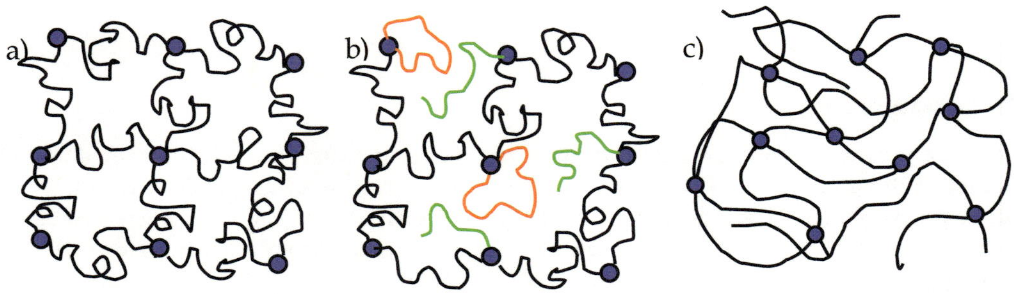

Fig. 1. Diagrams of polymer gel microstructures: a) ideal end-linked polymer gel; b) end-linked gel with dangling ends (in green) and loop defects (in orange); c) randomly crosslinked polymer gel.

The complex structure of a polymer gel dictates that a gel's response to external forces varies widely depending on the time scale of the application of the force. At a basic level, a gel is a collection of polymers. In dilute solution, a polymer chain has a spectrum of relaxation times that defines how quickly the polymer can relax from a deformation (Larson, 1988). Deformations stretch and align segments of the polymer, reducing the number of available conformations and hence reducing the polymer's entropy. Random Brownian motion drives the polymer to increase its conformational entropy and reduce the stored elastic stress. The ends of the polymer can rearrange very quickly but the middle of the polymer is constrained and must wait for the ends to relax before it can relax. The longest relaxation time of a free polymer, which determines the overall rheological behavior, is controlled by the molecular weight of the polymer and the viscosity of the surrounding fluid.

For polymer gels, the effective molecular weight is infinite, as is the longest relaxation time meaning the network will never completely relax from a deformation. But, unlike a purely elastic solid, polymer gels can still internally rearrange and dissipate energy resulting in a viscoelastic character. This is especially true of polymer gels where the length of the polymer chain, either in the solvent or as a part of the gel network, is large enough to allow the polymer to physically entangle with itself (Llorente & Mark, 1979; Patel et al., 1992). Figure 2 illustrates several relaxation mechanisms from an affine deformation. The polymers in the sol will relax the most quickly because their chain ends are unconstrained, followed by dangling polymers which have only one end attached to the network (c.f. Figure 2c). A polymer chain within the gel network will relax more slowly than a free polymer in solution because both ends of the chain are constrained at the crosslinks (c.f. Figure 2d). The final deformation of the gel is determined by its equilibrium modulus.

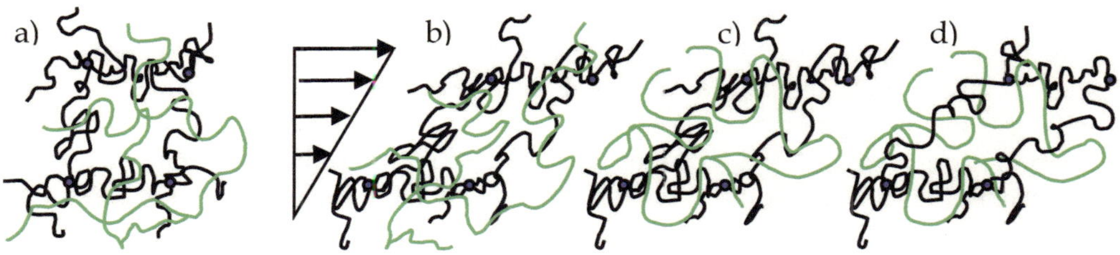

Fig. 2. Stages of relaxation in diluted polymer gel: a) Undeformed gel; b) short times: gel (black) and sol (green) polymers are stretched and aligned; c) intermediate times: sol polymers (in green) relax but gel network (black) is still deformed; d) long times: polymer in gel network relaxes but retains an equilibrium deformation.

The viscoelastic nature of polymer gels plays an important role in their adhesion properties. Adhesive properties of polymeric materials are fundamental to diverse industrial applications. Adhesives have been applied to sophisticated technologies such as nanotechnology, micro-electronics, and biotechnology (Moon et al., 2004). Newtonian liquids make poor adhesives because they flow under sustained forces such as gravity and would not stay in place. It can require a lot of work to peel two surfaces held together by a viscous Newtonian liquid apart due to the high internal friction which dissipates energy as the liquid flows. However, the liquid will leave a residue when the surfaces are separated which is not desirable. At the other extreme, stiff elastic rubbers tend to separate rapidly from a surface because they cannot deform and do not have many internal mechanisms to dissipate energy which leads to poor adhesive performance. Soft polymer gels can have excellent adhesion properties due to both the elastic and viscous properties (Zosel, 1991; Lenhart, 2006; Andrews & Jones, 2006). The gel does not flow or creep under small stresses allowing it to stay where it is applied. As surfaces bonded with a gel are peeled apart, the gel deforms, but because of the internal energy dissipation mechanisms (e.g., physical entanglements, network defects or solvent), only a fraction of the applied energy is stored as elastic energy in the gel network. Thus it requires more work to remove the gel from a surface than for an elastic material of the same equilibrium modulus. When the gel does separate, if the gel modulus is higher than the interfacial strength, it will not leave a residue on the surface.

2. Rheology of polymer gels

Today's modern rheometers allow the precise measurement of a complex material's response to an applied force (stress) or deformation (strain). Historically, rheometers were categorized as stress controlled (applies a force and measures the resulting deformation) or strain controlled (applies a deformation and measures the resulting force) (Macosko, 1994). Advances in instrument hardware and control have resulted in versatile instruments which can perform both types of tests. Strain controlled instruments are more expensive, but they can accurately probe higher oscillation frequencies and do not require frequent inertial calibration (Kavanagh & Ross-Murphy, 1998).

To demonstrate various aspects of gel rheology, a commercial fluorosilicone gel (Dow Corning DC4-8022) is used as an example. This platinum-catalyzed silicone gel is only lightly crosslinked and contains about half by weight of sol diluting the gel network.

Though quite soft, this is still considered a strong gel because the chemical crosslinks are permanent. The viscoelastic nature of these soft gels is readily apparent and contributes to a rich dynamic response. A series of gels varying from extremely soft to hard were examined (Table 1). The equilibrium modulus varied over almost an order of magnitude while keeping the soluble fraction fixed at 45 wt%. Each of these fluorosilicone gels has a glass transition temperature of -65°C indicating that the chemical backbones of the gel polymers are the same. For rheology testing, fluorosilicone gel is placed between parallel plates of a TA Instruments AR-G2 rheometer and cured at 82°C for 24 hours. The gel is then cooled to 25°C prior to further testing.

	G′ at 0.01Hz
High	1060 Pa
Medium	471 Pa
Medium-Low	229 Pa
Low	157 Pa

Table 1. Summary of fluorosilicone gel equilibrium modulus.

As a solution of polymer undergoes the crosslinking reaction either through chemical reaction or irradiation, the average molecular weight of the polymer grows. The gel point can be defined phenomenologically as the critical transition point between when the material is classified as a liquid before the gel point and when it is a solid afterwards. At the critical gel point, the liquid viscosity has diverged to infinity so it is no longer a liquid, but the equilibrium elastic modulus is zero so it is not yet a solid. There are several theories for the process of gelation which are described in detail elsewhere (Flory, 1941; Larson, 1999). Figure 3 illustrates the gelation process from a percolation theory perspective assuming a crosslinker with four reactive sites (tetrafunctional crosslinker). Before the gel point, the equilibrium response of the polymer is to flow. As the reaction progresses, the polymer molecular weight and viscosity increase, diverging as the polymer hits the gel point. At the critical gel point, the gel network just spans the entire sample and the molecular weight and relaxation time are both infinite. Both Flory's classical theory and percolation theory predict that the extent of reaction necessary to form a space filling network is $(f\text{-}1)^{-1}$ where f is the functionality of the crosslinker. In order to form a gel, the crosslinker must have a functionality of 3 or greater (i.e. the gel microstructure must be branched). Some gels are crosslinked to the point of being almost a solid. Others such as the lightly crosslinked silicone discussed here maintain a pronounced viscoelastic character.

2.1 Creep testing

The easiest way to identify if a polymer is a gel is to place the material under a constant stress and track the deformation with time (Anseth et al.,1996; Kavanaugh & Ross-Murphy, 1998). This type of test is referred to as a creep test. For a solid polymer (e.g., a hard epoxy or pencil eraser), the deformation will immediately reach a steady state value which is related to the material stiffness (c.f. Figure 4). Likewise, a Newtonian liquid will immediately reach a constant rate of deformation which is proportional to the reciprocal of the liquid's viscosity. A polymer liquid or other viscoelastic liquid will eventually reach a constant deformation rate. The amount of time it takes to reach a constant shear rate is determined by how long it takes for the polymer to reach its equilibrium deformation state and gives a measure of the longest relaxation time of the polymer. When a polymer gel or other viscoelastic solid is subjected to a constant stress τ, it will eventually reach a constant deformation. The equilibrium deformation γ scaled by the applied stress is called the creep compliance $J=\gamma/\tau$ and is indicative of the equilibrium modulus of the gel with stiffer gels deforming less (i.e. low compliance) than softer gels (c.f. Figure 4b). The creep compliance is approximately independent of the applied stress as indicated in Figure 4c showing that the deformation is linearly dependent on the applied stress.

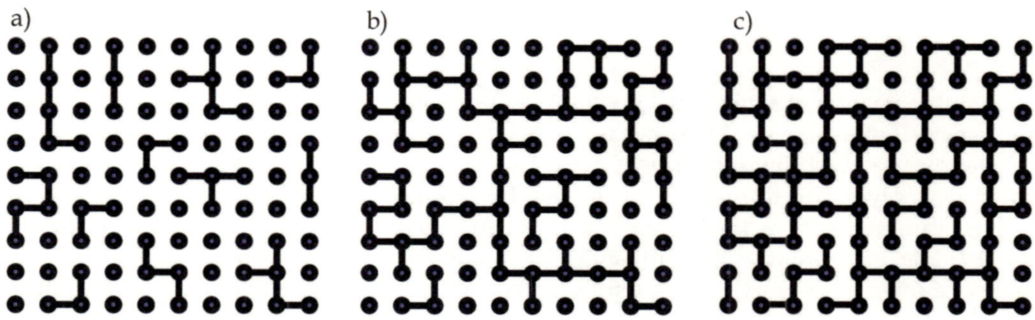

Fig. 3. Illustration of polymer gelation with crosslinker (dots) connecting polymer chains (lines): a) before gel point, isolated polymers increasing molecular weight; b) at critical gel point, network reaches percolation threshold; c) final gel network with defects.

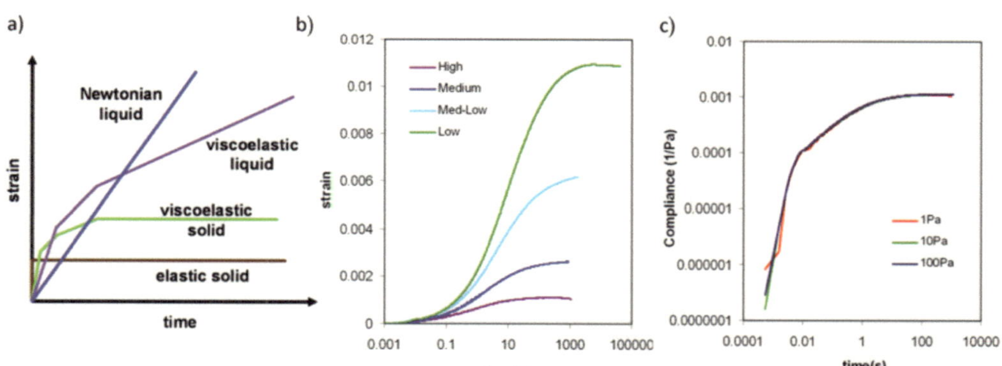

Fig. 4. Compliance curves for creep tests of dilute silicone gels showing the effect of gel stiffness on compliance. a) typical creep behavior for different classes of materials; b) time dependent strain of gels of varying stiffness under a constant stress of 1Pa; c) creep compliance J of a polymer gel under varying stresses.

2.2 Swelling and sol extraction

Gels will also behave differently than liquids when exposed to solvents. Viscoelastic polymer liquids will dissolve in a good solvent as the physical entanglements unravel. A chemically crosslinked gel will swell in the solvent, but the chemically crosslinked gel network will not completely dissolve. For the lightly crosslinked silicone polymers shown in Figure 5, 1 gram pieces of cured gel were placed in 100mL of methylethylketone (MEK) for 24 hours before being drained and weighed. Each of the gels swelled to more than ten times its original weight. The degree of swelling of the gel is related to the gel equilibrium modulus where stiffer gels will swell less than softer gels (Patel et al., 1992). The degree of swelling is presented as the volume fraction of gel in the swollen state (v_s). Assuming simple additivity of volumes, the volume fraction can be calculated as:

$$v_s = \frac{m_g / \rho_g}{m_g / \rho_g + m_{MEK} / \rho_{MEK}} \tag{1}$$

where m_g and m_{MEK} are the masses of the gel and MEK respectively and ρ_g & ρ_{MEK} are their densities.

The soluble fraction of a gel, or sol fraction, can be determined by using a Sohxlet extraction to remove all of the material which is not bound into the gel network (Gottleib et al, 1981). A Sohxlet extractor continuously rinses the gel with freshly condensed solvent allowing any unreacted polymer to diffuse out of the gel. For an ideal end-linked polymer gel, the sol fraction can be less than 1% (Patel et al, 1992). As discussed previously in order to form a space filling gel network, the probability that a given polymer chain is attached to the backbone must be at least 1/3 for tetrafunctional crosslinkers (f=4) (Flory, 1941) indicating that the maximum sol fraction can be 2/3. Below that degree of crosslinking, the polymer will still be a liquid and completely dissolve. However, once reacted, a gel can swell to many times its original size as shown in Figure 5 reaching much higher sol fractions.

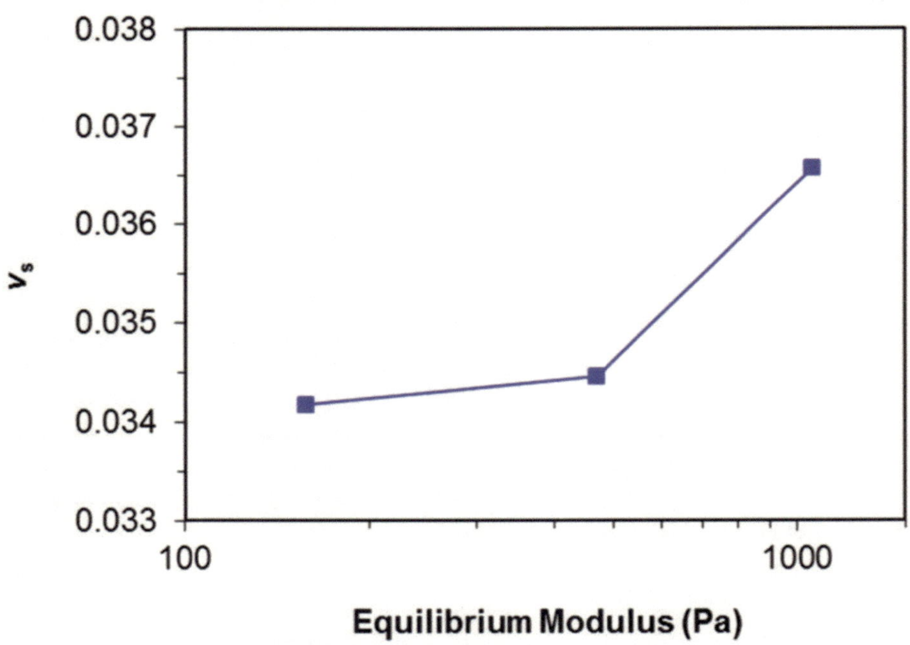

Fig. 5. Volume fraction of polymer in swollen gel as a function of elastic modulus.

2.3 Linear oscillatory rheology

While creep and extraction techniques can provide effective indications of when a material is a gel, oscillatory rheology provides the most sensitive measure of the critical gel point, the point when the material changes from a viscoelastic liquid to a viscoelastic solid. Linear oscillatory rheology subjects the material to a small oscillatory strain (or stress) of the form:

$$\gamma = A\sin(\omega t) \tag{2}$$

where γ is the strain, A is the amplitude of the oscillation and ω is the frequency of oscillation. The resulting response of the material is measured. For a purely elastic solid, the stress required to impose the deformation is proportional to the strain whereas for a viscous liquid, the stress is proportional to the strain rate

$$\dot{\gamma} = A\omega\cos(\omega t). \tag{3}$$

Viscoelastic solids such as gels will have a response that is somewhere between the two extremes. The complex shear modulus G^* can be separated into the fraction that is in-phase with the deformation and the part that is out-of-phase with the deformation. These are generally represented in terms of the elastic G' and viscous G'' shear moduli:

$$G' \propto \sin(\omega t), \quad G'' \propto \cos(\omega t), \quad G^* = \left(G'^2 + G''^2\right)^{1/2}, \quad \tan(\delta) = \frac{G''}{G'} \tag{4}$$

The phase angle δ shows the relative importance of the liquid-like viscous modulus G'' and the solid-like elastic modulus G'.

Oscillatory rheology is a powerful characterization tool because by varying the amplitude and frequency of the applied strain, a wide range of timescales and behaviors can be studied (Anseth et al., 1996). For this chapter, we will limit our discussion to small amplitude experiments within the linear viscoelastic regime which allows an investigation of the gel response without disruption of the gel structure. In the linear regime, the measured moduli are independent of the applied strain. Figure 6 shows an example of the shear moduli where the applied strain amplitude was varied at a fixed frequency of 1Hz. At very low strains (γ<0.01%), the signal is very weak and the data can be noisy. In Figure 6, the linear viscoelastic regime extends to strains of ~40%. At higher strains, changes occur in the gel structure (ruptured bonds or entanglements) resulting in a decrease in the measured moduli.

2.3.1 Determination of the critical gel point

To determine the gel point during a crosslinking reaction, the complex moduli are measured as a function of time as shown in Figure 7. At early times, both of the moduli are low and the elastic portion G' is much smaller than the viscous portion G''. This is characteristic of a polymer liquid at low frequencies. The presence of a small elastic contribution well before the critical gel point is due to the stretching of the polymers under deformation and potentially physical entanglements between the polymers. As the crosslinking reaction progresses, the molecular weight of the polymers increases, increasing both the viscosity and relative contribution of the elastic modulus G'. Longer polymers have longer relaxations times and more entanglements. At a time known as the cross-over point, the elastic modulus becomes larger than the viscous modulus. As the reaction progresses to completion, the elastic and viscous moduli approach their equilibrium values. Stiffer gels will have a higher elastic modulus and a smaller phase angle δ. In the limit of a very stiff gel, the viscous contribution may be negligible.

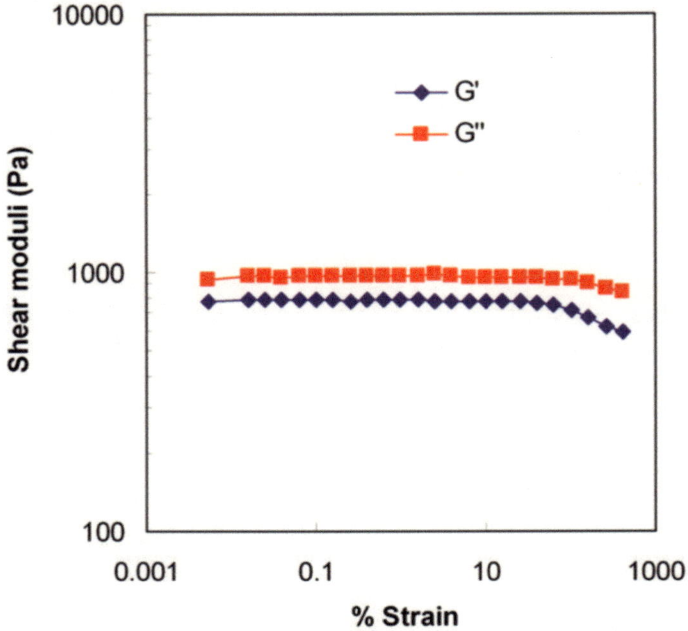

Fig. 6. In the linear viscoelastic regime, the measured shear moduli G' and G'' are independent of the applied strain or stress.

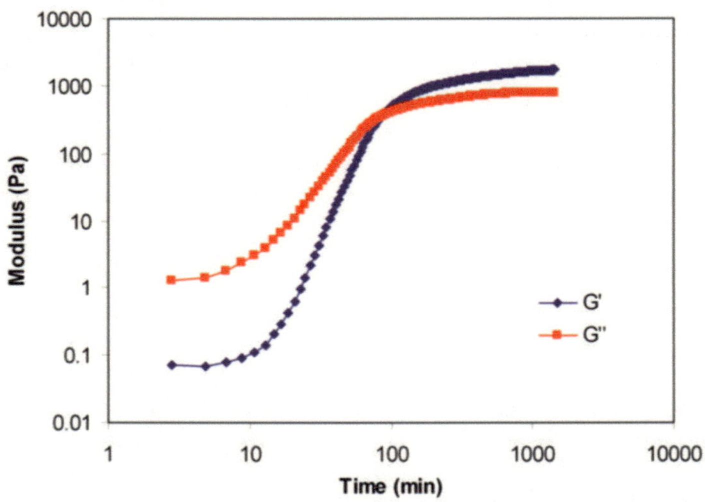

Fig. 7. Shear moduli as a function of time during crosslinking reaction of a silicone gel (ω=1Hz, stress=5Pa).

For ideal end-linked polymer gels, the critical gel point can be defined as the cross-over point where the elastic and viscous moduli are equal ($G'=G''$). However for non-ideal gels where network defects or physical entanglements are present, the cross-over point depends on the applied frequency. The critical gel point represents a physical transition from a liquid to a solid and hence should not depend on the measurement parameters. Chambon and Winter (1987) proposed what is now the definitive criterion for determination of the critical gel point. The critical gel point is when the two moduli exhibit a power law dependence on the applied frequency over a wide range of frequencies. Alternately at the critical gel point, the ratio of the shear moduli, tan(δ), is independent of frequency (Gupta, 2000)

$$G', G'' \propto \omega^n \quad \frac{G''}{G'} = \tan(\delta) = \tan\left(\frac{n\pi}{2}\right) \tag{5}$$

For ideal gels, the shear moduli at the critical gel point are equal and n=0.5. For gels which contain defects, the phase angle $\delta = n\pi/2$ will be independent of frequency with n in the range of 0.5 to 1, with gels containing more non-idealities having a larger value of n. An example is illustrated in Figure 8 for a lightly crosslinked silicone containing many defects. Winter & Mours (1997) summarize in detail other characteristics of a gel at the critical gel point.

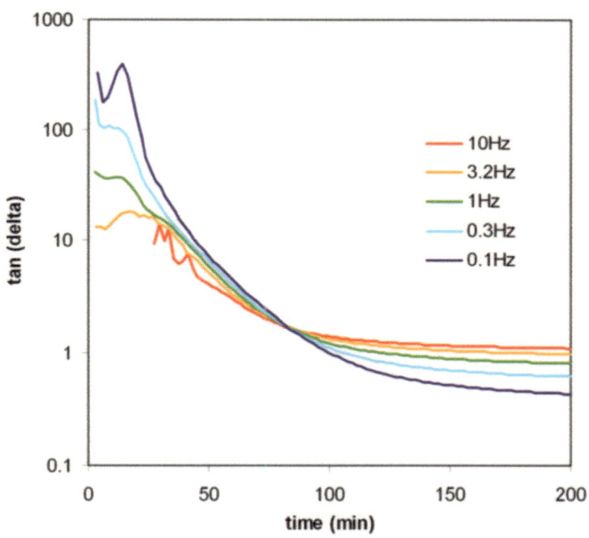

	n
Ideal	0.5
High	0.61
Medium	0.66
Low	0.7

Fig. 8. Determination of critical gel point and network quality from oscillatory rheology. The critical gel point is the time when the curves of tan(δ) at various frequencies coincide. The table shows the relaxation exponent n for gels of various equilibrium moduli.

2.3.2 Frequency dependence of gels

Ideal gels have an almost purely elastic response where the elastic modulus is much higher than the viscous modulus and is independent of frequency. In gel networks with imperfections, the response of the polymer gel will depend on frequency with both shear moduli increasing with frequency. Various time scales of a polymer gel can be investigated by adjusting the frequency of the applied oscillation to probe different relaxation times as illustrated in Figure 2. At low frequencies, both gel and sol polymers are rearranging due to Brownian motion so the measured properties are dominated by the equilibrium elastic deformation of the gel network (c.f. Figure 2d). Physical entanglements are created and broken quickly compared to the rate of deformation so they do not contribute drag or store elastic energy. At high frequencies the polymer does not have time to rearrange (c.f Figure 2b). Physical entanglements persist longer than the oscillation frequency so they physically constrain the polymers, store elastic energy and contribute to viscous dissipation (Patel et al., 1992).

This timescale dependence is demonstrated in Figure 9 for two silicone gels. At low frequencies, the stiffer gel (red curves) shows a higher elastic modulus and smaller phase angle relative to the softer gel (black curves). As the frequency approaches zero, the elastic modulus approaches a plateau value known as the equilibrium modulus of the gel network. The equilibrium modulus reflects only the chemical crosslinks in the gel because the lifetime of physical entanglements is much shorter than the oscillation period. At higher frequencies, the solvent and polymer entanglements begin to contribute to the material response increasing both the elastic and viscous moduli. At the highest frequencies, the shear moduli almost overlap because the response is dominated by local interactions between polymer chains and physical entanglements are indistinguishable from chemical crosslinks (Mrozek et al., 2011).

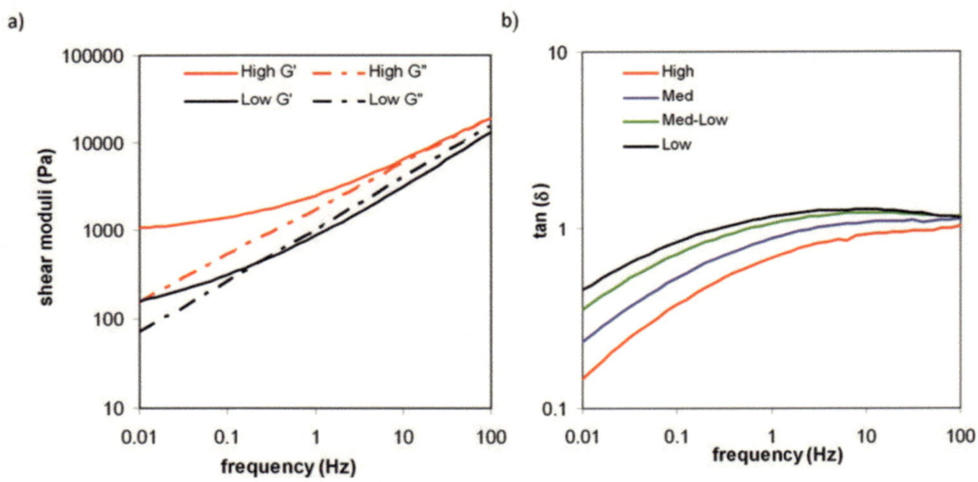

Fig. 9. Frequency dependence of (a) shear moduli and (b) phase angle tan(δ). The high modulus gel is shown in red and the low modulus gel in black..

3. Adhesion properties of polymer gels

When an uncrosslinked or lightly crosslinked polymer is brought into contact with the surface of another material at a temperature above its glass transition temperature, an adhesive bond of measurable strength is formed in most cases (Zosel, 1985). The adhesion of the polymer to the substrate is highly influenced by the viscoelasticity of the polymer as well as the surface and interfacial tensions of the polymer and substrate (Zosel, 1989). To function properly, polymeric adhesives must combine liquid-like characteristics to form good molecular contact under an applied pressure and solid-like characteristics to resist an applied stress once the bond has been formed. This combination of liquid-like and solid-like properties usually requires a high molecular weight polymer to form the backbone of the adhesive, and a low molecular weight fraction which favors flow and deformation (Roos et al., 2002). One common criterion for a material with good adhesive properties is an elastic modulus less than 10^5 Pa (Dahlquist criterion, Creton, 2003). Materials with elastic moduli exceeding the Dahlquist criterion have poor adhesion characteristics due to the inability to dissipate energy via viscous contributions or to deform to make good contact with a surface. Further, these materials have a high peak adhesive force, but fail quickly upon further strain (brittle failure with no fibril formation). However, not all polymer gels which meet the Dahlquist criterion are good adhesives. Many hydrogels have low equilibrium moduli, but have negligible viscous moduli. With no dissipative modes in the materials, even these soft materials undergo brittle failure and are poor adhesives. For the discussion in this section, only materials meeting the Dahlquist criterion ($G' < 10^5$ Pa) will be considered.

Adhesive bond formation also requires a sufficiently high segmental mobility in order to obtain contact at molecular dimensions between the adhesive and solid substrate during the possibly very short contact time. During the separation phase, the adhesive must be able to accommodate large deformations in order to store and dissipate a large amount of energy before fracture occurs (Zosel, 1991; Gay, 2002). When a surface comes in contact with the gel, initially there are only small contact zones where the polymer wets the surface. The number and size of the contacts increase with the contact time and contact force by wetting of the surface and deformation of the polymer to accommodate surface roughness. Contact formation is an important factor in determining the strength of an adhesive joint (Zosel, 1997).

Adhesion between soft polymeric materials and a substrate is typically measured in one of two ways: peel testing or probe (tack) testing. Peel testing is typically done by casting and/or curing a polymer film on a substrate. Once the polymer is cured, one edge of the film is gripped by a mechanical pulling device and subsequently peeled from the substrate at a constant velocity and at a constant peel angle (frequently 90°). During the peel test, the force required to peel the polymer from the substrate is recorded. A variety of analysis techniques can then be applied to the resulting data including recording the maximum force measured during the peel process. This maximum force can then be compared with maximum forces obtained for other polymer materials. The total peel energy may also be calculated by integrating the force versus displacement curve. To obtain meaningful data that can be compared to data for other materials or measurements made on different equipment, the data must be normalized by the width of the polymer film as the measured properties are highly dependent on film width. Peel testing results are difficult to interpret because the stress distribution near the advancing peel front greatly complicates the distribution of applied force. Further, the measured quantities will depend on the peel angle, peel velocity, and process of forming the interface (Crosby, 2003; Gent, 1969).

Probe tests to measure the tack adhesion are accomplished by bringing a probe into contact with the surface of the polymer material being tested under a given force for a specified period of time. The probe is then raised at a constant velocity while measuring the force required to do so. The resulting force versus distance curve provides valuable information on the adhesion properties of the material. The measured adhesion depends on the probe speed, contact time and force as well as the probe shape and surface characteristics. Probe testing eliminates several of the complicating factors associated with peel testing and is the focus of this section.

3.1 Tack adhesion measurement

In a typical tack adhesion test for polymer gels, a rigid probe is brought into contact with a polymer gel film at a given rate (c.f. Figure 10a). Once contact between the probe and gel is established, a holding period is performed where a constant force is applied to the gel for a given period of time. The probe and gel film are then separated at a constant rate while measuring the force (normal force) required for separation (c.f. Figure 10b). The adhesion energy, or work of adhesion, is then determined from the integral of the resulting stress versus strain curve (c.f. Figure 10c). Other useful information obtained from the resulting curve includes the peak adhesive force and the strain to failure. Critical variables impacting the measured adhesion energy are the contact force, contact time, and separation speed (Gent, 1969; Zosel, 1985, 1997, 1998; Hui, 2000). Researchers have reported that the roughness of the probe surface also plays an important role in the tack adhesion measured, with rough surfaces resulting in poorer molecular contact between the polymer and the probe (Gay, 2002; Hui, 2000). The overall shape of the force versus distance curve is determined by the viscoelastic and molecular properties of the gel as well as the microscopic debonding mechanisms (Derks, 2003; Lakrout et al., 2001; O'Connor & Willenbacher, 2004).

For the measurements reported here, fluorosilicone films were prepared by casting uncured fluorosilicone gel onto aluminum plates (50 mm in diameter) and then curing at 82°C for 24 hours. The samples were then cooled to room temperature before use. The cured gel films were 0.9 ± 0.05 mm thick. Tack measurements were performed on a TA Instruments ARES G2 rheometer. The probe used was an 8 mm diameter, flat plate fixture.

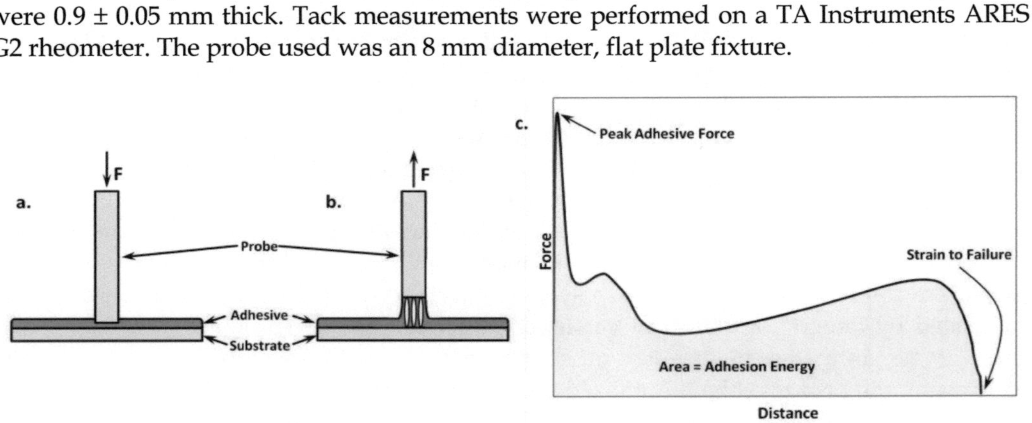

Fig. 10. Schematic representation of a typical tack adhesion measurement apparatus during (a) contact and (b) separation steps. (c) A typical force vs. distance curve obtained from a tack measurement.

3.2 Contact force and contact time

For soft materials such as polymer gels, adhesion is largely dictated by two factors. First, the ability of the material to achieve intimate contact with the substrate. If the material is cured on the substrate, then intimate contact is determined by the ability of the uncured material to wet the substrate surface (Lenhart & Cole, 2006). If the gel is brought into contact with the substrate after curing, the contact between the gel and the substrate is influenced by the contact force, contact time, and rheology of the gel. Generally, increasing the contact force results in better contact and, thus, better adhesive strength (Zosel, 1997). Obtaining intimate molecular contact between the gel and the substrate greatly determines the strength of the adhesive joint (Zosel, 1997). Second, gel adhesion is largely dictated by the ability of the bulk material to dissipate energy effectively. The energy dissipating ability of the material is directly related to its viscoelastic properties.

For lightly crosslinked polymer gels, the adhesion energy is observed to increase with the contact force while holding contact time constant for low to moderate contact forces (Figure 11a). However, the adhesion energy reaches a plateau value as the contact force is further increased. The contact force has a pronounced effect on the shape of the force versus displacement curves as well (Figure 11b). As the contact force is increased, increases in both the peak adhesive force and the strain to failure are also observed. The increased adhesion energy, peak adhesive force, and strain to failure indicate that better molecular contact is achieved between the polymer gel and the probe when the contact force is increased.

The peak at small distances in the force versus distance curve is related to the onset of cavitation, or the formation of small air pockets between the polymer gel and the probe (c.f. Figure 11b). Once cavitation occurs, the measured force decreases significantly. The air pockets then grow as the polymer gel is stretched further and fibrils are formed. The force increases as the fibrils are stretched and strain harden. Finally, the fibrils detach cleanly from the probe and the measured force returns to a value of zero. The debonding mechanisms are discussed further in Section 4.

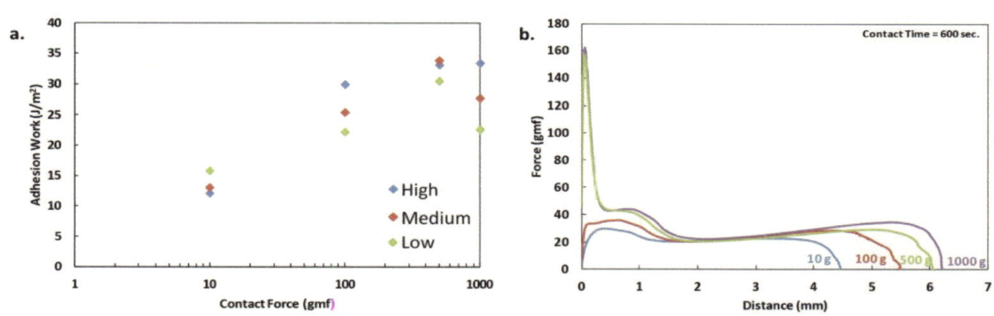

Fig. 11. (a) Work of adhesion as a function of contact force for three different polymer gels of varying equilibrium moduli (low, medium, high). (b) Force versus distance curves measured for a polymer gel (medium) for several different contact forces.

The contact between the gel and substrate is also influenced by the contact time. If the contact force is held constant, increasing the contact time can result in better adhesive strength as intimate molecular contact is enhanced. The correlation between contact time and adhesive strength is dependent on the rheology of the polymer gel. Gels with a lower modulus are better able to relax under the applied force and conform to the substrate which results in better contact. On the other hand, gels with a higher modulus take much longer to relax and may not be able to completely conform to the substrate because of network limitations. The inability to conform to the substrate results in a poorer contact and, thus, lower adhesion energy.

For lightly crosslinked polymer gels, a slight dependence on contact time is observed for differing values of the contact force (c.f. Figure 12). For a contact force of 10 grams, the adhesion energy increases by 25% when the contact time is increased from 30 seconds to 1000 seconds. The increase in adhesion energy at long contact times indicates that under a force of 10 grams, contact between the probe and the polymer gel is enhanced with time. However, for a contact force of 1000 grams, there is very little change in adhesion energy over the same range of contact times for the same material indicating that contact between the polymer gel and the probe is independent of time over the range reported here. Further, the magnitude of the adhesion energy is higher (180%) for the higher contact force indicating that better molecular contact between the probe and the polymer gel was achieved. At high contact forces, the measured work of adhesion reaches a plateau where the best possible contact between the gel and the probe has been achieved.

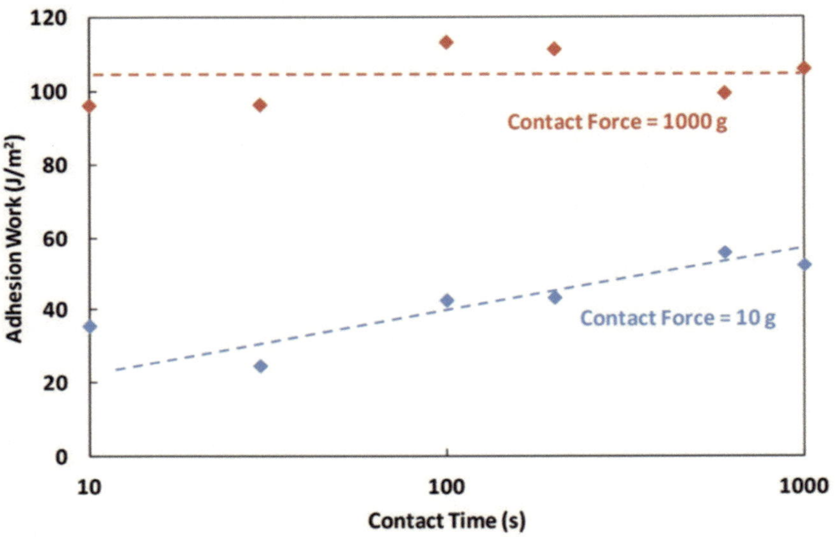

Fig. 12. Work of adhesion as a function of contact time for two different values of contact force for a polymer gel with an equilibrium modulus of 470 Pa.

3.3 Separation speed

The separation speed is important in the tack behavior of polymer gels because changing the separation speed changes the time scale of the deformation of the polymer. Based on the rheological behavior of the polymer gel, changes in the time scale of the deformation can result in significantly different viscoelastic behavior which, in turn, results in significantly different adhesion behavior. For lightly crosslinked polymer gels, the adhesion energy and the peak adhesive force are observed to increase with increasing separation velocity (c.f. Figure 13a). At separation velocities below 0.01 mm/s, a plateau is observed where the adhesion energy is independent of separation velocity. As the separation velocity increases, there is a power law relationship between adhesion energy and separation velocity. At high separation velocities, the adhesion energy again plateaus and becomes independent of separation velocity.

The trends in adhesion energy observed for a lightly crosslinked polymer gel can be better understood by examining the rheological behavior of the gel over a wide range of deformation time scales or oscillatory frequencies. The ratio of the viscous modulus (G'') to the elastic modulus (G') provides insight into the behavior of the gel at various time scales. At low frequencies (long time scales), the value of tan(δ) approaches zero as the elastic response dominates the viscous modulus (c.f. Figure 13b) indicating that, at very long time scales, the gel response is dominated by the equilibrium behavior of the gel network. Thus, most of the energy applied to the polymer gel is stored elastically in the equilibrium deformation of the network. The plateau in the adhesion energy curve at low separation velocities corresponds to the behavior of the polymer gel at long time scales. At very long deformation time scales, the polymer chains rapidly rearrange releasing physical entanglements and are able to maintain an equilibrium configuration while the gel network is being deformed. All of the applied force is stored elastically in the gel network since the viscous contributions (friction between polymer chains, chain disentanglement and rearrangement, etc.) are small.

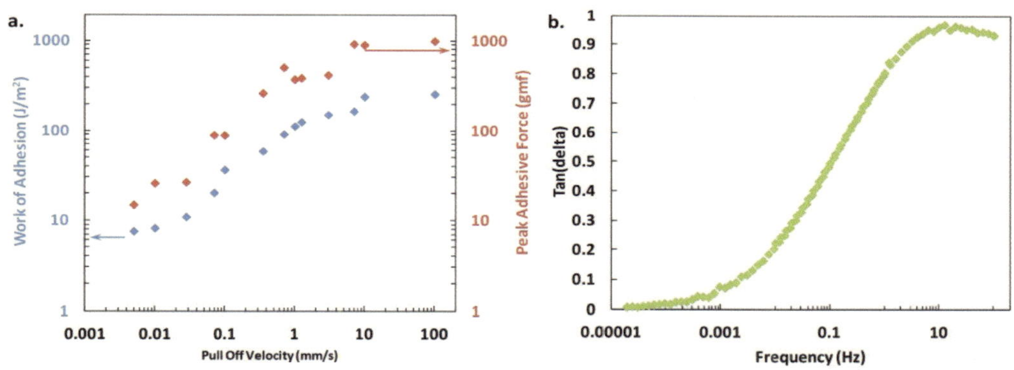

Fig. 13. Correlation in time scale dependence between adhesion and linear rheology.
a) Adhesion energy (blue points) and peak adhesive force (red points) as a function of separation velocity for a polymer gel with an equilibrium modulus of 470 Pa; b) tan(δ) as a function of oscillation frequency.

A plateau is also observed in the tan(δ) function at high frequencies (short time scales, Figure 13b). Above an oscillation frequency of 10 Hz, tan(δ) maintains a constant value that is independent of the applied frequency. Although the magnitude of the moduli change above 10 Hz, the ratio of the moduli remains constant. At short time scales, the polymer chains do not have sufficient time to rearrange or disentangle when a stress is applied. Thus, the physical entanglements between chains persist contributing both to the viscous dissipation and also increasing the effective number of crosslinks. The force applied is now distributed almost equally between the viscous and elastic modes, thus much more force is required to deform the gel. The work of adhesion for short time scales (high separation velocities) is about a factor of 100 higher than at long time scales.

Further insight into the polymer gel behavior during the tack test can be gained by examining the stress versus strain curves for several different separation velocities (c.f. Figure 14). At a separation velocity of 0.01 mm/s relatively low forces are required because only the gel backbone is being deformed (c.f. Figure 2d) while most of the polymers are able to relax. For separation velocities between 0.1 mm/s and 10.0 mm/s, much of the increase in adhesion energy can be attributed to the development and enhancement of the peak adhesive force observed at low strain values. The development of the peak adhesive force greatly increases the area beneath the stress versus strain curve (adhesion energy). For separation velocities above 10.0 mm/s, the stress versus strain curves become very similar in both shape and values. This similarity corresponds to the plateau in tan(δ) at high oscillation frequencies (i.e., the polymer gel behaves similarly at 10.0 mm/s and 100.0 mm/s). At low oscillation frequencies and slow tack measurement speeds, the response is dominated by the equilibrium modulus due to deformations of the gel network. At high frequencies or fast probe speeds, the physical entanglements dissipate energy through internal friction requiring a larger peak force to deform the gel and a larger overall work of adhesion.

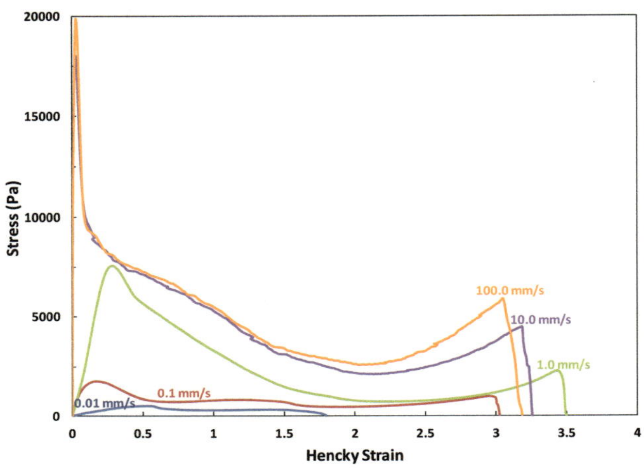

Fig. 14. Stress versus strain curves for a polymer gel with an equilibrium modulus of 470 Pa (medium) at several different pull off velocities (0.01 mm/s – 100 mm/s).

4. Failure modes

The mechanism of failure of an adhesive bond is complicated because not only interfacial interactions, but also bulk rheology, can play a significant role (Moon et al., 2004). There are two main failure mechanisms for adhesive applications; adhesive separation and cohesive failure. Adhesive separation for materials below the Dahlquist criterion is generally characterized by the formation, extension, and eventual failure of fibrillar structures within the polymer material. In cohesive failure, the failure is due to a fracture within the polymer film rather than separation at an interface. Typically, if a cohesive fracture occurs both of the resulting surfaces will be covered in the polymer material. For materials to have high tack and peel strength, the material must be able to dissipate a large amount of deformation energy during separation. Studies indicate that the large degree of energy dissipation is connected to formation of fibrillar structures during separation (Zosel, 1998).

4.1 Adhesive separation

Adhesive separation of soft polymer gels ($G'<10^5$ Pa) is generally characterized by the formation of fibrillar structures during the separation process. The molecular conditions necessary for fibril formation have been discussed by Zosel, who argued that a high molecular weight between crosslinks (i.e., a low equilibrium modulus) is a necessary condition for the formation of the fibrillar structure (Zosel, 1989, 1991, 1998). A slight degree of branching and crosslinking is beneficial for the stability of the fibrils but excessive crosslinking can lead to a premature failure of the fibrils, therefore significantly reducing the adhesion energy (Lakrout et al., 1999). For a fibril forming polymer gel, it can be clearly seen that the material is split into separate filaments or fibrils which are anchored on both the fixed bottom substrate and the moving probe surface. These fibrils are increasingly stretched as the probe is raised from the gel causing the storage and dissipation of energy (Zosel, 1989). The microscopic mechanisms of adhesive separation are commonly divided into 4 parts (c.f. Figure 15a).

1. Homogeneous deformation
2. Cavitation
3. Rapid lateral growth of cavities
4. Fibrillation

First, there is a homogeneous deformation of the polymer gel where the stress is distributed throughout the material. Upon further deformation, cavitation occurs and small air pockets form near the probe surface. Next, the air pockets grow and the fibrils are formed. At this point in the deformation, the stress is no longer supported by the entire volume of polymer material. The bulk of the stress is supported by the newly formed fibrils. As the fibrils are stretched, strain hardening may occur, which causes a slight upturn in the stress versus strain curve at high strain values (Figure 13b). Upon further deformation the fibrils either break cohesively or detach adhesively from the probe surface, causing complete debonding (Roos et al., 2002).

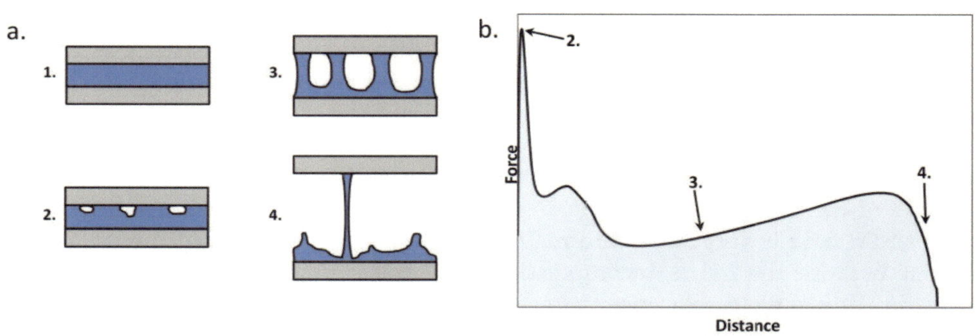

Fig. 15. (a) Schematic representation of the separation of a fibril forming adhesive and the substrate in a tack adhesion experiment. (b) Typical curve of force versus distance for a fibril forming polymeric material.

4.2 Cohesive failure

Cohesive failure is observed when a crack propagates in the bulk of the polymeric material which leads to failure (Figure 16a). In most cases, the surfaces of the adherents (substrate and probe in the case of a tack experiment) will be covered with the polymeric material following separation. The crack may originate and propagate near the center of the polymer material or near an interface. For a material exhibiting cohesive failure, the measured force during separation quickly reaches a maximum then gradually decreases to zero (Figure 16b). The debonding process is generally governed by the viscous nature of the polymer (Zosel, 1989). Studies of debonding mechanisms show that yield stress fluids exhibit cohesive failure where air enters (crack propagation) the center of the fluid layer (Derks et al., 2003).

While cohesive failure is interesting and desired in many applications, materials that fail cohesively do not generally exhibit a high degree of tack adhesion. For the polymer gel materials discussed here, cohesive failure is not observed in any case.

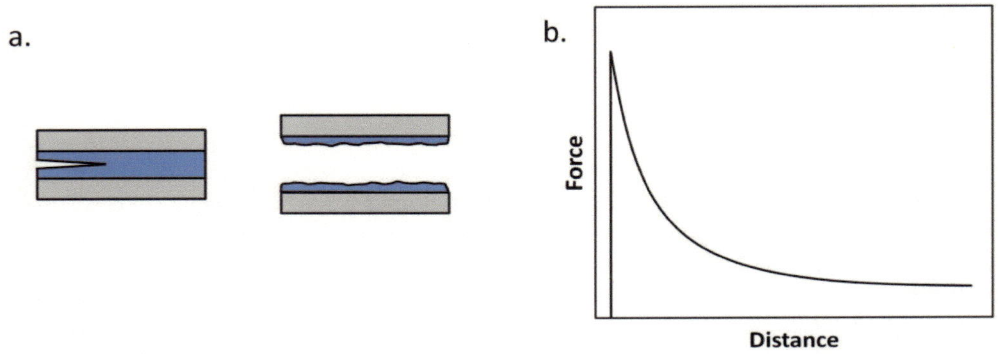

Fig. 16. (a) Schematic representation of a material exhibiting cohesive failure. (b) Typical curve of force versus distance for a cohesively failing material.

5. Hardness

Another useful parameter in characterizing the stiffness of a gel network is hardness. Hardness is measured by indenting a probe into the gel at a specified velocity while measuring the force required for the indentation. The force required to indent the gel to a certain depth is the hardness. While the measured hardness does depend strongly on the modulus of the gel, it also depends on many other measurement parameters such as the size and shape of the gel sample, probe size, speed, and indentation depth. Additionally, the applied strain field is very non-uniform. The strain and strain rate near the probe can be high, but because the gel is incompressible, the entire volume of the gel experiences deformation due to the displacement of gel by the probe. Thus hardness is at best a relative measure of gel material properties. However, since hardness is used by some common gel manufacturers to specify their materials, it is important to understand it in the context of other rheological characterization methods.

For the measurements here, a 1.27 cm diameter probe with a hemispherical cap was used to indent 50g of polymer gel cured in a 5cm diameter glass jar on a Texture Technologies TA.XT Plus Texture Analyzer. Figure 17 shows results for four lightly crosslinked polymer gels of varying equilibrium modulus. As with other characterization methods we have discussed, the measured hardness is strongly dependent on the measurement speed.

At very low speeds, the indentation probes primarily the response of the gel network and hence the measured hardness shows a strong dependence on the equilibrium modulus. Defects cause both the gel equilibrium modulus as well as the hardness to decrease. At fast measurement speeds, physical entanglements and viscous friction between polymer chains contribute substantially to the measured response. The measured hardness increases and becomes less dependent on the quality of the polymer network as quantified by the equilibrium modulus.

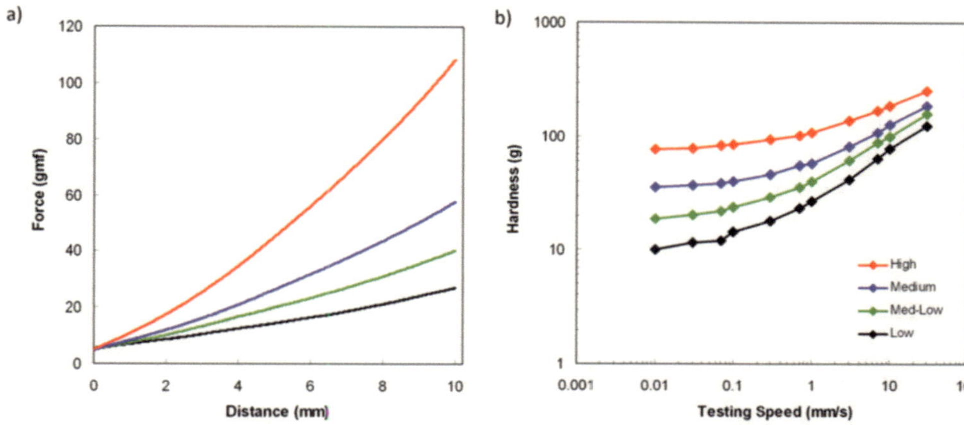

Fig. 17. Hardness test results: a) for four different polymer gels, b) as a function of speed.

Figure 18 shows a direct comparison between the measured hardness and other gel characteristics. As mentioned above, the hardness correlates well with the measured equilibrium modulus and the inverse of the creep compliance. The correlation with adhesion properties is more complex. There is a striking similarity between the dependence

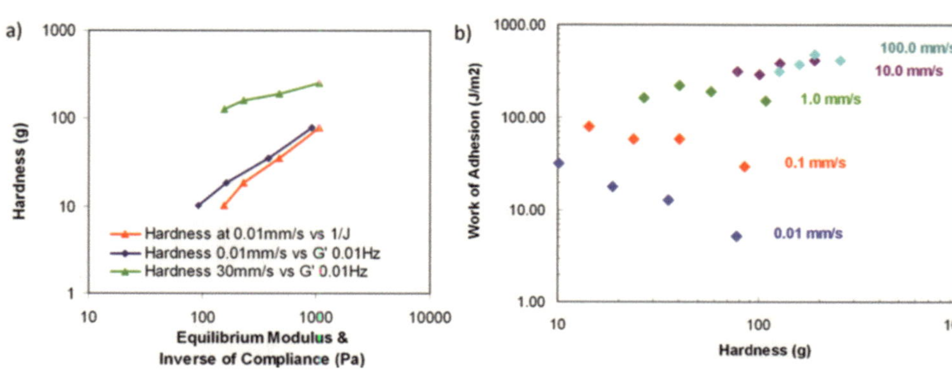

Fig. 18. Gel rheological properties as a function of hardness. a) Hardness at 2 speeds as a function of equilibrium modulus (G' at 0.01Hz) and the inverse of the compliance ($1/J$); b) Work of adhesion as a function of hardness.

of the hardness as a function of indentation speed (Figure 17b) and the work of adhesion as a function of pull off speed (Figure 13a). Figure 18b shows the relationship between the work of adhesion and the measured hardness for a variety of probe indentation and pull-off speeds. For slow speeds where the hardness is primarily probing the response of the gel network, the measured work of adhesion depends strongly on the hardness. Higher hardness gels have a lower work of adhesion. At high speeds, the work of adhesion becomes almost independent of hardness of the gel suggesting that hardness testing retains more dependence on the equilibrium modulus of the gel at higher speeds than the tack testing. Whereas tack testing at high speed imposes a fairly uniform high strain rate on the gel, hardness testing imposes a non-uniform strain field such that a large fraction of the gel experiences only small strains and strain rates due to displacement of the gel by the probe.

6. Conclusions

Polymer gels are viscoelastic solids which exhibit a wide variety of dynamic rheological behavior demonstrated here using a series of lightly crosslinked fluorosilicone gels with a range of equilibrium moduli. At long time scales, the elastic response of the gel network dominates and the measured equilibrium parameters are highly dependent on the gel microstructure and the presence of defects. The equilibrium gel response is almost purely elastic and shows a plateau at low frequencies where the gel response becomes independent of frequency. That plateau is more apparent for stiffer polymer gels. In non-ideal gels which have entanglements, network defects and polymer solvent, viscoelastic effects become more important as the frequency is increased. At short time scales, physical entanglements and solvent effects become dominant and the measured gel response becomes almost independent of the gel network quality. The importance of time scales is readily apparent in all forms of characterization including linear oscillatory rheology, adhesion and hardness testing.

Linear oscillatory rheology provides a precise tool for determining the critical gel point and for understanding gel behavior by manipulating both the applied strain and frequency. The viscoelastic properties are critically important for good adhesive properties. The presence of dissipation mechanisms (as evidenced by the viscous modulus) are important for creating a large work of adhesion whereas the presence of an equilibrium modulus is required for keeping the adhesive in place. The frequency dependence of the work of adhesion closely mirrors the linear oscillatory rheology. At low frequencies, the work of adhesion is strongly dependent on the equilibrium modulus of the gel. At high frequencies, the adhesive response becomes independent of the gel network quality and instead is dominated by physical entanglements. Hardness testing can also provide useful qualitative information about gel dynamic response, though because of the non-uniform strain field applied, the frequency dependence is not as pronounced.

7. Acknowledgements

The authors would like to acknowledge many insightful discussions with Joseph Lenhart of the U.S. Army Research Laboratory.

Sandia National Laboratories is a multi-program laboratory managed and operated by Sandia Corporation, a wholly owned subsidiary of Lockheed Martin Corporation, for the U.S. Department of Energy's National Nuclear Security Administration under contract DE-AC04-94AL85000

8. References

Andrews, G. P. & Jones, D. S. (2006). Rheological Characterization of Bioadhesive Binary Polymeric Systems Designed as Platforms for Drug Delivery Implants. *Biomacromolecules*, Vol. 7, pp. 899-906, ISSN 1525-7797

Anseth, K. S.; Bowman, C. N. & Brannon-Peppas, L. (1996). Mechanical Properties of Hydrogels and their Experimental Determination. *Biomaterials*, Vol. 17, No.17, pp. 1647-1657, ISSN 0142-9612

Chambon, F. & Winter, H. H. (1987). Linear Viscoelasticity at the Gel Point of a Crosslinking PDMS with Imbalanced Stoichiometry. *Journal of Rheology*, Vol. 31, No. 8, pp. 683-697, ISSN 0148-6055

Creton, C. (2003). Pressure-Sensitive Adhesives: An Introductory Course. *MRS Bulletin*, Vol. 28, No. 6, pp. 434-439, ISSN 0883-7694

Crosby, A. J. (2003). Combinatorial Characterization of Polymer Adhesion. *Journal of Materials Science*, Vol. 38, pp. 4439-4449, ISSN 0022-2461

Derks, D., Lindner, A., Creton, C. & Bonn, D. (2003). Cohesive Failure of Thin Layers of Soft Model Adhesives Under Tension. *Journal of Applied Physics*, Vol. 93, No. 1, pp. 1557-1566, ISSN 0021-8979

Gay, C. (2002). Stickiness - Some Fundamentals of Adhesion. *Integrative and Comparative Biology*, Vol. 42, pp. 1123-1126, ISSN 1540-7063

Gent, A. N. & Petrich, R. P. (1969). Adhesion of Viscoelastic Materials to Rigid Substrates. *Proceedings of the Royal Society A*, Vol. 310, pp. 433-448, ISSN 1471-2946

Gottlieb, M.; Macosko, C. W.; Benjamin, G. S.; Meyers, K. O. & Merrill, E. W. (1981). Equilibrium Modulus of Model Poly (dimethylsiloxane) Networks. *Macromolecules*, Vol. 14,, pp. 1039-1046, ISSN 0024-9297

Gupte, R. K. (2000) *Polymer and Composite Rheology*, Marcel Dekker, ISBN 9780824799229, New York, NY.

Hui, C. Y., Lin, Y. Y & Baney, J. M. (2000). The Mechanics of Tack: Viscoelastic Contact on a Rough Surface. *Journal of Polymer Science: Part B: Polymer Physics*, Vol. 38, pp. 1485-1495, ISSN 1099-0488

Kavanagh, G. M. & Ross-Murphy, S. B. (1998). Rheological Characterization of Polymer Gels. *Progress in Polymer Science*, Vol. 23, pp. 533-562, ISSN 0079-6700

Lakrout, H., Sergot, P. & Creton, C. (1999). Direct Observation of Cavitation and Fibrillation in a Probe Tack Experiment on Model Acrylic Pressure-Sensitive-Adhesives. *Journal of Adhesion*, Vol. 69, pp. 307-359, ISSN 0021-8464

Lakrout, H., Creton, C., Ahn, D. & Shull, K. R. (2001). Influence of Molecular Features on the Tackiness of Acrylic Polymer Melts. *Macromolecules*, Vol. 34, pp. 7448-7458, ISSN 0024-9297

Larson, R. G. (1988) *Constitutive Equations for Polymer Melts and Solutions*, Butterworths, ISBN 0409901199, Boston, MA.

Lenhart, J. L. & Cole, P. J. (2006). Adhesion Properties of Lightly Crosslinked Solvent-Swollen Polymer Gels *The Journal of Adhesion*, Vol. 82, pp. 945-971, ISSN 0021-8464

Llorente, M. A. & Mark, J. E. (1979). Model Networks of end-Linked Polydimethylsiloxane chains. IV Elastomeric Properties of the Tetrafunctional Networks Prepared at Different Degrees of Dilution. *Journal of Chemical Physics*, Vol. 71, No. 2, pp. 682-689, ISSN 0021-9606

Macosko, C. W. (1994) *Rheology: Principles, Measurements, and Applications*, Wiley-VCH, Inc., ISBN 1560815795, New York, NY.

Moon, S., Swearingen, S. & Foster, M. D. (2004). Scanning Probe Microscopy Study of Dynamic Adhesion Behavior of Polymer Adhesive Blends. *Polymer*, Vol. 45, pp. 5951-5959, ISSN: 0032-3861

Mrozek, R. A.; Cole, P. J.; Otim, K. J.; Shull, K. R. & Lenhart, J. L. (2011). Influence of Solvent Size on the Mechanical Properties and Rheology of Polydimethylsiloxane-based Polymeric Gels. *Polymer*, Vol. 52, pp. 3422-3430, ISSN 0032-3861

O'Connor, A. E. & Willenbacher, N. (2004). The Effect of Molecular Weight and Temperature on Tack Properties of Model Polyisobutylenes. *International Journal of Adhesion & Adhesives*, Vol. 24, pp. 335-346, ISSN 0143-7496

Patel, S. K.; Malone, S.; Cohen, C.; Gillmor, J. R. & Colby, R. H. (1992). Elastic Modulus and Equilibrium Swelling of Poly(dimethylsiloxane) Networks. *Macromolecules*, Vol. 25, pp. 5241-5251, ISSN 0024-9297

Richter, S. (2007). Recent Gelation Studies on Irreversible and Reversible Systems with Dynamic Light Scattering and Rheology – A Concise Summary. *Macromolecular Chemistry and Physics*, Vol. 208, pp. 1495-1502, ISSN 1022-135

Roos, A., Creton, C., Novikov, M. B. & Feldstein, M. M. (2002). Viscoelasticity and Tack of Poly(Vinyl Pyrrolidone)-Poly(Ethylene Glycol) Blends. *Journal of Polymer Science: Part B: Polymer Physics*, Vol. 40, pp. 2395-2409, ISSN 1099-0488

Ross-Murphy, S. B. (1995). Structure-Property Relationships in Food Biopolymer Gels and Solutions. *Journal of Rheology*, Vol. 39, No. 6, pp. 1451-1463, ISSN 0148-6055

Solomon, M. J. & Spicer, P. T. (2010). Microstructural Regimes of Colloidal Rod Suspensions, Gels and Glasses. *Soft Matter*, Vol. 6, pp. 1391-1400, ISSN 1744-683X

Tunick, M. H. (2010). Small-Strain Dynamic Rheology of Food Protein Networks. *Journal of Agricultural and Food Chemistry*, Vol. 59, pp. 1481-1486. ISSN 0021-8561

Winter, H. H. & Chambon, F. (1986). Analysis of Linear Viscoelasticity of a Crosslinking Polymer at the Gel Point. *Journal of Rheology*, Vol. 30, No. 2, pp. 367-382, ISSN 0148-6055

Winter, H. H. & Mours, M. (1997). Rheology of Polymer Near Liquid-Solid Transitions. *Advances in Polymer Science*. Vol. 34, pp. 165-234

Zosel, A. (1985). Adhesion and Tack of Polymers: Influence of Mechanical Properties and Surface Tensions. *Colloid & Polymer Science*, Vol. 263, pp. 541-553, ISSN 0303-402X

Zosel, A. (1989). Adhesive Failure and Deformation Behaviour of Polymers. *Journal of Adhesion*, Vol. 30, pp. 135-149, ISSN 0021-8464

Zosel, A. (1991). Effect of Cross-Linking on Tack and Peel Strength of Polymers. *Journal of Adhesion*, Vol. 34, pp. 201-209, ISSN 0021-8464

Zosel, A. (1997). The Effect of Bond Formation on the Tack of Polymers. *Journal of Adhesion Science and Technology*, Vol. 11, No. 11, pp. 1447-1457, ISSN 0169-4243

Zosel, A. (1998). The Effect of Fibrilation on the Tack of Pressure Sensitive Adhesives. *International Journal of Adhesion & Adhesives*, Vol. 18, pp. 265-271, ISSN 0143-7496

4

Fracture Behaviour of Controlled-Rheology Polypropylenes

Alicia Salazar and Jesús Rodríguez
Department of Mechanical Technology, School of Experimental Sciences and Technology, University of Rey Juan Carlos, Madrid, Spain

1. Introduction

The combination of economy, unique properties and ease of recycling offered by polypropylene (PP) leads to its widespread use as a commodity in the food and medical packing, automobile, furniture and toy industries (Karger-Kocsis, 1995; Pasquini, 2005). Some of these applications put stringent demands on PP properties to achieve the final commercial end-uses. PP produced in conventional reactors by fourth generation Ziegler Natta catalyst system presents relatively high molecular weight (M_w) and broad molecular weight distribution (MWD). These features cause high melt viscosity which makes them unsuitable for commercial end-uses such as fiber spinning, blown film and extruded and injection-molded thin-walled products.

The MWD largely determines the rheological properties of polypropylene melts and many physical properties. Therefore, this parameter must be controlled to improve the material response during processing and to achieve the diversity in polymer grades suitable for the different applications of polypropylene. Control of MWD of PP can be done in the conventional reactor through improvement in the polymerization technology or via a post-reactor operation in the original PP with high M_w and broad MWD by means of different degradation methods. The former is not profitable because it requires the addition of chain terminators and transfer agents which normally leads to decrease the output and increase the cost. Instead, the latter technique, such as reactive extrusion, consists of incorporating peroxides to induce controllable degradation of PP chains which efficiently results in polymers with tailor made properties. The polypropylenes prepared this way are termed "Controlled-Rheology Polypropylenes" (CRPP) and they have superior processing properties due to reduced M_w and narrowing MWD (Asteasuain et al., 2003; Azizi & Ghasemi, 2004; Azizi et al., 2008; Baik & Tzoganakis, 1998; Barakos et al., 1996; Berzin et al., 2001; Blancas & Vargas, 2001; Braun et al., 1998; Do et al., 1996; Gahleitner et al., 1995, 1996; Graebling et al., 1997; Pabedinkas et al., 1989; Ryu et al., 1991a, 1991b; Tzoganakis et al., 1989; Yu et al., 1990). The advantages of CRPPs versus reactor made PPs are less shear sensitivity, high elongation at break and heat distortion temperature, surface smoothness and better physical properties such as clarity and gloss (Pabedinkas et al., 1989; Baik & Tzoganakis, 1998). Moreover, the method of reactive extrusion is a simply operation with low cost and high productivity.

The main drawback of PP is its low fracture toughness, especially at low temperatures and/or high strain rates (Hodgkinson et al., 1983; Prabhat & Donovan, 1985). For that reason, a dispersed rubbery phase is incorporated in PP matrix inducing the appearance of toughening mechanisms via either physical blending or by copolymerization with other polyolefins with lower Tg than PP, such as the polyethylene, PE (Karger-Kocsis, 1995; Pasquini, 2005). The copolymerization with ethylene is preferable instead of blend systems due to the strong incompatible nature of polyethylenes and polypropylenes (Teh et al., 1994). The resulting block copolymers are heterophasic materials with a two phase structure where an elastomeric phase in form of spherical domains, usually ethylene-propylene copolymer rubber (EPR), is dispersed uniformly within the PP homopolymer matrix (Sun & Yu, 1991; Sun et al., 1991). As PP homopolymers, these ethylene-propylene block copolymers, EPBCs, also termed as Heterophasic Copolymers (HECO), can be also peroxide degraded to attain different grades. With controlled-rheology (CR) grades of copolymer the M_w is not only reduced by chain cleavage but the ethylene containing block of the copolymer also experiences chain growth. This can lead to a very small decrease in impact strength but the improved flow properties counteract this.

Reactor-made PPs and EPBCs are semicrystalline polymers which have been extensively investigated. The mechanical and fracture response of PPs produced via catalyzers is strongly influenced by the microstructure. There are many reports on the effect of crystallinity degree, size, distribution and shape of spherulites, the lamellar thickness and the crystalline orientation of reactor-made PPs (Allen & Bevington, 1989; Avella et al., 1993; Chen et al., 2002; Dasari et al, 2003b, 2003b; Fukuhara, 1999; Grein et al., 2002; Ibhadon, 1998; Ogawa, 1992; Pasquini, 2005; Sugimoto et al., 1995; Tjong et al., 1995; Van der Wal et al., 1998) and the ethylene content in case of EPBCs (Chen et al., 2002; Doshev et al., 2005; Fukuhara, 1999; Grein et al., 2002; Grellmann et al., 2001; Kim et al. 1996; Lapique et al., 2000; Starke et al., 1998; Sun & Yu, 1991; Van der Wal et al., 1999; Yokoyama & Riccò, 1998) on the mechanical properties.

In case of PP, crazes are the main mechanism of deformation and failure. They are formed in the weak points of the microstructure, normally nearby the intercrystalline regions. Craze structure and extension are closely related to the molecular weight. Several authors have proven that the crack initiation and growth, as well as the breakdrown, are controlled by the amorphous interconnections among the spherulites, the entanglement density of which improves as the molecular weight increases (Allen & Bevington, 1989; Azizi, 2004; Dasari et al, 2003a, 2003c; Fukuhara, 1999; Ibhadon, 1998; Sugimoto et al., 1995; Yokoyama & Riccò, 1998). Gahleitner et al., 1995, 1996, Xu et al., 2001, 2002, and Yu et al., 2003 have also shown that the impact strength is increased for polymers with small spherulite size and low crystallinity.

Failure in EPBCs is related to the dispersed elastomeric particles. Upon loading, small cavities are nucleated at the weak points of the copolymer such as the boundary between the EPR particles and the PP matrix or the intercrystalline zones in the PP matrix (Dasari et al., 200a; Grellmann et al., 2001; Starke et al., 1998; Van der Wal et al., 1999; Yokoyama & Riccò, 1998). These voids are stabilised by fibrillar bridges of PP filaments which are plastically deformed and orientated. As in case of PPs, these fibrillar bridges are stronger as the molecular weight is higher (Fukuhara, 1999; Ibhadon, 1998; Kim et al. 1996; Van der Wal et al., 1999; Yokoyama & Riccò, 1998).

Design engineers need fracture parameters for the construction of structural elements made of PPs and EPBCs in many technological and industrial applications. Although extensive research exists into the reactive mechanism, the rheological and crystallization behaviour as well as the mechanical properties of CRPPs (Asteasuain et al., 2003; Azizi & Ghasemi, 2004; Azizi et al., 2008; Baik & Tzoganakis 1998; Barakos et al., 1996; Berzin et al., 2001; Blancas & Vargas, 2001; Gahleitner et al., 1995, 1996; Pabedinkas et al., 1989; Ryu et al., 1991; Tzoganakis et al., 1989; Yu et al., 1990) and blends of PP and PE (Braun et al., 1998; Do et al., 1996; Graebling et al., 1997; Yu et al., 1990), there is little information on the fracture behaviour of peroxide degraded PP and EPBCs (Sheng et al., 2008) and even less concerned with the influence of supramolecular characteristics as the spherulite size and distribution on the fracture parameters of these materials. In Sheng et al., 2008, the effect on the fracture behaviour with the Post-Yield Fracture Mechanics approach was analyzed through the evaluation of essential work of fracture parameters of controlled-rheology EPBC films. The molecular weight of which was adjusted by reactive extrusion with the incorporation of dicumylperoxide. The results revealed the same tendency as in the case of reactor-made PPs or EPBCs with the same ethylene content but different molecular weight: the fracture toughness decreases as the molecular weight decreases as a result of the addition of peroxide. In addition, there are no explicit researches which tackle with the influence of the spherulite size and distribution on the fracture mechanics parameters of CRPPs, which allow us to extrapolate the results attained in reactor-made PPs (Allen & Bevington, 1989; Avella et al., 1993; Chen et al., 2002; Dasari et al., 2003a, 2003b; Fukuhara, 1999; Ibhadon, 1998; Van der Wal et al., 1998; Xu et al., 2001, 2002; Yu et al., 2002). Even more, most of the results found in the literature have been performed on thermally treated specimens with controlled spherulitic architectures and sizes (Avella et al., 1993; Chen et al., 2002; Dasari et al., 2003a, 2003b, 2003c; Gahleitner et al., 1995, 1996; Ibhadon, 1998; Wang et al., 2008; Xu et al., 2001, 2002; Yu et al., 2002) but not on real injected or extruded samples, ready for engineering or industrial applications.

For all those reason, the present chapter is focussed on:

1. Analyzing the evolution of the fracture toughness parameters of different grades of PP homopolymer and EPBCs prepared controlling the addition of peroxide in reactive extrusion. The fracture strength of these CRPPs and controlled-rheology EPBCs is determined through J-integral methodologies, paying special attention to the analysis of the micromechanisms of failure via optical microscopy and scanning electron microscopy. An analysis of the rheological, thermal and mechanical response is also provided.
2. Discussing the effect of the peroxide content, the structural parameters (MW, MWD) and the supramolecular characteristics such as the spherulite size and distribution on the fracture toughness parameters and the failure mechanisms. The analysis is carried out on real injected CRPPs in which neither thermal treatment was applied nor nucleating agents were used.
3. Evaluating the goals and limitations of controlled-rheology grades over conventional ones through the analysis of the fracture toughness parameters. For that, a controlled-rheology EPBC and a reactor-made EPBC with similar structural properties were examined.

2. Methodologies for the fracture characterization

In many in-service applications, PPs and EPBCs present a pronounced non linear mechanical response brought about by contained plastic deformation. Elastic-Plastic Fracture Mechanics (EPFM) approach through the J-integral methodologies is to be used to achieve the fracture toughness at crack initiation, J_{IC}, which is determined from the crack resistance curve, J-R curve, where J is plotted versus the crack extension, Δa. For the J-R curves construction of polymers, ESIS (Hale & Ramsteiner, 2001) and ASTM D6068 recommend the multiple specimen method. This methodology, though straightforward and effective, is time and material intensive, as at least a minimum of seven specimens have to be tested to generate the R-curve. For that reason, indirect methods have been developed to obtain J-R curves with fewer specimens and, thus, less time requirements. The single specimen methods are based on the load separation criterion (Ernst et al., 1981), and offer an easy and effective alternative approach to obtain J-R curves. Among the single specimen methods, the normalization method (Baldi & Riccò, 2005; Bernal et al., 1996; Landes and Zhou, 1993; Morhain & Velasco, 2001; Rodríguez et al., 2009; Varadarajan et al., 2008) and the load separation parameter method have been successfully applied to polymeric materials (Rodríguez et al., 2009; Salazar & Rodríguez, 2008; Wainstein et al., 2007). Next, a brief description of the methods utilized is presented.

2.1 Multiple specimen method

The multiple specimen method, first proposed by Landes and Begley, 1974, is the most common approach for deducing J-R curves. The basic guidelines of ESIS TC4 Protocol "J-Fracture toughness of polymers at slow speed" are applied (Hale & Ramsteiner, 2001). A set of identical specimens is loaded to various displacements, unloaded, cooled at low temperatures and finally fractured. The initial and final stable crack lengths are measured physically from the broken surfaces, while J is calculated from the total energy required to extend the crack, U, which is determined from the area under the load versus load-point displacement curve up to the line of constant displacement corresponding to the termination of the test:

$$J = \frac{2U}{B(W - a_0)} \tag{1}$$

where B is the specimen thickness, W is the specimen width and a_0 is the initial crack length.

2.2 Normalization method

The J-R curves obtained via the normalization method is focussed on determining accurate crack length predictions using the load (P)-displacement (δ) data alone. The instructions given by ASTM E1820-06 were taken as a guide. Although this test method covers the procedures for the determination of fracture toughness of metallic materials, it has proven its applicability to polymeric materials (Baldi & Riccò, 2005; Bernal et al., 1996; Landes and Zhou, 1993; Morhain & Velasco, 2001; Rodríguez et al., 2009; Varadarajan et al., 2008).

The first step for the determination of the J-R curve is an optical crack-length measurement of the initial, a_o, and final, a_f, crack lengths. Subsequently, each value of the load P_i up to, but not including P_{max} is normalized using the following expression:

$$P_{Ni} = \frac{P_i}{WB\left[\frac{W - a_{bi}}{W}\right]^{\eta pl}} \tag{2}$$

with η_{pl} = 2 for three point bending specimens (SENB). a_{bi} is the blunting corrected crack length given by:

$$a_{bi} = a_o + \frac{J_i}{2m\sigma_{ys}} \tag{3}$$

$$J_i = \frac{K_i^2\left(1-\nu^2\right)}{E} + J_{pli} \tag{4}$$

where σ_{ys} is the yield stress and m is a dimensionless constant that depends on stress state and materials properties, named as crack tip constraint factor. m=1 has been reported for PP based materials (Morhain & Velasco, 2001; Rodríguez et al., 2009). In turn, K_i is the stress intensity factor, E is the Young´s modulus, ν is the Poisson´s ratio and J_{pl} is the plastic part of the J-integral according to ASTM E1820-06.

Each corresponding load line displacement, δ_i, is normalized to give a normalized plastic displacement:

$$\delta'_{pli} = \frac{\delta_{pli}}{W} = \frac{\delta_i - P_i C_i}{W} \tag{5}$$

being C_i the specimen elastic load line compliance, based on the crack length a_{bi}.

Thus, data points up to maximum force are normalized. In order to obtain the final point, the same equations are employed, but instead of the initial crack length, the final crack length is used. The normalized plastic displacement values above 0.001 up to maximum force, excluding P_{max} value itself, and the points obtained with the use of the final crack length are used for the normalization function fit. The normalization function can be analytically expressed:

$$P_N = \frac{a + b\delta'_{pl} + c\delta^{2'}_{pl}}{d + \delta'_{pl}} \tag{6}$$

where a, b, c and d are searched fitting coefficients. When the fitting parameters are determined, an iterative procedure is further applied to force all P_{Ni} data to lie on the fitted curve by ai adjustment. Knowing P_i, δ_i and a_i values the construction of the J-R curve is followed with J-integral values given by:

$$J = \frac{2U_i}{B(W - a_0)} \tag{7}$$

where U_i is determined from the area under the load versus load-point displacement curve up to (P_i, δ_i); versus $\Delta a_i = a_i - a_o$.

2.3 Determination of the crack growth initiation energy, J_{IC}

Once the J-R curve is constructed either by the multiple specimen or the normalization method, it should be described by a power law $J=C\cdot\Delta a^N$, with $N\leq1$. The crack initiation resistance or fracture toughness, J_{IC}, is calculated as a pseudo-initiation value $J_{0.2}$, which defines crack resistance at 0.2 mm of the total crack growth (Hale & Ramsteiner, 2001).

The size requirements for plane strain J_{IC} are given by (Williams, 2001):

$$B, a, W-a > 25\frac{J}{\sigma_{ys}} \tag{8}$$

3. Materials and sample preparation

The materials under study were an ISPLEN polypropylene homopolymer, PP0, and two ethylene-propylene block copolymers, EPBC0 and EPBC0-2, with an ethylene content of ~8 wt%, supplied by REPSOL Química. They were manufactured using a Spheripol process with a fourth generation Ziegler-Natta catalyst. The peroxide used was di-tert-butylperoxide (DTBP).

The reactive experiments and sample preparation were carried out using a twin-screw extruder, Werner & Plfeiderer (model ZSK-30), with a length/diameter ratio of the dies, L/D, of 25. In case of PP, the peroxide and the PP0 were premixed to prepare master batches using 0, 154, 402 and 546 ppm of peroxide content. Instead, for the copolymers, the starting reactor-made copolymer for the reactive extrusion experiments was EPBC0. So, the peroxide was added to the EPBC0 to obtain master batches with peroxide contents of 0, 101, 332 and 471 ppm. Independently of the grade, both components were inserted into the extruder. Experiments were performed at a profile temperatures of 190, 220, 240, 220, 200 °C for the PP and of 190, 220, 240, 220 °C for the EPBC with a screw rotation speed of 150 rpm in both cases. Extrudates were cooled through a water bath and were granulated. Samples for mechanical and fracture properties were injection moulded.

4. Experimental procedure

4.1 Melt flow rate and molecular weight characterization

Melt flow rate (MFR) was measured following the ISO 1133 standard using a Ceast 6932 extrusion plastometer at 230°C/2.16 kg.

The molecular weight distributions were determined with Gel Permeation Chromatography (GPC) using a Polymer Labs PL220 equipment and taking ISO 14014 as a guide. The samples were dissolved at 143 °C in 1,2,4 trichlorobenzene at a polymer concentration of 1.3 mg/ml. A phenolic antioxidant Irganox 1010 was added to the solution to prevent any degradation.

4.2 Thermal analysis

The apparent melting temperature, T_m, the crystallinity temperature, T_c, and the crystallinity index, α, of all the samples were measured via Differencial Scanning Calorimetry (DSC) using a Mettler-Toledo (model DSC822) equipment. Two scans were done at 10 °C/min, from 0 to 200 °C under nitrogen atmosphere, in aluminum spans with 10 mg of sample. The values of T_c were obtained from the maxima of the crystalline peaks meanwhile the values of T_m and the apparent enthalypy, ΔH, were calculated from the maxima and the area of the melting peaks, respectively. The crystallinity index via this technique was determined using

$$\alpha = \frac{\Delta H}{\Delta H^0} \tag{9}$$

where the enthalpy of fusion of 100% pure crystalline α-PP, ΔH^0, was taken as 190 J/g (Pasquini, 2005).

4.3 Supramolecular characterization: Determination of spherulite size

Films with 3 µm in thickness were sectioned from the centre of bulk injected specimens with a microtome (Rotary Microtome Leica RM2265) as the fracture process in the notched specimens occurs within the sample. The films were immersed in a solution containing 1.3 wt% potassium permanganate, 32.9 wt% dry H3PO4 and 65.8 wt% concentrated H2SO4 for 36 h at room temperature (Thomann et al., 1995; Wei et al., 2000). The etched samples were subsequently washed in a mixture of 2 parts by volume of concentrated sulphuric acid and 7 parts of water for 15 minutes in an ultrasonic bath. The resulting sections were picked up and mounted on microscope slides to be analyzed via transmitted light microscopy (Leica DMR) or Au-Pd sputter coated for scanning electron microscopy (Hitachi S-3400N).

Ten scanning views were chosen randomly under enlargement at 1000x for quantitative determination. The image analysis software Image Pro-Plus 4.5 was used to obtain the distribution and spherulite size, taken as the maximum length inside the spherulite. Percentage of porosity was also determined from the microstructural images.

4.4 Mechanical characterization

Tensile tests were carried out, following the ISO527-2:1997 standard, in order to measure the yield stress, the stress at break and the strain at break. Specimens with 10x115x4 mm in the narrow section were tested on an electromechanical testing machine (MTS Alliance RT/5) under displacement control at a cross-head speed of 50 mm/min. Strain values were measured with a high strain extensometer (model MTS DX2000) attached to the sample.

Flexure tests were performed following the guidelines described in ISO178:2003 standard to determine the Young´s modulus. Samples with 10x80x4 mm were tested on an electromechanical testing machine (INSTRON 4465) under stroke control at a cross-head speed of 2 mm/min.

4.5 Fracture characterization

At room temperature and under low loading rates, PPs and EPBCs present a pronounced nonlinear mechanical response. The ESIS TC4 Protocol entitled "The determination of J-fracture toughness of polymers at slow speed" (Hale & Ramsteiner, 2001), was followed to achieve the J-R curves from multiple specimens of CRPPs. In turn, the guidelines given by

ASTM E1820-06 were ensued to achieve via the normalization method the J-integral response of controlled-rheology EPBCs.

Fracture toughness tests were carried out at 23 °C on single edge notch bending (SENB) specimens obtained directly from the mould with 6x18x79 mm in size and an initial notch length of 8.1 mm. A sharp crack was created by sliding a razor blade into the root of the mechanized notch. The resulting initial crack length was ~ 9.0 mm. Some recent results question the validity of this notch sharpening procedure because some damage can be produced ahead of the crack tip (Salazar et al., 2010a, 2010b, 2010c). Therefore, the fracture toughness obtained from this type of specimens should be considered as an upper limit.

The tests were conducted for each material at room temperature and under displacement control at a cross-head speed of 1 mm/min using a three-point bend fixture of 72 mm loading span. An electromechanical testing machine (MTS Alliance RF/100) with a load cell of ± 5 kN was utilized.

After the tests, the fracture surfaces of the broken specimens were examined using light microscopy (Leica DMR) and scanning electron microscopy (Hitachi S-3400N) to analyze the extension of the stress-whitened region due to plastic deformation as well as the micromechanisms of failure. For scanning electron microscopy analysis, the samples were Au-Pd sputter coated.

5. Results and discussion

5.1 MWD and MFR

5.1.1 Controlled-rheology PPs

Table 1 collets the MFR and the GPC results for the polypropylene with 0, 154, 402 and 546 ppm of peroxide content termed as PP0, PP-CR154, PP-CR402 and PP-CR546, respectively. Several conclusions can be drawn from the analysis of these data. Firstly, MFR increased remarkably with increasing the peroxide content. Secondly, controlled-rheology polypropylene drops more abruptly the weight average molecular weight, M_w, than the number average molecular weight, M_n. As a consequence, the MWD, described by the ratio M_w/M_n, got narrower. The reduction in the MWD reached up to ~ 40% and was in agreement with the decrease observed by other researchers for similar peroxide contents (Baik & Tzoganakis, 1998; Barakos et al., 1996; Berzin et al., 2001; Gahleitner et al., 1995, 1996). Thirdly, the main differences in M_w were found for a small concentration of peroxide, PP-CR154, while there were no pronounced changes among the controlled-rheology polypropylenes. Finally, all these results confirm that increasing the amount of peroxide leads to more chain scission, reducing efficiently the length of the chains.

	MFR (g/10min)	M_w (kg/mol)	M_n (kg/mol)	MWD (M_w/M_n)
PP0	11	319.1	51.7	6.16
PP-CR154	26	238.6	49.5	4.81
PP-CR402	48	179.4	45.6	3.93
PP-CR546	59	172.0	44.0	3.90

Table 1. Melt flow rate (MFR), weight and number average molecular weights (M_w and M_n) and molecular weight distribution (MWD) of the PP, PP-CR154, PP-CR402 and PP-CR546

5.1.2 Controlled-rheology EPBCs

Table 2 summarizes the ethylene content, MFR and the GPC results for the ethylene-propylene block copolymers with 0, 101, 332 and 471 ppm of DTBP content termed as EPBC0, EPBC-CR101, EPBC-CR332 and EPBC-CR471, respectively. As expected, the MFR increased noticeably with increasing the peroxide content, decreasing also the average molecular weights of EPBC-CRs. However, the MWD reduced as much as 30%. This diminish contrasted with those obtained for controlled-rheology polypropylenes homopolymers, which underwent much more degradation for similar peroxide contents (Table 1). Berzin et al., 2001 noted the presence of high residual masses, which indicates the presence of long chains, for the copolymers even for the more degraded products. PP/PE mixtures are known to show opposite effects during peroxide-initiated scission reactions (Berzin et al., 2001; Braun et al., 1998; Do et al., 1996; Graebling et al., 1997). While for the PP matrix, the free radicals formed from the thermal decomposition of the organic peroxide lead to β-scissions because of the low stability of the tertiary hydrogen atoms of macroradicals, for the PE, peroxide attack leads to chain branching and to crosslinking by macroradical recombination. These two competing mechanisms could be the reason for the much lesser reduction in MWD for copolymers than in homopolymers during peroxide degradation.

	Et (wt%)	MFR (g/10min)	M_w (kg/mol)	M_n (kg/mol)	MWD (M_w/M_n)
EPBC0	7.78	12	287.0	56.4	5.09
EPBC0-2	8.62	18	264.1	56.0	4.72
EPBC-CR101	8.19	20	253.0	53.3	4.75
EPBC-CR332	8.36	41	199.4	53.5	3.73
EPBC-CR471	7.91	63	178.2	50.3	3.54

Table 2. Ethylene content (Et), melt flow rate (MFR), weight and number average molecular weights (M_w and M_n) and molecular weight distribution (MWD) of the EBPC0, EPBC0-2, EPBC-CR101, EPBC-CR332 and EBPC-CR471.

This table also shows the basic rheological properties of another reactor-made EPBC, EPBC0-2, with similar characteristics to the peroxide degraded copolymer with the lowest peroxide content, EPBC-CR101.

5.2 Thermal properties

5.2.1 Controlled-rheology PPs

Table 3 shows the apparent melting temperature, T_m, the crystallinity temperature, T_c, the apparent enthalpy, ΔH, and the crystallinity index, α, obtained from DSC measurements for polypropylenes PP0, PP-CR154, PP-CR402 and PP-CR546. All the values measured are in accordance with data reported in the literature (Karger-Kocsis, 1995; Pasquini, 2005). T_m decreased with the peroxide content while T_c and α remained constant. The evolution of T_m with the peroxide content underwent the same trend as that of the M_w with peroxide content among the controlled-rheology polypropylenes (Table 1). However, the present results stood up for the no enhancement of either the crystalline degree or the crystallinity temperature with the peroxide content. Therefore, the total mass fraction of lamellae must be constant and the major expected difference in structure must be an increase in entanglement density of the amorphous zone with molecular weight attending to the tendency shown by T_m with molecular weight. This behaviour was also reported by Tzoganakis et al., 1989 for polypropylene homopolymers.

	T_m (ºC)	T_c (ºC)	ΔH (J/g)	α (%)
PP0	165	118	112	56
PP-CR154	163	118	104	55
PP-CR402	161	118	104	55
PP-CR546	161	118	104	55

Table 3. Melting temperature, T_m, crystallinity temperature, T_c, apparent enthalpy, ΔH, and crystallinity index, α, obtained from DSC measurements of the polypropylenes PP0, PP-CR154, PP-CR402 and PP-CR546

5.2.2 Controlled-rheology EPBCs

The results of the thermal analysis carried out on EPBCs with different DTBP content are listed in Table 4. As in case of the PP homopolymers, the melting peak temperature reduced with the peroxide content while the crystallization temperature was almost constant. However, the main difference was found in the crystallinity degree, which slightly decreased with the peroxide content. During the degradation of the copolymer, both chain scissions in the PP matrix and crosslinks in the elastomeric phase were produced. Berzin et al., 2001, Braun et al., 1998 and Do et al., 1996 have reported that extensive crosslinks of the elastomeric phase should disturb the crystallinity, reducing not only the size but even the amount. This might be the reason of the slight reduction in the crystalline behaviour of the EPBC-CRs and was confirmed when comparing the crystallinity degree between the reactor made EPBC0-2 and the controlled-rheology EPBC-CR101. Although these reactor-made and controlled-rheology grades presented similar structural properties (Table 2), the reactor-made showed higher crystallinity degree than the peroxide degraded EPBC. This clearly responds to the crosslinking disturbance in the spherulite architecture of the latter.

	T_m (ºC)	T_c (ºC)	ΔH (J/g)	α (%)
EPBC0	164	123	95	53
EPBC0-2	164	118	95	53
EPBC-CR101	164	122	94	51
EPBC-CR332	163	122	93	50
EPBC-CR471	162	123	91	49

Table 4. Melting temperature, T_m, crystallinity temperature, T_c, apparent enthalpy, ΔH, and crystallinity index, α, obtained from DSC measurements of the various EPBCs

5.3 Spherulite size and distribution characterization

The crystal morphology of PP0 and controlled-rheology PPs, PP-CR154, PP-CR402 and PP-CR546, is shown in Figures 1a, 1b, 1c and 1d, respectively. The spherulitic morphology is composed of an aggregate of lamellae that radiate from the centre outward, typical of monoclinic α-form of isotactic PP, iPP (Karger-Kocsis, 1995; Pasquini, 2005; Thomann et al., 1995). The spherulite size was uneven and it seemed to increase with the addition of peroxide content. Moreover, small pores with ~ 1 μm in diameter are also discernible in those images associated to controlled-rheology polypropylenes with high peroxide content. These pores were originated during the sample manufacture mainly due to poorly residual left peroxide removal as no pores were found in any of the analyzed PP0 images. The percentage of porosity is collected in Table 5. As shown, porosity increased with peroxide content, attaining the highest increase for the materials with 402 and 546 ppm of peroxide.

	Porosity (%)	$\bar{D}$ (μm)	σ^2 (μm^2)
PP0	0	21.8	15.8
PP-CR154	4.8 ± 0.1	27.5	13.7
PP-CR402	7.2 ± 0.2	33.3	7.8
PP-CR546	13.6 ± 1.1	35.2	8.4

Table 5. Porosity and normal distribution parameters of the spherulite size of propylene homopolymer, PP0, and the controlled-rheology-polypropylenes, PP-CRs: mean spherulitic size, $\bar{D}$, and variance, σ^2.

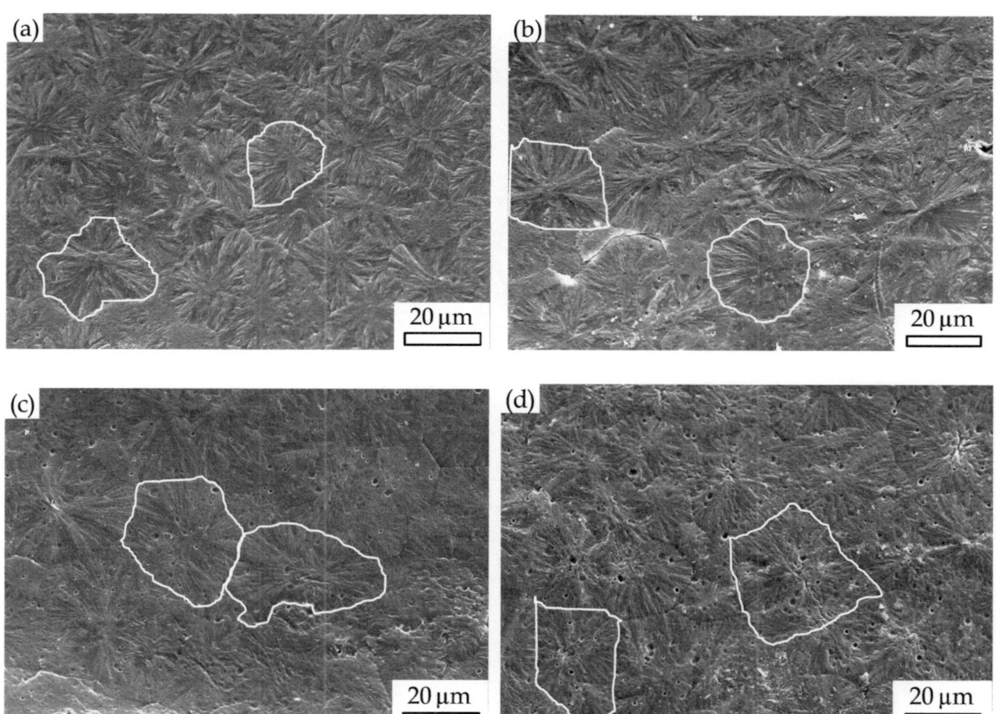

Fig. 1. SEM images at a magnification of x1000 showing the spherulitic mofphology of (a) polypropylene homopolymer, PP0, and controlled-rheology polypropylenes, CRPPs, with (b) 154, (c) 402 and (d) 546 ppm of peroxide content. By way of example some spherulites have been outlined with white lines.

The resulting analysis of the spherulite size led to histograms that were approximated to a normal or Gaussian distribution as the kurtosis for all the materials reached values of utmost 0.15, whereas the perfect Gaussian distribution takes values of 0 (Sheskin, 2000). Thus, the normal distribution in spherulite size for all the materials under study is a number average and are combined in Figure 2, whereas the statistical parameters as the mean spherulitic size, $\bar{D}$, and the variance, σ^2, are collected in Table 5. Firstly, the addition of peroxide increased the spherulite size but not the crystalline percentage (Table 3). The rise in peroxide content drives more chain scission and therefore, reduces efficiently the length of the chains. Although some works are found in the literature reporting the increase in the crystalline fraction as the decrease of chain length promotes more nucleating sites and so easier forming of crystals (Gahleitner et al., 1995, 1996; Azizi & Ghasemi, 2004; Wang et al., 2008), the present results stand up for the no enhancement of the crystalline degree with the peroxide content because the total mass fraction of lamellae maintains constant with the only growth in the spherulite size at the expense of the reduction in the density of tie molecules (Tzoganakis et al., 1989; Ryu et al., 1991). Secondly, the distribution width, represented by σ^2, tended to decrease with the peroxide content. This indicates that the addition of peroxide not only seems to increase the spherulite size but also to a slightly more uniform distribution (Xu et al., 2002).

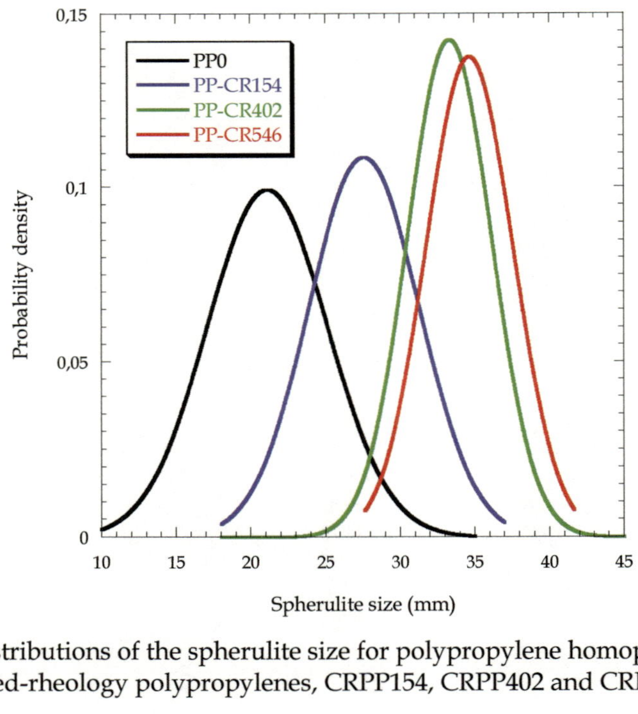

Fig. 2. Normal distributions of the spherulite size for polypropylene homopolymer, PP0, and controlled-rheology polypropylenes, CRPP154, CRPP402 and CRPP546.

5.4 Mechanical characterization

5.4.1 Controlled-rheology PPs

The average Young´s modulus, E, the yield stress, σ_{ys}, the tensile strength, σ_t, and the elongation at break, ε_r, together with their corresponding standard deviations, are summarized in Table 6. Regarding the elastic properties, the Young´s modulus decreased with the addition of peroxide. The elastic modulus is highly dependent on the amount of stiffest part of the structure, that is, the crystalline region (Avella et al., 1993; Dasari et al., 2003a, 2003c; Gahleitner et al., 1995; Van der Wal et al., 1998). The crystallinity percentage reduced scarcely with a small amount peroxide and held constant for higher additions of peroxide. The evolution of the Young´s modulus with the peroxide content does not correlate with the trend shown by the percentage of crystalline with the peroxide addition. Nevertheless, apart from the crystallinity degree, the Young´s modulus is highly sensitive to the presence of defects as pores. The morphological analysis performed to reveal the spherulites evidenced the presence of porosity due to the pore peroxide removal (Figure 1). The percentage of porosity increased with the peroxide content, reaching values up to ~ 14% for the PP-CR546 (Table 5). Therefore, the elastic modulus evolves with the peroxide content just the opposite way as the porosity does.

Concerning the strength properties, the yield stress showed a slight decrease with a small addition of peroxide but seemed not to be influenced by the amount included. On the contrary, the stress and strain at break presented a significant change with the peroxide content. Both magnitudes tended to diminish as the amount of peroxide increased and this tendency is in accordance with the relationships between the mechanical properties and the molecular weight reported for reactor-made grades previously (Dasari et al., 2003a, 2003b, 2003c; Ogawa, 1992; Sugimoto et al., 1995). The reduction in the molecular weight with the peroxide content involves changes above all in the amorphous part and the results obtained for the tensile stress at break and the strain at break evidence that they are predominantly governed by this amorphous region. However, an exception is noted. The strain at break of CRPP154 diverted from this tendency and the value obtained was the highest. This observation was also shown by Azizi & Ghasemi, 2004, who assumed that at low levels of peroxide content the high molecular weight tail of PP matrix was degraded while the low molecular weight remained unchanged and therefore the slippage of polymer chains could be easier and greater.

	E (MPa)*	σ_{ys} (MPa)	σ_t (MPa)	ε_r (%)
PP0	1336	33.9 ± 0.3	20.4 ± 0.3	145 ± 50
PP-CR154	1231	31.5 ± 0.4	19.0 ± 0.5	320 ± 110
PP-CR402	1245	31.9 ± 0.3	17.5 ± 0.9	90 ± 60
PP-CR546	1241	31.1 ± 0.4	16 ± 2	44 ± 12

Table 6. Mechanical properties, such as the Young´s modulus, E, the yield stress, σ_{ys}, the tensile strength, σ_t, and the elongation at break, ε_r, of the propylene homopolymer, PP0, and the controlled-rheology polypropylenes, PP-CR154, PP-CR402 and PP-CR546.*The standard deviations of the Young´s modulus were not available.

5.4.2 Controlled-rheology EPBCs

The mechanical properties of the reactor-made copolymers, EPBC0 and EPBC0-2, and controlled-rheology EPBCs are presented in Table 7. Regarding the elastic properties, the Young´s modulus decreases with the peroxide content, which is not surprising since the cristallinity is reduced with the peroxide treatment (Table 4). As already discussed for the polypropylenes, this phenomenon can be explained with the help of the degree of cristallinity. The elastic modulus is highly dependent on the stiffest part of the structure, that is, the crystalline region (Avella et al., 1993; Dasari et al., 2003a; Van der Wal et al., 1998); thus, the lower the latter, the lower the former.

The yield stress reduces with increasing peroxide content and this tendency is in accordance with the relationships between the mechanical properties and the molecular weight reported elsewhere (Ogawa, 1992; Dasari et al., 2003b; Van der Wal et al., 1999). However, in this case two exceptions are remarkably. Firstly, the elongation at break increases with the addition of DTBP up to 101 ppm concentration and decreases after that. This behaviour has been also shown for the polypropylenes (Table 6) and a possible explanation to this fact was given by Azizi & Ghasemi, 2004. It is assumed that at low levels of peroxide content the high molecular weight tail of PP matrix was degraded while the low molecular weight remained unchanged and therefore the slippage of polymer chains could be easier and greater. Secondly, the other parameter that shows an anomalous trend is the tensile strength, which initially decreased with the addition of peroxide but for greater amounts of DTBP, a sudden increase occurs. At this degree of copolymer degradation, probably the elastomeric phase could be constituted by either solid crosslinked particles or high molecular weight highly branched species and the failure properties were transferred to the minority phase (Berzin et al., 2001).

	E (MPa)	σ_{ys} (MPa)	σ_t (MPa)	ε_r (%)
EPBC0	1165 ± 20	25.2 ± 0.5	17.5 ± 0.5	92 ± 50
EPBC0-2*	1291	23.4	16.5	161
EPBC-CR101	1061 ± 15	23.9 ± 0.3	17.2 ± 0.3	175 ± 42
EPBC-CR332	1022 ± 31	22.6 ± 0.1	15.1 ± 0.5	49 ± 23
EPBC-CR471	995 ± 30	22.2 ± 0.1	17.5 ± 0.3	58 ± 4

Table 7. Mechanical properties, such as the Young´s modulus, E, the yield stress, σ_{ys}, the tensile strength, σ_t, and the elongation at break, ε_r, of the reactor-made copolymers, EPBC0 and EPBC0-2, and the controlled-rheology EPBCs, EPBC-CR101, EPBC-CR332 and EPBC-CR471. *The standard deviations of the mechanical properties of EPBC0-2 were not available.

With the aim of evaluating the influence of the reactive extrusion on the mechanical properties, the comparison of the mechanical properties between the reactor-made EPBC0-2 and the controlled-rheology EPBC-CR101 with similar structural properties draws some information. The elastic modulus of the reactor-made copolymer is higher than that of the peroxide degraded copolymer and this is in accordance with the crystallinity degree (Table 4). The percentage of crystallinity of the reactor-made copolymer is slightly higher than that of the controlled-rheology copolymer as in the former there are no crosslinkings in the elastomeric phase which lead to obstruct the crystalline architecture. In contrast, the yield stress, tensile strength and elongation at break of EPBC0-2 are a bit lower than those of the EPBC-CR101. The crosslinks produced in the elastomeric phase during the peroxide degradation give rise to copolymers with the capacity to absorb more energy than the reactor-made copolymer.

5.5 Fracture characterization

5.5.1 Fracture behavior of controlled-rheology PPs

Figure 3 shows the load (P) - load line displacement (δ) records obtained from fracture tests at room temperature and at quasi-static conditions (low loading rates) of propylene homopolymer, PP0, and thee controlled-rheology PPs. The mechanical response for all the materials presented clearly elastic-plastic behaviour and this justifies the use of the EPFM multiple specimen method to evaluate the fracture behaviour. In addition, all the curves deviated from linearity and at a certain deflection level, sudden instability occurred and the specimen broke in two halves. The difference in stiffness is due to the different initial crack lengths.

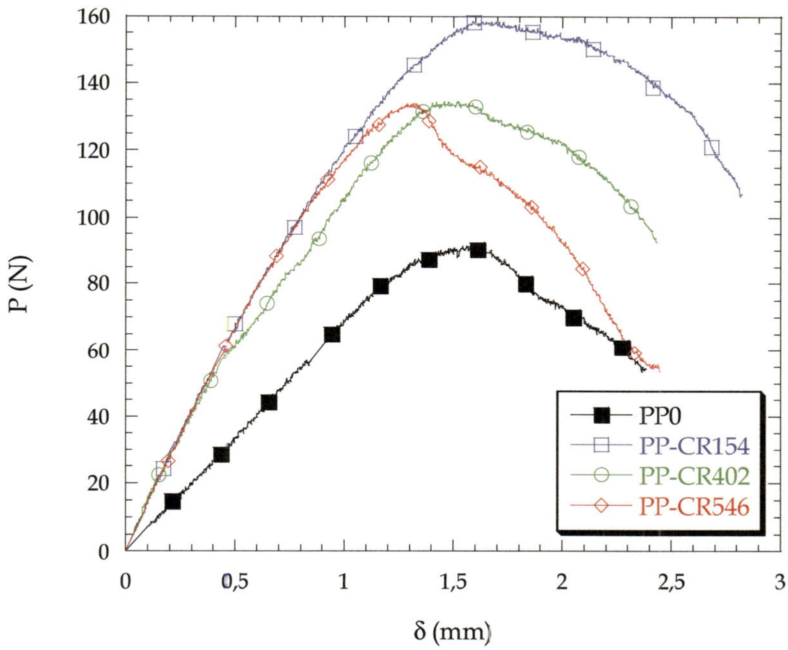

Fig. 3. Load (P) versus load-line displacement (δ) of the polypropylene homopolymer, PP0, and controlled-rheology polypropylenes, CRPP154, CRPP402 and CRPP546.

Figure 4 presents the J-R curves constructed for the propylene homopolymer, PP0, and the controlled-rheology PPs, PP-CR154, PP-CR402 and PP-CR546. These plots also include the fit of the J-crack growth resistance curve to the power law $J=C \cdot \Delta a^N$, with $N \leq 1$. The reactor-made polypropylene was the toughest material and the addition of peroxide acts in detriment of the toughness. This is indicated more clearly in Figure 5 where the crack growth initiation energy, J_{IC}, determined following the guidelines described by Hale & Ramsteiner, 2001, is plotted. All the fracture toughness values verified the size criterion described by equation 8, the fulfilment of which guarantees the plane strain state. As expected, the fracture toughness values drop abruptly with a small addition of peroxide and this decrease continues gradually with peroxide content.

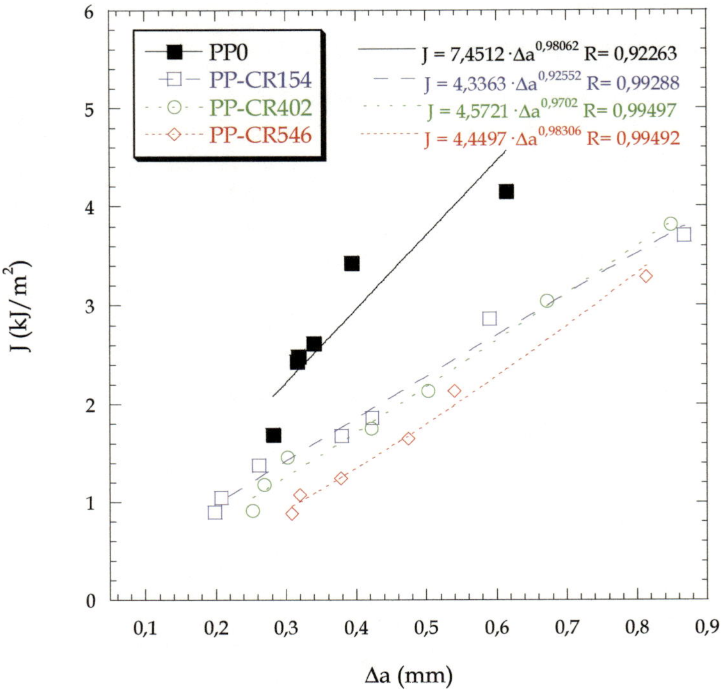

Fig. 4. J-R curves of polypropylene homopolymer, PP0, and controlled-rheology polypropylenes, PP-CR154, PP-CR402 and PP-CR546.

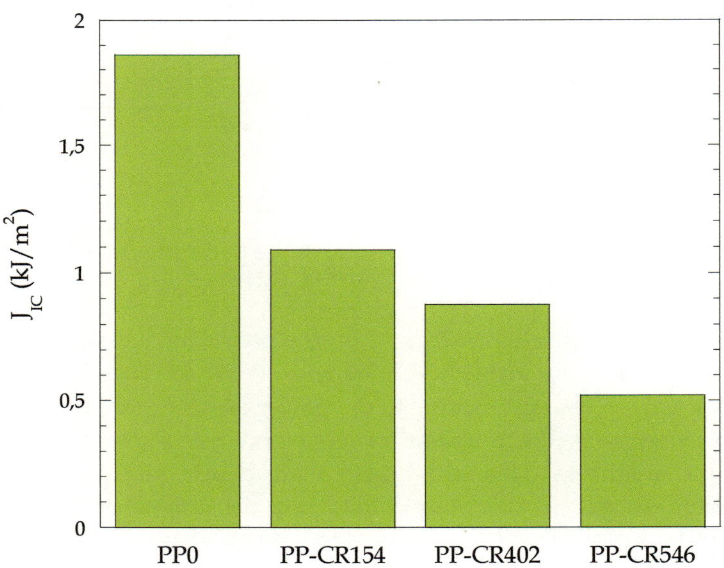

Fig. 5. Evolution of the crack growth energy, J_{IC}, with peroxide content of the polypropylenes under study.

The loss of ductility with the addition of peroxide can be easily observed in Figure 6, which shows the micrographs obtained via light microscopy and performed on the fracture surfaces of the tested specimens broken afterwards at liquid nitrogen and high loading rates to reveal the stable crack length. All the polypropylenes underwent stable crack length with ~ 0.3 mm in size. Independently of the material, three regions were distinguishable. The first one, close to the notch tip, is attributed to stable crack propagation. This zone is followed by a stress whitened area and ahead of it; the remainder of the fracture surface is characterized by a rough, un-whitened and uniform area related to the virgin material. Two main features can be drawn from the analysis of these fracture surfaces. Firstly, the stable crack length is more and more diffuse with the peroxide content. Indeed, to minimise the risk of incorrect measurements in the crack length, scanning electron microscopy means were to be used in the peroxide degraded polypropylenes instead of light microscopy. Secondly, the intensity and the extension of the stress- whitened region ahead of the stable crack length were

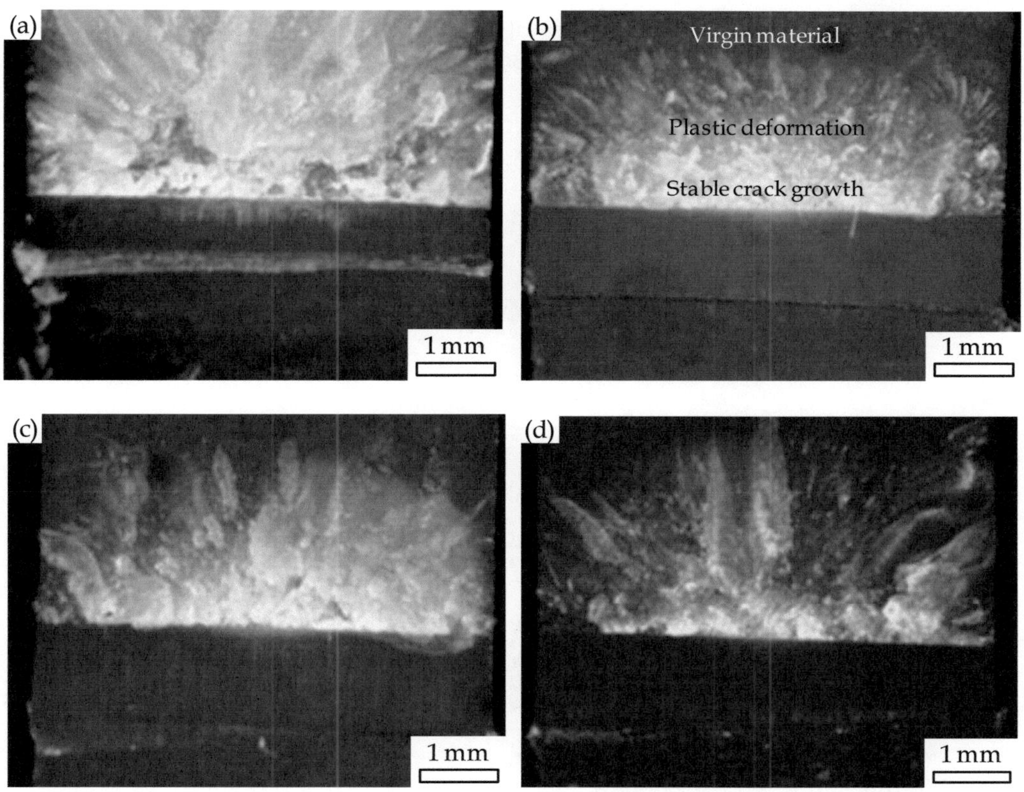

Fig. 6. Fracture surfaces obtained via light microscopy of (a) PP0, (b) PP-CR154, (c) PP-CR402 and (d) PP-CR546. All the specimens showed a stable crack length ~ 0.3 mm. By way of example, the three characteristic regions appearing in the fracture surfaces are indicated in CRPP154 micrograph.

reduced with the addition of peroxide. This very tendency has been also observed previously (Avella et al., 1993; Fukuhara, 1999; Sugimoto et al., 1995). They focused their research on understanding the effect of Mw on the fracture toughness of semicrystalline polymers, and observed that the fracture toughness increased strongly with increasing molecular weight. The key factor were the tie molecules which join the lamellae bundles together. As the molecular weight lowers, the number of tie molecules decreases, the material becomes less interconnected and the fracture occurs at lower stress and strain levels.

Finally, the fractographic study revealed little differences among the four materials with different rheological properties (Figure 7). The analysis of the stable crack region close to the notch displayed spherulites interconnected by amorphous regions, being the interspherulitic links breakage the dominant mode of failure (Lapique et al., 2000). Consequently, this type of morphology indicates that crazes are the main micromechanism of deformation. Upon loading, spherulites are separated from one another and sustained by tie molecules, the density and strength of which are diminished with the peroxide content. With increasing loading, craze area spreads out and propagates within the interspherulitic and amorphous regions. The cracks are formed when these fibrillar bridges fail (Avella et al., 1993; Botsis et al., 1989; Chen et al., 2002; Dasari et al., 2003a, 2003b, 2003c; Ibdhadon, 1998; Ogawa, 1992; Sugimoto et al., 1995).

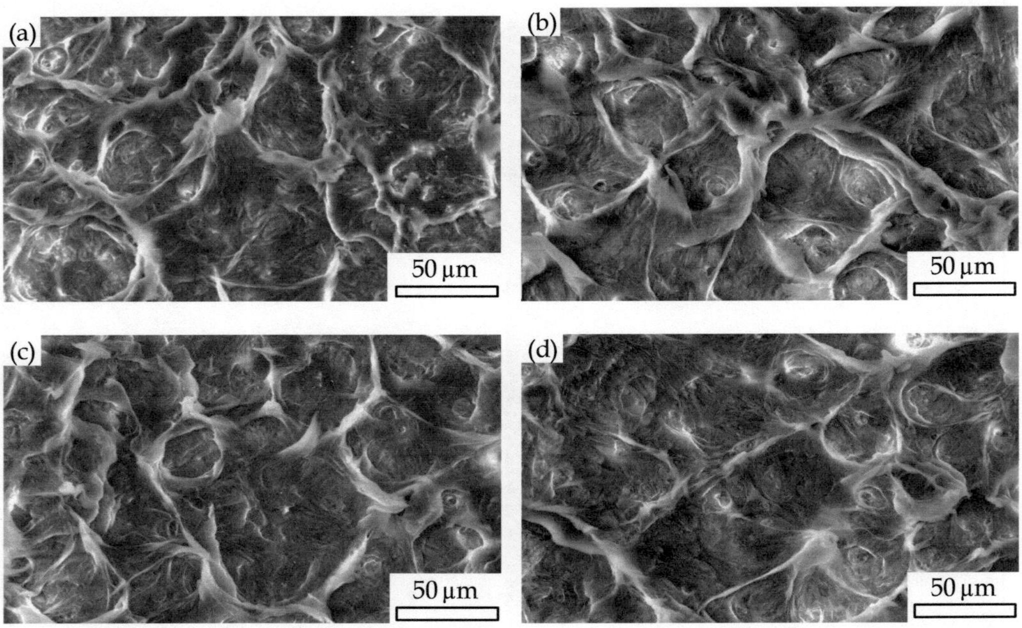

Fig. 7. Fracture surfaces obtained via scanning electron microscopy of the stable crack growth region close to the notch: (a) PP0, (b) CR-PP154, (c) CR-PP402 and (d) CR-PP546.

5.5.2 Fracture behavior of controlled-rheology EPBCs

Figure 8 presents the typical load (P) – load line displacement (δ) diagrams obtained from fracture tests at room temperature and at low loading rates for the reactor-made copolymer, EPBC0, and the controlled-rheology copolymers taking as the starting copolymer EPBC0, EPBC-CR101, EPBC-CR332 and EPBC-CR471. The mechanical response for all the copolymers under study was clearly nonlinear with complete stable fracture. Therefore, the appropriate fracture mechanics procedure to characterize the fracture behaviour of these copolymers was the EPFM normalization method. Although all the materials showed not only the same geometry and size but also similar initial crack lengths, it is evident that the area under the curve, that is, the energy absorbed by the specimen, differs from that of the non-peroxide treated copolymer, EPBC0, to that with a small content of peroxide, EPBC-CR101, and even more to those with higher peroxide content as EPBC-CR332 and EPBC-CR471. The load relaxation, related to stable crack growth, occurred at lower load values as the peroxide content increases and the degree of fracture stability is also lesser prominent.

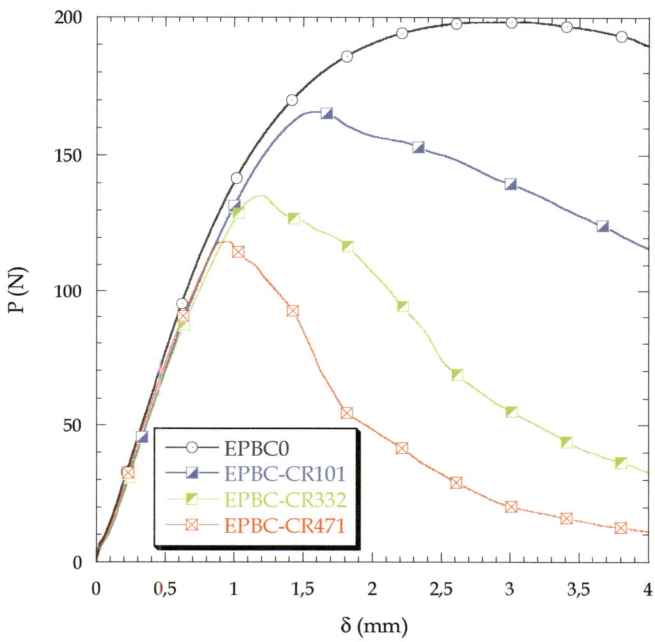

Fig. 8. Load (P) versus load-line displacement (δ) of the copolymer, EPBC0, and controlled-rheology copolymers, EPBC-CR101, EPBC-CR332 and EPBC-CR471

The J-R curves constructed for the non-degraded copolymer, EPBC0, and the peroxide-degraded copolymers, EPBC-CR101, EPBC-CR332 and EPBC-CR471, are shown in Figure 9. These plots also include the fit of the J-crack growth resistance curve to the power law $J=C\cdot\Delta a^N$, with $N\leq 1$. Analysis of the curves reveals that the reactor-made copolymer was the toughest material and the addition of peroxide acts in detriment of the toughness. Interestingly, the copolymers with the highest peroxide content, EPBC-CR332 and EPBC-CR471, showed almost flat resistance curves, indicating the remarkable loss of ductility with the peroxide content. This is clearly observed in Figure 10, where the crack initiation resistance, J_{IC}, of all the copolymers under study is presented. The fracture toughness of the peroxide-degraded copolymers verified the size requirement specified in equation 8, the fulfilment of which guarantees plane strain state, but the value of the reactor-made copolymer was not in plane strain state. Even so, the values are comparable as all the fracture specimens are of the same thickness. As expected, the fracture values drop abruptly with a small addition of peroxide and this reduction continues gradually with the peroxide content. This behaviour is similar to that shown by polypropylene homopolymers (Figure 5) and can aso be explained in terms of the molecular weight. The fracture toughness increases strongly with increasing molecular weight due to the linking molecules that join the crystalline blocks together. As the molecular weight decreases, the number of linking molecules also decreases, the material becomes less interconnected and fracture occurs at lower stress and strain levels (Avella et al., 1993; Fukuhara, 1999; and Sugimoto et al., 1995).

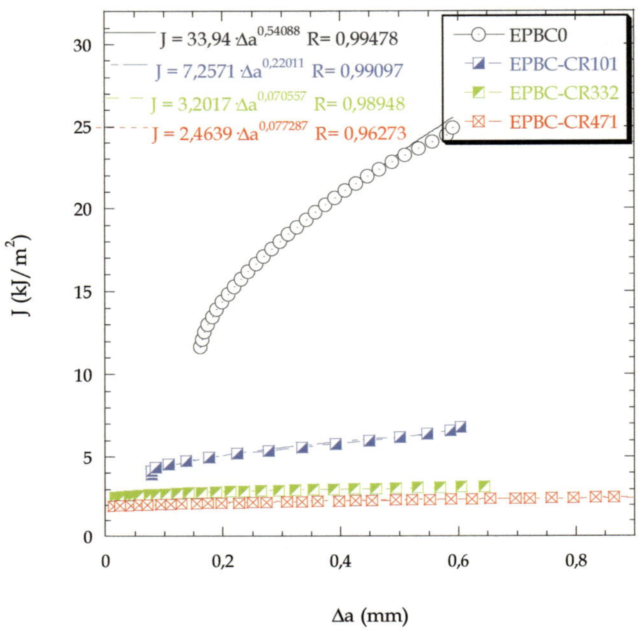

Fig. 9. J-R curves of the non-degraded copolymer, EPBC0, and the controlled-rheology copolymers, EPBC-CR101, EPBC-CR332 and EPBC-CR471.

The reduction of toughness and the loss of ductility with the addition of peroxide is easily observed in Figure 11, which shows the micrographs obtained via light microscopy and performed on the fracture surfaces of the tested specimens broken afterwards at liquid nitrogen and high loading rates to reveal the stable crack length. All the copolymers underwent stable crack length with ~ 0.6 mm in size. Independently of the material, three regions are distinguishable. The first one, close to the notch tip, is attributed to the stable crack propagation. This zone is followed by a stress whitening area and, ahead of it, the remainder of the fracture surface is characterized by a rough, un-whitened and uniform area related to the virgin material. Two main features can be drawn from the analysis of these fracture surfaces. Firstly, the stable crack length is more and more diffuse with the peroxide content. Indeed, to minimise the risk of incorrect measurements in the crack length, scanning electron microscopy means was used for the peroxide degraded copolymers instead of light microscopy. Secondly, the intensity and the extension of the stress whitening region ahead of the stable crack length reduced with the addition of peroxide. These two factors explain the decrease of the slope of the R-curves with the addition of peroxide (Figure 9).

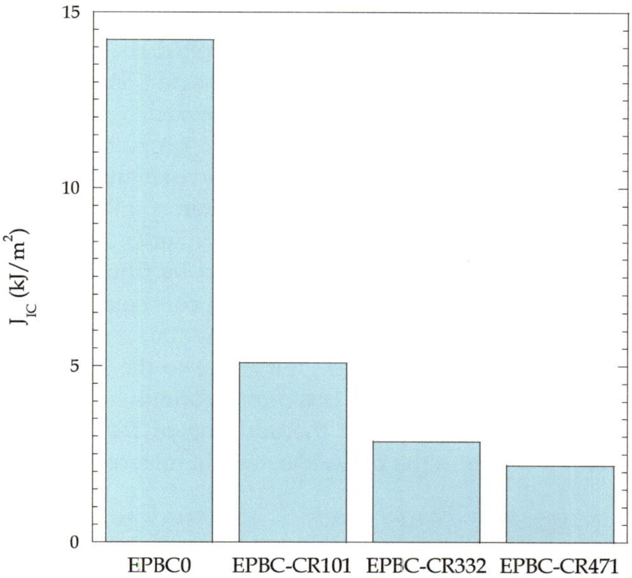

Fig. 10. Evolution of the crack growth energy, J_{IC}, with peroxide content of the copolymers under study.

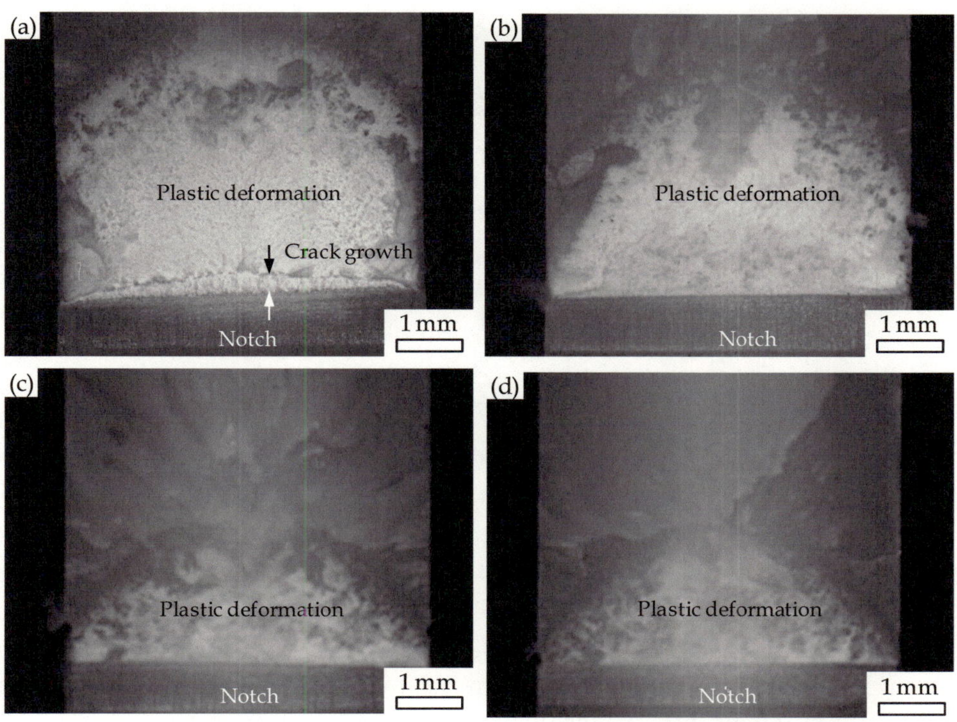

Fig. 11. Fracture surfaces obtained via light microscopy: (a) EPBC0, (b) EPBC-CR101, (c) EPBC-CR332 and (d) EPBC-CR471. All specimens show a stable crack length of ~ 0.6 mm.

The fracture surfaces were also examined via scanning electron microscopy with the aim of analyzing more deeply the fracture behaviour of controlled-rheology-copolymers. Figure 12 shows the morphology of the stable crack growth region for every copolymer. All the fracture surfaces displayed a macroductile tearing formed by broken stretched filaments oriented in the perpendicular direction to the crack propagation. Upon loading, small cavities were nucleated at the weak points of the copolymer such as the boundary between the elastomeric particles and the PP matrix or the intercrystalline zones in the PP matrix (Dasari et al.; 2003a; Doshev et al., 2005; Grellmann et al., 2001;. Starke et al., 1998; Yokoyama & Riccò, 1998). These voids were stabilised by fibrillar bridges of PP filaments which were plastically deformed and orientated. With consequent increase in deformation, excessive plastic flow occurred at these PP filaments and a stable crack nucleated and propagated through the closely PP bridges giving rise to the fibrillated morphology with ductile pulling of ligaments left behind. It is worth mentioning that the extension of such a ductile tearing was less pronounced with the addition of DTBP as the strength of these fibrillar PP bridges was reduced as the molecular weight lowered.

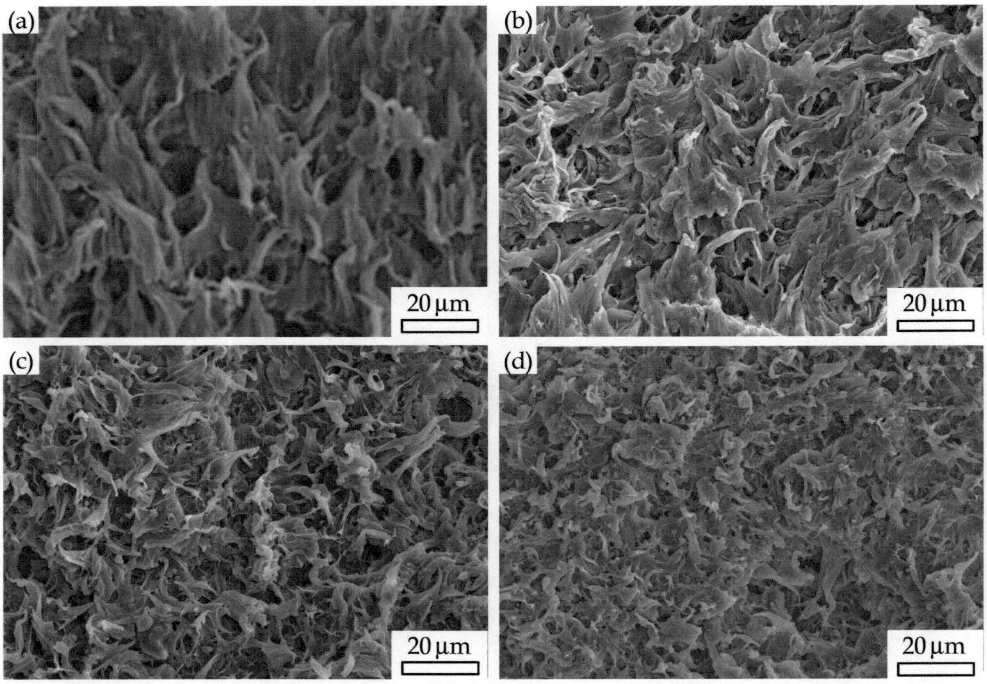

Fig. 12. Fracture surfaces obtained via scanning electron microscopy of the stable crack growth zone close to the notch: (a) EPBC0, (b) CR-EPBC101, (c) CR-EPBC332 and (d) CR-EPBC471.

The nucleation and growth of this mechanism of failure was clearly observed in the stress whitening region, which occurred ahead of the stable crack growth zone, of the post mortem fracture surfaces (Figure 13). The micrographs reveal the way the voids grow around the elastomeric phase. Despite the different content in peroxide, no appreciable differences could be evidenced in this region.

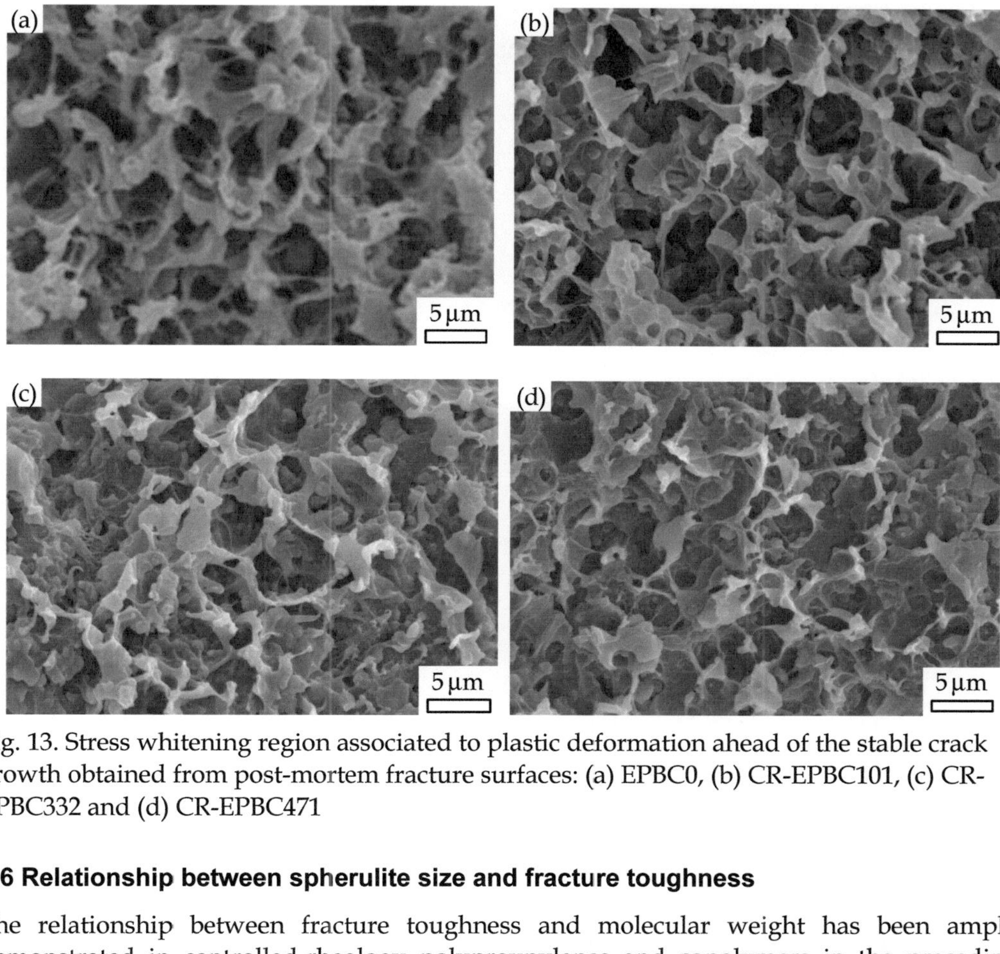

Fig. 13. Stress whitening region associated to plastic deformation ahead of the stable crack growth obtained from post-mortem fracture surfaces: (a) EPBC0, (b) CR-EPBC101, (c) CR-EPBC332 and (d) CR-EPBC471

5.6 Relationship between spherulite size and fracture toughness

The relationship between fracture toughness and molecular weight has been amply demonstrated in controlled-rheology polyproypylenes and copolymers in the preceding sections. Besides, it has been also shown that the decrease in the molecular weight is closely correlated with an increase in the spherulite diameters (Figures 1 and 2). Figure 14 represents the relationship between the fracture toughness and the spherulite size of the reactor-made polypropylene, PP0, and the controlled-rheology polypropylenes, PP-CR154, PP-CR402 and PP-CCR546. As observed, the fracture toughness is highly dependent not only on the molecular characteristics but also on supramolecular characteristics as the spherulites (Ibhadon, 1998): the bigger the spherulite diameter the lower the fracture toughness. The increase of the crystalline size declines the number and flexibility of the molecular chains which control the fracture toughness. The contribution to the fracture toughness is determined by the energy required to strain and break the tie molecules which hold together the lamellae bundles. Upon loading, the lamellae bundles separation is accompanied by massive voiding with the simultaneous onset of a craze like a microporous structure. As the molecular weight is decreased, the spherulite size is raised and hence, the density of tie molecules is suppressed impeding the formation of this microporous structure within the spherulitic and interspherulitic regions (Chen et al., 2002). This favours the brittleness and therefore the reduction in the fracture toughness with the increase of the spherulite diameter/addition of peroxide content.

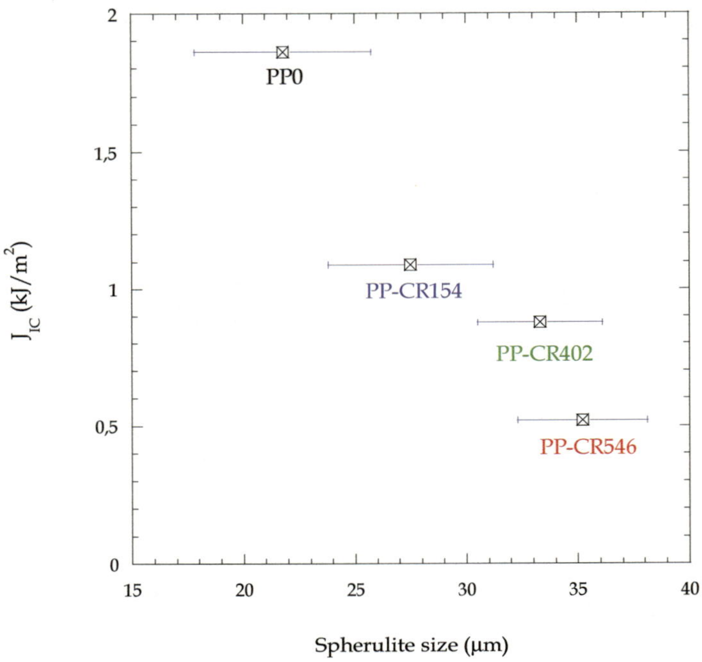

Fig. 14. Frature toughness, J_{IC}, versus the spherulite size of the reactor-made polypropylene, PP0, and the controlled-rheology polypropylenes, PP-CR154, PP-CR402 and PP-CR546.

It is worth mentioning that the controlled-rheology polypropylenes PP-CR402 and PP-CR546 are very similar from the structural (analogous molecular weight characteristics) (Table 1) and morphological (spherulite size) (Table 5) points of view. However, their fracture toughness values are quite different (Fig. 14). This disparity does not correlate properly with the small differences observed in the size of the spherulites but does with the variation in the porosity (Table 5). PP-CR546 presented a ~ 14% of porosity versus a ~ 8% for CRPP402.

5.7 Fracture behaviour of controlled-rheology versus conventional grades

Although the reactor-made copolymer EPBC0-2 presents analogous structural properties to the peroxide degraded copolymer EPBC-CR101 (molecular weights, MWD and melt flow rate) (Table 2), it has been shown some distinctness in both thermal and mechanical properties. In general, the thermal and mechanical properties of EPBC0-2 were more similar to the other reactor-made copolymer EPBC0 than to the controlled-rheology copolymer EPBC-CR101. With the aim of evaluating if fracture behaviour sustained this trend, Figure 15 shows the resistance curves of EPBC0-2 and EPBC-CR101 obtained at room temperature and low loading rates. These plots also include the fit of the J-crack growth resistance curve to the power law $J=C\cdot\Delta a^N$, with $N\leq1$. Despite the likeness in the structural properties the J-R curves are completely different. Moreover, the fitting parameters of EPBC0-2 are similar to those of the EPBC0 (Fig. 9). This is more clearly seen when comparing the crack growth initiation energy of these copolymers, which were 14.2, 14.0 and 5.09 kJ/m² for EPBC0, EPBC0-2 and EPBC-CR101, respectively. Both EPBC0 and EPBC0-2 were not in plane strain state as neither of them verified the size criterion displayed in equation 8.

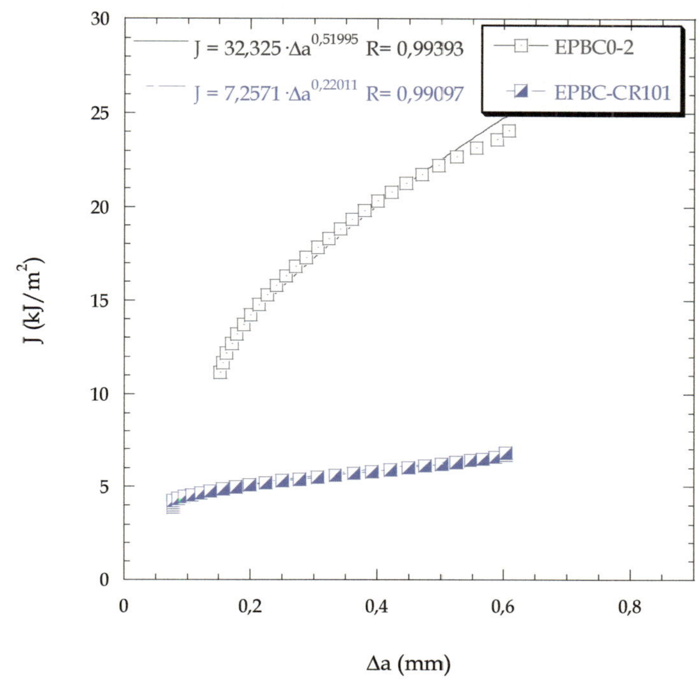

Fig. 15. J-R curves of a rector-made copolymer, EPBC0-2, and rheology controlled copolymer, EPBC-CR101, with similar structural characteristics (molecular weights and melt flow rate) (Table 5).

Interestingly, not only the measurements of J-fracture toughness values but also the fractographic analysis performed via optical microscopy indicated that the reactor-made EPBC0-2 copolymer was tougher than the reactive extrusion copolymer EPBC-CR101. Figure 16 shows the optical micrographs of the fracture surfaces broken after the test at liquid nitrogen temperature and high loading rates of EPBC0-2 and EPBC-CR101. The stable crack growth in both materials was ~ 0.6 mm. It is evident that the stable crack length is more easily distinguishable in EPBC0-2 than in EPBC-CR101. Besides, both the intensity and the extension of the stress-whitening region ahead of the stable crack growth are more pronounced in EPBC0-2 than in EPBC-CR101. Furthermore, when comparing the optical fracture surface of EPBC0 (Fig. 11a) with that of EPBC0-2 (Fig. 16a), scarce differences can be observed.

The fractographic analysis carried out via scanning electron microscopy on EPBC0-2 showed the same micromechanism of failure as that described for the controlled-rheology copolymers, that is, macroductile tearing formed by broken stretched filaments oriented in the direction perpendicular to the crack propagation (Fig. 17). However, the ductile tearing of EPBC0-2 is more marked than that of EPBC-CR101. Once more, there is more similarity between the fracture surfaces of reactor-made copolymers, EPBC0 (Fig. 12a) and EPBC0-2 (Fig. 17a), than between the copolymers with analogous structural properties but processed by distinct procedures (Fig. 17).

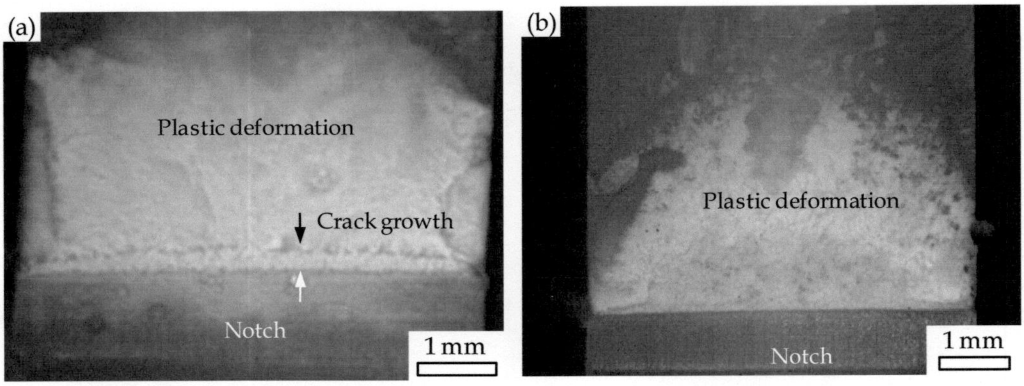

Fig. 16. Fracture surfaces obtained via light microscopy of (a) the reactor-made copolymer EPBC0-2 and (b) rheology controlled copolymer EPBC-CR101. Both materials are characterized by similar structural properties. The specimens presented a stable crack length of ~ 0.6 mm.

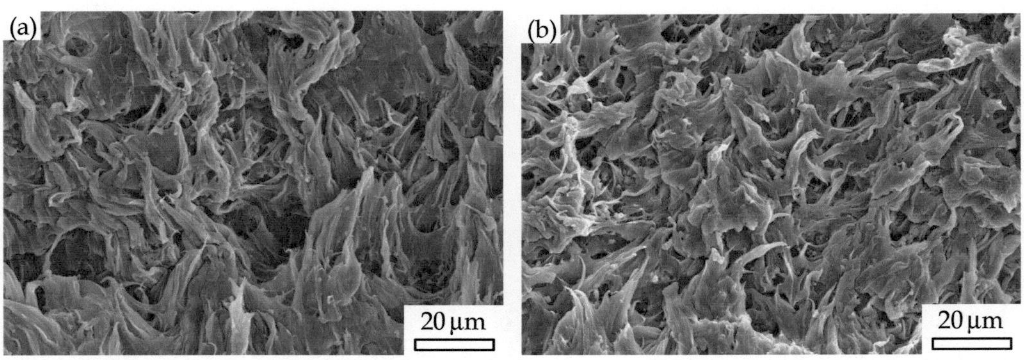

Fig. 17. Fracture surfaces obtained via scanning electron microscopy of (a) the reactor-made copolymer EPBC0-2 and (b) rheology controlled copolymer EPBC-CR101. Both materials are characterized by similar structural properties.

Finally, the reactor-made copolymer, EPBC0-2, results to be a very tough material with excellent mechanical properties and above all, superior processing properties, analogous to those of peroxide-degraded copolymers.

6. Conclusions

A set of controlled-rheology propylene homopolymers and ethylene-propylene block copolymers have been analyzed. The investigation has been focussed on the influence of the DTBP addition on the supramolecular characteristics as the spherulite size and distributuion of these semicrystalline polypropylenes and above all, on the fracture behaviour under EPFM. Besides, an in-depth and systematic analysis of the MWD, thermal and mechanical properties has also been performed. Finally, controlled-rheology and conventional grades with similar structural properties were compared to evaluate the goals and limitations of the reactive extrusion process. The main conclusions of each of these aspects are:

- For either the propylene homopolymer or copolymers, the peroxide reduced the molecular weight and narrowed the MWD leading to an increase of the MFR. This reduction was more pronounced in the propylene homopolymers than in the copolymers because in the latter, two competing effects were present: β-scissions of the long PP chains and chain branching and crosslinking of the PE phase.
- Concerning the thermal properties, the melting temperature reduced with the peroxide content for both the propylene homopolymers and the copolymers. However, the crystalline degree remained constant with the peroxide addition for the propylene homopolymers but reduced slightly for the copolymers. The crosslinks produced in the elastomeric phase may disturb the crystallinity architecture of the copolymers.
- Regarding the evolution of the supramolecular properties of the propylenes homopolymers, the average spherulite size seems to enhance and the distribution width seems to reduce, attaing a more unifrom distribution, as the DTBP content increases. The reduction of M_w, promoted by peroxide addition, leads to the maintenace of the crystalline degree, with the only growth of the spherulite size at the expense of the reduction in the density of tie molecules. However, for very high peroxide content, this trend is not followed because of the presence of pores.
- In general, the mechanical properties tends to decrease with the peroxide content. The strength parameters as the yield stress, tensile strenght and elongation at break, are highly dependent on the molecular weight. The increase in peroxide content entails a decrease in the molecular weight which preferentailly affects the amorphous regions. However, some exceptions were observed in the evolution of the tensile strength and elongation at break with the peroxide addition of the copolymers. This anomalous behavious can be also explained in the light of the presence of the long residual chains of PE which were probably branched.
- Concerning the fracture behaviour, the fracture toughness values of either the reactor-made polypropylenes or copolymers were the highest of all the materials under study. Independently of the material, the fracture toughness dropped abruptly with a small addition of peroxide content and a continuous reduction with further addition of DTBP. In addition, the copolymer experienced an appreciable loss of ductility with the peroxide content not only evident form the load-displacement records and the naked-eye examination ofo the fracture surfaces but also from the slope of the J-R curves.
- In case of PPs, the mechanism of failure of all the materials was crazing and no differences were found in the fracture surfaces among the different grades of polypropylene. All the results point at the influence of molecular weight which is closely related to the amorphous region. On the other hand, the mechanism of failure of all the copolymers analyzed involved growth of voidds around the elastomeric particles which resulted in a macroductile tearing. The degree of ductile tearing was less marked as the polymer degradation was more accentuated. Therefore, the reduction of the fracture parameters with the decrease in the molecular weight in semi-crystaline

propylene homopolymers and copolymers can be explained by the reduction in number and strength of the linking molecules that join the crystalline blocks.

- Finally, the controlled-rheology and the reactor-made EPBCs with similar structural characteristics presented diferences in the mechanical properties, being specially remarkable in the fracture behaviour. The micromechanisms of failure of both types of copolymers were similar although more likeness was found between the fracture surfaces of the reactor-made copolymers than between the copolymers with analogous structural parameters processed by distinct procedures.

7. Acknowledgment

The author is indebted to Ministerio de Educación of Spain for its financial support through project MAT2009-14294 and to REPSOL for the materials supply.

8. References

Allen, G. & Bevington, J.C. (1989). *Comprehensive Polymer Science. The synthesis, characterization, reactions and applications of polymers; Vol. 2 Mechanical properties.* Colin booth & Colin Price Eds. Pergamon Press, ISBN 0080325157, Oxford, UK

Asteasuain, M.; Sarmoria, C. & Brandolin, A. (2003). Controlled rheology of polypropylene: modelling of molecular weight distributions. *Journal of Applied Polymer Science,* Vol.88, No.7, (May 2003), pp. 1676-1685, ISSN 1097-4628

Avella, M.; dell´Erba, R.; Martuscelli, E. & Ragosta, G. (1993). Influence of molecular mass, thermal treatment and nucleating agent on structure and fracture toughness of isotactic polypropylene. *Polymer,* Vol.34, No.14, (1993), pp. 2951-2960, ISSN 0032-3861

Azizi, H. & Ghasemi, I. (2004). Reactive extrusion of polypropylene: production of controlled-rheology polypropylene (CRPP) by peroxide-promoted degradation. *Polymer Testing,* Vol.23, No.2, (April 2004), pp. 137-143, ISSN 0142-9418

Azizi, H.; Ghamesi, I. & Karrabi, M. (2008). Controlled-peroxide degradation of polypropylene: rheological properties and prediction of MWD from rheological data. *Polymer Testing,* Vol.27, No.5, (August 2008), pp. 548-554, ISSN 0142-9418

Baik, J.J. & Tzoganakis, C. (1998). A study of extrudate distorsion in controlled-rheology polypropylenes. *Polymer Engineering and Science,* Vol.38, No.2, (February 1998), pp: 274-281, ISSN 1548-2634

Baldi, F. & Riccò, T (2005). High rate J-testing of toughened polyamide 6/6: applicability of the load separation criterion and the normalization method. *Engineering Fracture Mechanics,* Vol.72, No.14, (September 2005), pp. 2218–2231, ISSN 0013-7944

Barakos, G.; Mitsoulis, E.; Tzoganakis, C. & Kajiwara, T. (1996). Rheological characterization of controlled-rheology polypropylenes using integral constitutive equation. *Jourrnal of Applied Polymer Science,* Vol.59, No.3, (January 1996), pp. 543-556, ISSN 1097-4628

Bernal, C.R.; Montemartini, P.E. & Frontini P.M. (1996) The use of load separation criterion and normalization method in ductile fracture characterization of thermoplastic polymers. *Journal of Polymer Science Part B - Polymer Physics,* Vol.34, No.11, (August 1996), pp. 1869–1880, ISSN 0887-6266

Berzin, F.; Vergnes, B. & Delamare L. (2001). Rheological behaviour of controlled-rheology polypropylenes by peroxide-promoted degradation during extrusion: comparison between homopolymer and copolymer. *Journal of Applied Polymer Science*, Vol.80, No.8, (May 2001), pp. 1243-1252, ISSN 1097-4628

Blancas, C. & Vargas, L. (2001). Modeling of the industrial process of peroxide initiated polypropylene (homopolymers) controlled degradation. *Journal of Macromolecular Science Part B-Physics*, Vol.40, No.3-4, (2001), pp. 315-326, ISSN 0022-2348

Botsis, J.; Oerter, G. & Friedrich, K. (1999). Fatigue fracture in polypropylene with different spherulitic sizes. In: *Imaging and image analysis application for plastics*, B. Pourdeyhimi, (Ed.), 289-298, ISBN 978-1-884207-81-5, North Carolina State University, NC, USA

Braun, D.; Richter, S.; Hellmann, G.P. & Rätzsch, M. (1998). Peroxy-initiated chain degradation, crosslinking and grafting in PP-PE blends. *Journal of Applied Polymer Science*, Vol.68, No.12, (June 1998), pp. 2019-2028, ISSN 1097-4628

Chen, H.B.; Karger-Kocsis, J.; Wu, J.S. & Varga, J. (2002). Fracture toughness of α- and β-phase polypropylene homopolymers and random- and block-copolymers. *Polymer*, Vol.43, No.24, (November 2002), pp. 6505-6514, ISSN 0032-3861

Dasari, A.; Rohrmann, J. & Misra R.D.K. (2003). Microstructural evolution during tensile deformation of polypropylenes. *Materials Science and Engineering A*, Vol.351, (June 2003), pp. 200-213, ISSN 0921-5093

Dasari, A.; Rohrmann, J. & Misra R.D.K. (2003). Microstructural aspects of surface deformation processes and fracture of tensile strained high isotactic polypropylene. *Materials Science and Engineering A*, Vol.358, (October 2003), pp. 372-383, ISSN 0921-5093

Dasari, A.; Rohrmann, J. & Misra R.D.K. (2003). Surface microstructural modification and fracture behavior of tensile deformed polypropylene with different percentage crystallinity. *Materials Science and Engineering A*, Vol.360, (November 2003), pp. 237-248, ISSN 0921-5093

Do, I.H.; Yoon, L.K.; Kim, B.K. & Jeong, H.M. (1996). Effect of viscosity ratio and peroxide/coagent treatment in PP/EPR/PE ternary blends. *European Polymer Journal*, Vol.32, No.12, (December 1996), pp. 1387-1393, ISSN 0014-3057

Doshev, P.; Lach, R.; Lohse, G.; Heuvelsland, A.; Grellmann, W. & Radusch, H.J. (2005). Fracture characteristics and deformation behavior of heterophasic ethylene-propylene copolymers as a function of the dispersed phase composition. *Polymer*, Vol.46, No.22, (October 2005), pp. 9411-9422, ISSN 0032-3861

Ernst H.A.; Paris, P.C. & Landes J.D. (1981). Estimations on J-integral and Tearing Modulus T from a single specimen test record. *Fracture Mechanics: Thirteenth Conference ASTM STP 743*, Roberts R., (Ed.), pp. 476–502.

Fukuhara, N. (1999). Influence of molecular weight on J-integral testing of polypropylene. *Polymer Testing*, Vol.18, No.2, (April 1999), pp. 135-149, ISSN 0142-9418

Gahleitner, M.; Bernreitner, K.; Neiβl, W.; Paulik, C. & Ratajski. (1995). Influence of molecular structure on crystallization behavior and mechanical properties of polypropylene. *Polymer Testing*, Vol.14, No.2, (1995), pp. 173-187, ISSN 0142-9418

Gahleitner, M.; Wolfschwenger, J.; Bachner, C; Bernreitner, K. & Neiβl, K. (1996). Crystallinity and mechanical properties of PP-homopolymers as influenced by molecular structure and nucleation. *Journal of Applied Polymer Science*, Vol.61, No.4, (July 1996), pp. 649-657, ISSN 1097-4628

Graebling, D.; Lambla, M. & Wautier, H. (1997). PP/PE blends by reactive extrusion: PP rheological behavior changes. *Journal of Applied Polymer Science,* Vol.66, No.5, (October 1997), pp. 809-819, ISSN 1097-4628

Grein, C.; Plummer, C.J.G.; Kausch, H.H.; Germain, Y. & Béguelin, Ph. (2002). Influence of β nucleation on the mechanical properties of isotactic polypropylene and rubber modified isotactic polypropylene. *Polymer,* Vol.43, No.11, (May 2002), pp. 3279-3293, ISSN 0032-3861

Grellmann, W.; Seidler, S.; Jung, K. & Kotter, I. (2001). Crack-resistance behavior of polypropylene copolymers. *Journal of Applied Polymer Science,* Vol.79, No.13, (January 2001), pp. 2317-2325, ISSN 1097-4628

Hale, G.E. & Ramsteiner, F. (2001). J-fracture toughness of polymers at slow speed. In: *Fracture mechanics testing methods for polymers, adhesives and composites,* Moore, D.R.; Pavan, A. & Williams J.G., editors, pp 123-157, Elsevier Science Ltd. and ESIS, ISBN 0 08 043689 7, The Netherlands.

Hodgkinson, J.M.; Savadori, A. & Williams, J.G. (1983). A fracture mechanics analysis of polypropylene/rubber blends. *Journal of Materials Science,* Vol.18, No.8, (1983), pp. 2319-2336, ISSN 1573-4803

Ibhadon, A.O. (1998). Fracture mechanics of polypropylene: effect of molecular characteristics, crystallization conditions, and annealing on morphology and impact performance. *Journal of Applied Polymer Science,* Vol.69, No.13, (September 1998), pp. 2657-2661, ISSN 1097-4628

Karger-Kocsis, J. (1995). *Polypropylene: structures, blends and composites: copolymers and blends.* Chapman and Hall, ISBN 0-412-61420-0, London, UK

Kim, G.M.; Michler, G.H.; Gahleitner, M. & Fiebig, J. (1996). Relationship between morphology and micromechanical toughening mechanisms in modified polypropylenes. *Journal of Applier Polymer Science,* Vol.60, No.9, (May 1996), pp. 1391-1403, ISSN 1097-4628

Landes, J.D. & Begley, J.A. (1974). The results from J-integral studies: an attempt to establish a J_{IC} testing procedure. *Fracture Analysis, ASTM STP 560,* American Society for Testing and Materals, pp. 170-186.

Landes, J.D. & Zhou, Z. (1993). Application of load separation and normalization methods for polycarbonate materials. *International Journal of Fracture,* Vol. 63, No.4, (1993), pp. 383–393, ISSN 0376-9429

Lapique, F.; Meakin, P.; Feder, J. & Jossang, T. (2000). Relationships between microstructure, fracture-surface morphology, and mechanical properties in ethylene and propylene polymers and copolymers. *Journal of Applied Polymer Science,* Vol.77, No.11, (September 2000), pp. 2370-2382, ISSN 1097-4628

Morhain, C. & Velasco J.I. (2001). Determination of J-R curve of polypropylene copolymers using the normalization method. *Journal of Materials Science,* Vol.36, No.6, (March 2001) pp. 1487–1499, ISSN 1573-4803

Ogawa, T. (1992). Effects of molecular weight on mechanical properties of polypropylene. *Journal of Applied Polymer Science,* Vol.44, No.10, (April 1992), pp. 1869-1871, ISSN 1097-4628

Pabedinkas, A.; Cluett, W.R. & Balke, S.T. (1989). Process control for polypropylene degradation during reactive extrusion. *Polymer Engineering and Science,* Vol.29, No.15, (August 1989), pp. 993-1003, ISSN 1548-2634

Pasquini, N. (Ed.) (2005). *Polypropylene Handbook*. Hanser Publishers, ISBN 3-446-22978-7, Munich, Germany

Prabhat, K. & Donovan, J.A. (1985). Tearing instability in polypropylene. *Polymer*, Vol.26, No.13, (December 1985), pp. 1963-1970, ISSN 0032-3861

Rodríguez, C.; Maspoch, M.Ll. & Belzunce F.J. (2009). Fracture characterization of ductile polymers through methods based on load separation. *Polymer Testing*, Vol.28, No.2, (April 2009), pp. 204–208, ISSN 0142-9418

Ryu, S.H.; Cogos, C.G. & Xanthos, M. (1991). Parameters affecting process efficiency of peroxide-initiated controlled degradation of polypropylene. *Advances in Polymer Technology*, Vol.11, No.2, (July 1991), pp. 121-131, ISSN 1098-2329

Ryu, S.H.; Cogos, C.G. & Xanthos, M. (1991). Crystallization behavior of peroxide modified polypropylene. *Society of the Plastics Industry Annual Technical Conference ANTEC 1991*, pp. 886-888, Montreal, Canada

Salazar, A. & Rodríguez, J. (2008). The use of load separation parameter S_{pb} method to determine the J-R curves of polypropylenes. *Polymer Testing*, Vol.27, No.8, (December 2008), pp. 977–984, ISSN 0142-9418

Salazar, A.; Rodríguez, J.; Segovia, A. & Martínez, A.B. (2010). Influence of the notch sharpening technique on the fracture toughness of bulk ethylene-propylene block copolymers. *Polymer Testing*, Vol.29, No.1, (February 2010), pp. 49-59, ISSN 0142-9418

Salazar, A.; Rodríguez, J;. Segovia, A. & Martínez A.B. (2010). Relevance of the femtolaser notch sharpening to the fracture of ethylene-propylene block copolymers. *European Polymer Journal*, Vol.46, No.9, (June 2010), pp. 1896-1907, ISSN 0014-3057

Salazar, A.; Segovia, A.; Martínez, A.B. & Rodríguez, J. (2010). The role of notching damage on the fracture parameters of ethylene-propylene block copolymers. *Polymer Testing*, Vol.29, No.7, (October 2010), pp. 824-831, ISSN 0142-9418

Sheng, B.R.; Li, B.; Xie, B.H.; Yang, W.; Feng, J.M. & Yang, M.B. (2008). Influences of molecular weight and crystalline structure on fracture behaviour of controlled-rheology-propylene prepared by reactive extrusion. *Polymer Degradation and Stability*, Vol.93, No.1, (January 2008), pp. 225-232, ISSN

Sheskin, D.J. (2007). *Handbook of parametric and nonparametric statistical procedures*. 4th edition Chapman & Hall/CRC, ISBN 9781584888147, Florida (USA)

Starke, J.U.; Michler, G.H.; Grellmann, W.; Seidler, S.; Gahleitner, M.; Fiebig, J. & Nezbedova, E. (1998). Fracture toughness of polypropylene copolymers: influence of interparticle distance and temperature. *Polymer*, Vol.39, No.1, (January 1998), pp. 75-82, ISSN 0032-3861

Sugimoto, M.; Ishikawa, M. & Hatada, K. (1995). Toughness of polypropylene. *Polymer*, Vol.36, No.19, (1995), pp. 3675-3682, ISSN 0032-3861

Sun, Z. & Yu, F. (1991). SEM study on fracture behaviour of ethylene/propylene block copolymers and their blends. *Macromolecular Chemistry and Physics*, Vol.192, No.6, (June 1991), pp. 1439-1445, ISSN 1022-1352

Sun, Z.; Yu, F. & Qi, Y. (1991). Characterization, morphology and thermal properties of ethylene-propylene block copolymers, *Polymer*, Vol.32, No.6, (1991), pp. 1059-1064, ISSN 0032-3861

Teh, J.W.; Rudin, A. & Keung, J.C. (1994). A review of polyethylene–polypropylene blends and their compatibilization. *Advances in Polymer Technology*, Vol.13, No.1, (April 1994), pp. 1-23, ISSN 1098-2329

Thomann, R.; Wang, C.; Kressler, J.; Jüngling, S. & Mülhaupt, R. (1995). Morphology of syndiotactic polypropylene. *Polymer,* Vol.36, No.20, (1995), pp. 3795-3801, ISSN 0032-3861

Tjong, S.C.; Shen, J.S. & Li, R.K.Y. (1995). Impact fracture toughness of β-polypropylene. *Scripta Materialia,* Vol.33, (1995), pp. 503-508, ISSN 1359-6462

Tzoganakis, C.; Vlachopoulos, J.; Hamielec, A.E. & Shinozaki, D.M. (1989). Effect of molecular weight distribution on the rheological and mechanical properties of polypropylene. *Polymer Engineering and Science,* Vol.29, No.6, (March 1989), pp. 390-396, ISSN 1548-2634

Van der Wal, A.; Mulder, J.J. & Gaymans, R.J. (1998). Fracture of polypropylene: 2. The effect of crystallinity. *Polymer,* Vol.39, No.22, (October 1998), pp. 5477-5481, ISSN 0032-3861

Van der Wal, A.; Nijhof, R. & Gaymans R.J. (1999). Polypropylene–rubber blends: 2. The effect of the rubber content on the deformation and impact behavior. *Polymer,* Vol.40, No.22, (October 1999), pp. 6031-6044, ISSN 0032-3861

Varadarajan, R.; Dapp, E.K. & Rimnac C.M. (2008). Static fracture resistance of ultra high molecular weight polyethylene using the single specimen normalization method. *Polymer Testing,* Vol.27, No.2, (April 2008), pp. 260–268, ISSN 0142-9418

Wang, S.H.; Yang, W.; Xu, Y.J., Xie, B.H.; Yang, M.B. & Peng, X.F. (2008). Crystalline morphology of β-nucleated controlled-rheology polypropylene. *Polymer Testing,* Vol.27, No.5, (August 2008), pp. 638-644, ISSN 0142-9418

Wainstein, J.; Fasce, L.A.; Cassanelli, A. & Frontini P.M. (2007). High rate toughness of ductile polymers. *Engineering Fracture Mechanics,* Vol.74, No.13, (September 2007), pp. 2070–2083, ISSN 0013-7944

Wei, G.X.; Sue, H.J.; Chu, J.; Huang, C. & Gong, K. (2000). Toughening and strengthening of polypropylene using the rigid–rigid polymer toughening concept. Part I. Morphology and mechanical property investigations. *Polymer,* Vol.41, No.8, (April 2000), pp. 2947-2960, ISSN 0032-3861

Williams, J.G. (2001). Introduction to elastic-plastic fracture mechanics. In: *Fracture mechanics testing methods for polymers, adhesives and composites,* Moore, D.R.; Pavan, A. & Williams J.G., (Eds.), pp 119-122, Elsevier Science Ltd. and ESIS, ISBN 0 08 043689 7, The Netherlands.

Xu, T.; Yu, J. & Jin, Z. (2001). Effects of crystalline morphology on the impact behavior of polypropylene. *Materials & Design,* Vol.22, No.1, (February 2001), pp. 27-31, ISSN 0261-3069

Xu, T.; Lei, H. & Xie, C. (2002). The research on aggregation structure of PP materials under different condition and the influence on mechanical properties. *Materials & Design,* Vol.23, No.8, (December 2002), pp. 709-715, ISSN 0261-3069

Yokoyama, Y. & Riccò T. (1998). Toughening of polypropylene by different elastomeric systems. *Polymer,* Vol.39, No.16, (June 1998), pp. 3675-3681, ISSN 0032-3861

Yu, D.W.; Xanthos, M. & Gogos, C.G. (1990). Peroxide modified polyolefin blends: Part I. Effects on ldpe/pp blends with components of similar initial viscosities. *Advances in Polymer Technology,* Vol.10, No.3, (October 1990), pp. 163-172, ISNN 1098-2329

Yu, J.; Xu, T.; Tian, Y. Chen, X. & Luo, Z. (2002). The effects of the aggregation structure parameters on impact-fractured surface fractal dimension and strain-energy release rate for polypropylene. *Materials & Design,* Vol.23, No.1, (February 2002), pp. 89-95, ISSN 0261-3069

5

Poisson's Ratio and Mechanical Nonlinearity Under Tensile Deformation in Crystalline Polymers

Koh-hei Nitta and Masahiro Yamana
Division of Material Sciences, Graduate School of Natural Science and Technology, Kanazawa University, Kakuma, Kanazawa, Japan

1. Introduction

Semicrystalline polymers contain liquid-like amorphous and ordered crystalline phases. When solidified from the pure melt, these polymers show a spherulitic structure in which crystalline lamellae composed of folded chain crystallites radiate from the center of the spherulite in such a way that a constant long period or crystallinity is approximately maintained. The amorphous regions reside in the interlamellar regions in the form of tie chains, whose ends are attached to adjacent lamellae; loop chains, whose ends are attached to the same lamella; cilia chains with only one end attached to a lamella (or dangling chain ends), and floating chains which are not attached to any lamellae. This hierarchical structure is illustrated in Figure 1.

Nitta-Takayanagi have introduced the idea of stacked lamellae running parallel forming clusters, the sizes of which are in the range of the end-to-end distance of single Gaussian chains, to explain the tensile yielding and necking phenomena (Nitta & Takayanagi, 2003). Similar concepts to the lamellar cluster have been proposed by other investigators. Based on transmission electron microscopy (TEM) observations of a uniaxially deformed PE sample, Kilian et al. (Kilian & Pietralla, 1978) have identified the presence of stacks of several lamellae running parallel and forming clusters, the sizes of which are in the range of several hundreds of angstroms. They showed that the elastic behavior for semicrystalline polymers can be explained using a cluster-network model in which the clusters consisting of mosaic blocks are connected by means of tie molecules to the neighboring ones and are considered to operate as junction points. In addition, the microindentation hardness was demonstrated to be governed by the cooperative shearing of the mosaic blocks within the clusters. Separately from Kilian's work, the direct observation of the stacked lamellar units containing three to ten lamellar crystals has been confirmed by Tagawa et al.(Tagawa, 1980) from high-resolution scanning electron micrographs of blown PE films. They emphasized that the stacks of lamellae containing three to ten lamellae act as one unit and do not separate into single lamellae during deformation. A similar structural unit has been already proposed by Peterlin et al. (Peterlin, 1971, 1975) to describe the necking process.

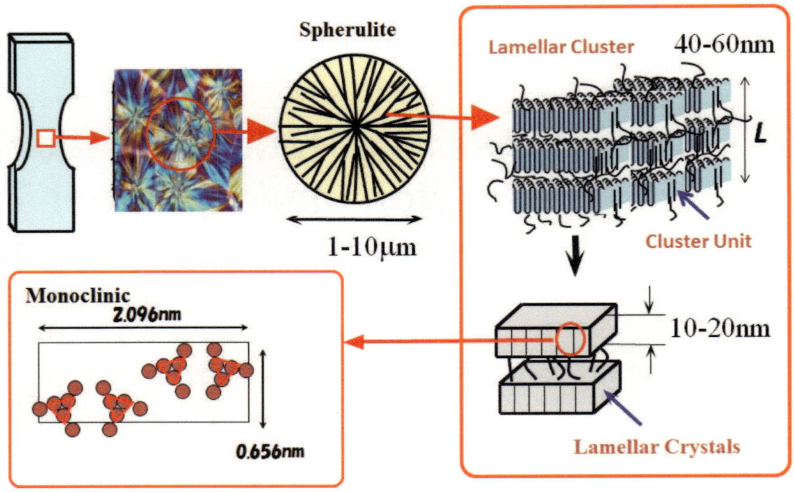

Fig. 1. Supermolecular structure of isotactic polypropylene.

The tensile test is one of the most popular and important mechanical tests. A typical nominal stress-strain curve of semicrystalline solids measured in the temperature range between the glass transition of the amorphous phase and the melting of crystalline phase is illustrated in Figure 2. The stress-strain curve is stepwise and divided into four zones. In the initial strain region, the stress is almost proportional to the applied strain; the deformation proceeds homogeneously and the sample specimen recovers to its original size after the stress is released. Young's modulus can be conventionally estimated from the slope of the line. This region is called the "elastic region". After the initial deformation, the material shows a clear yield point as a maximum point on the nominal stress-strain curve. The maximum point; i.e. yield point is associated with the onset of *temporal* plastic deformation and is referred to as the "failure point". Beyond the yield point, a concave contraction suddenly initiates on the specimen and coalesces into a well-defined neck at which is the onset of the *permanent* plastic deformation. This point is more clearly observed as the second yield point in the case of polyethylene-based materials, but this double yield point seems to be a general feature of semicrystalline polymers as suggested by Séguéla et al. (Séguéla, 1994). Finally, after the necking boundaries have propagated throughout the entire length of the specimen, an upsweep in the stress–strain curve, termed *strain hardening*, occurs, and the stress continues to increase up to the break point. Carothers et al. (Carothers & Hill, 1932) called this phenomenon "cold drawing". In the necking process, the macroscopic morphological transformation from isotropic spherulitic to anisotropic fibril structures takes place accompanied by the destruction and/or rearrangement of parts of the stacked crystalline lamellae. The extension ratio in the neck propagation region is generally defined as the natural draw ratio (Séguéla, 2007).

In general, the transverse strain (perpendicular to the stretching direction under uniaxial elongation), decreases at a constant rate with deformation. The Poisson's ratio is defined as the ratio of the transverse strain to the applied strain. The Poisson's ratio of rubbery materials is in the range of 0.4 to 0.5 (compared to 0.3 for hard materials such as metal or glass). As described above, most semicrystalline polymeric materials, however, show inhomogeneous deformation accompanied with neck formation. The evaluation of the transient change in Poisson's ratio under a uniaxial extension is essential to characterize the inhomogeneous deformation and the stress-strain relationship. The aim of this chapter is to clarify phenomenologically the relationship between tensile mechanical nonlinearity and geometrical nonlinearity resulting from the transient change in Poisson's ratio in crystalline polymers. For the purpose, we examined the transient Poisson's ratio and tensile properties using a series of typical semicrystalline polymers such as isotactic polypropylene (iPP), high density polyethylene (HDPE), and low density polyethylene(LDPE) samples.

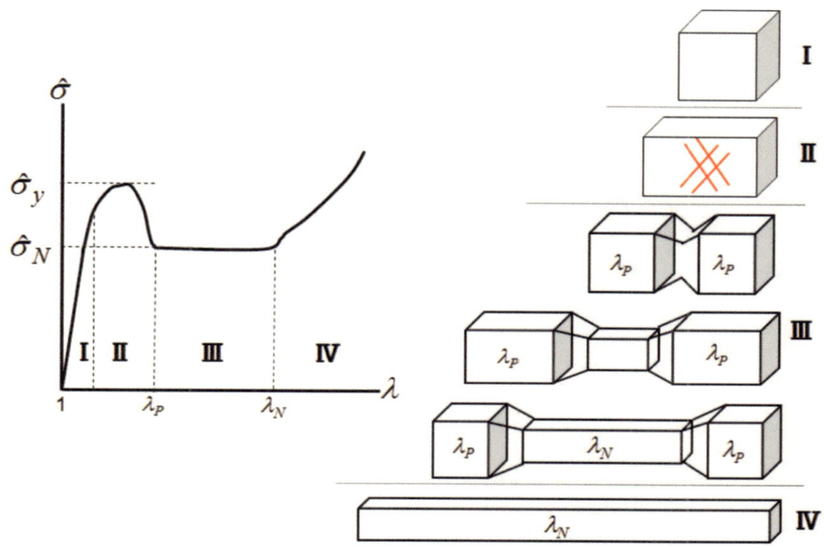

Fig. 2. A typical nominal stress-elongation curve corresponding schematics of deformed specimen during tensile deformation.

2. Evaluation of Poisson's ratio

Polymer pellets of isotactic polypropylene (iPP) were melted in a laboratory hot press for 15 min at 230 °C and 10 MPa, and then quenched in ice water (0 °C) or boiling water (100 °C). To control the crystallinity of the iPP sample, we annealed the sample sheets prepared at 0 °C for 2 hr at various temperatures from 40 to 140 °C. Square pillar shaped test specimens were cut out from a 1 mm thick sheet and used for the measurements. Double-edge notched sample specimens with the gauge length of 2 mm were used for the temperature dependence and the dumbbell shaped specimen with gauge length of 5 mm for the elongation speed dependence.

The tensile elongation was performed using a tensile tester which was specially designed with two clamps that move symmetrically with respect to the center point of the specimen. A laser sizer KeyenceLS3934 is mounted on the tensile tester to evaluate the transient thickness of the center of the specimen during tensile deformation. The resolution of the thickness was within ±2 μm. In addition, the three-dimensional change in specimen size during uniaxial elongation was estimated by monitoring successive images of the test specimen with a video camera every 0.5 sec. The experimental setup is shown in Figure 3. The measurements were performed under the following conditions: (1) the nominal strain rate was changed between 1.25 and 5.00 min^{-1}; (2) the temperature was varied within the range of 25 °C to 70 °C; and (3) the degree of crystallinity was changed from 40% to 70% by annealing the iPP sheets quenched at 0 °C in an oven for 2 hr at different temperatures from 40 to 140 °C. Poisson's ratio values were calculated from the thickness data of each deformed specimen under these conditions. It is noted here that the transient sample dimension in the initial strain region below 0-0.5 sec could not be evaluated.

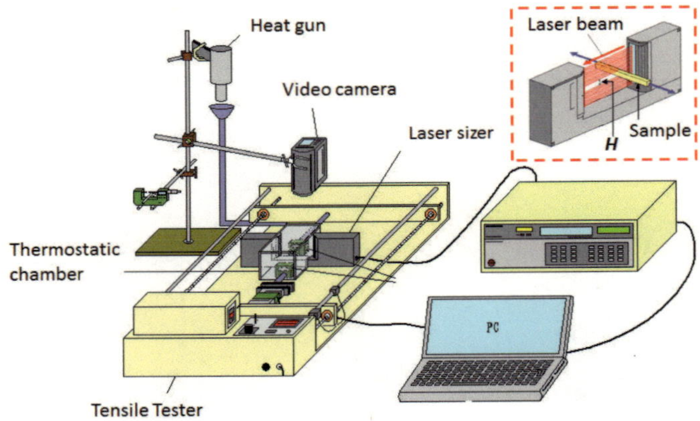

Fig. 3. Experimental setup for simultaneous measurements of the dimensions of the specimen and the external load under a constant elongation speed.

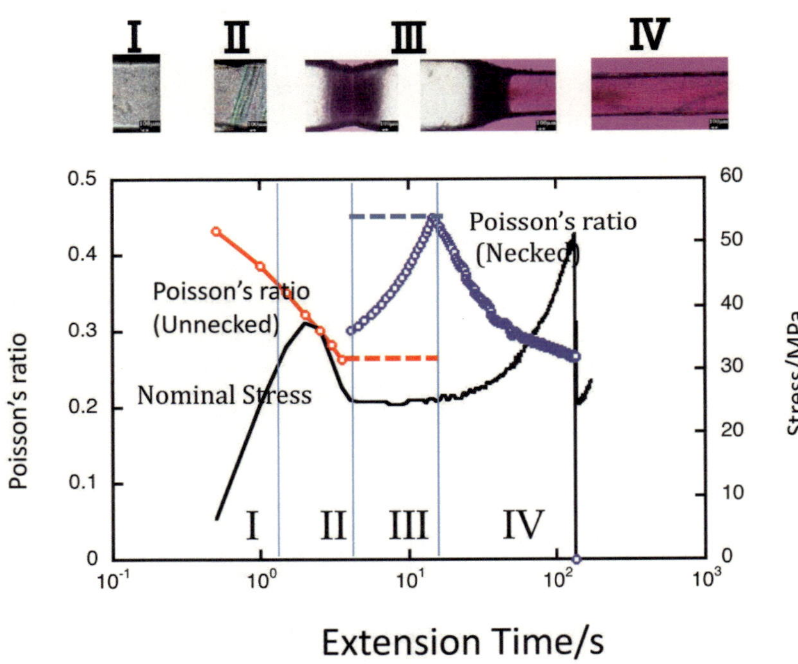

Fig. 4. Poisson's ratio and nominal stress plotted against the extension time under a tensile elongation at a fixed elongation speed of 20 mm/min.

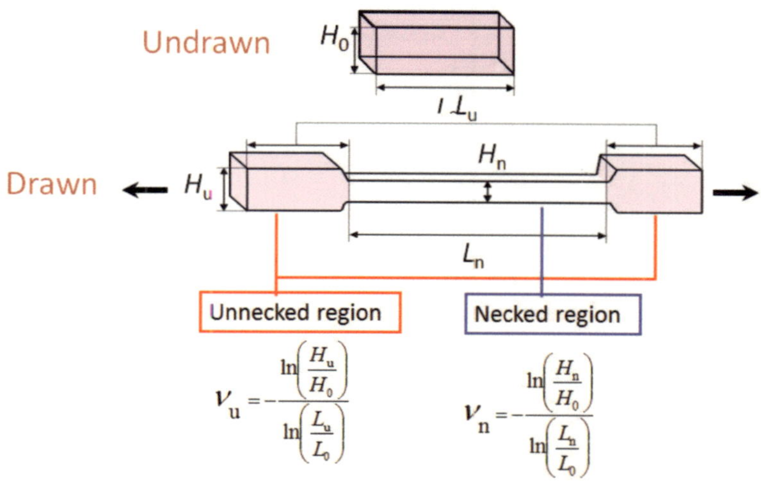

Fig. 5. Poisson's ratio at the necked and unnecked portions of a deformed specimen, where ν_n and ν_u are the Poisson's ratios at the necked (white box) and unnecked portions (red box), respectively.

Figure 4 exemplifies the nominal stress-strain curves and the Poisson's ratio-strain curves for a iPP specimen simultaneously measured at a constant extensional speed (20 mm/min) and room temperature (25 °C). Photographs showing the deformation of the samples are shown above the figure. As seen in this figure, the iPP specimen shows a clear necking phenomenon above the yield point and the strains for both the necking and unnecked regions are almost constant during the neck propagation. The Poisson's ratio values in this figure were estimated from the thickness of the center portion of the deformed specimen under the assumption that the longitudinal strain is given by the extension of the gauge region of the sample specimen. In the pre-yielding region, the value of Poisson's ratio gradually decreases from around 0.5 to zero. The line has been broken at the neck initiation. On the other hand, in the post-yielding region, the Poisson's ratio increases to 0.5 during neck propagation as the necked part grows to encompass the entire specimen. Poisson's ratio has a maximum at the onset of strain-hardening and then monotonically decreases with further increase in applied strain. Since the strains in the unnecked region and the necked region are almost constant, the Poisson's ratio during necking is considered to be an averaged Poisson's ratio.

For the specimens which show clear necking, it is necessary to estimate the Poisson's ratio separately for the necking and unnecked regions. The Poisson's ratios in the unnecked and necked portion were separately estimated from each thickness value for sample specimens with clear necking phenomena, as shown in Figure 5.

3. Poisson's ratio under tensile tests

3.1 Elongation speed dependence

To examine the effects of elongation speed, we investigated Poisson's ratio in the range of 5.0 to 20 mm/min (nominal strain rate of 1 to 4 /min) at room temperature, 25 °C.

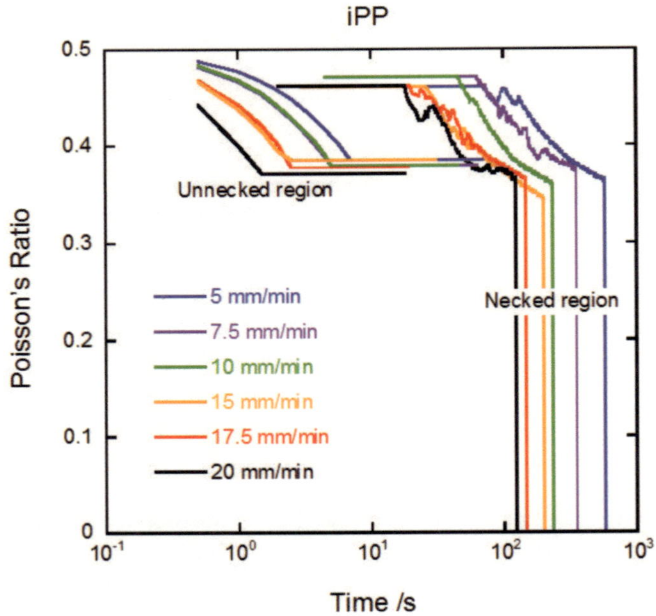

Fig. 6. Elongation speed dependence of Poisson's ratio of iPP at the unnecked and necked regions plotted against elongation time.

Figure 6 shows the Poisson's ratio-elongation time curves in the necked and unnecked regions for iPP. As seen in this figure, in the unnecked region, the value of Poisson's ratio gradually decreases from around 0.5 to around 0.35-0.37 and becomes constant during necking. This indicates that the deformation proceeds mostly through neck propagation after the neck initiation. In the necking region, the Poisson's ratio of the necked portion was almost constant at around 0.46, and the Poisson's ratio started to decrease again after the termination of necking. It is interesting to note that the samples were broken when the Poison's ratio reached about 0.33, independent of the elongation speed. In addition, it was found that the Poisson's ratio-time curves simply shift to longer times as the elongation speed increases; the curves can be superimposed on the elongation time axis. The stress-time curves can also be superposed. The data in the necked region could be superposed by merely shifting along the logarithmic time axis for the entire elongation range. Figure 7 shows the shift factor of the Poisson's ratio vs elongation time curves at a reference elongation speed of 20 mm/min. There is a linear relationship between the shift factor and the elongation speed. Using this relationship, we can determine the Poisson's ratio-elongation time curves at any elongation speed.

HDPE also showed clear necking. The necking initiation time decreases with increasing elongation speed, similarly to iPP, but the level of Poisson's ratio monotonically decreases with increasing elongation speed, as shown in Figure 8. It is interesting to note that the Poisson's ratio in the necking region is strongly dependent on the elongation speed. In contrast, the Poisson's ratio is relatively insensitive to the elongation speed in the post-necking region, with no clear differences in the Poisson's ratio-elongation time curves observed within experimental error (unlike the behavior observed for iPP). The Poisson's ratio at the break point was about 0.36 similar to that of iPP, and independent of the elongation speed.

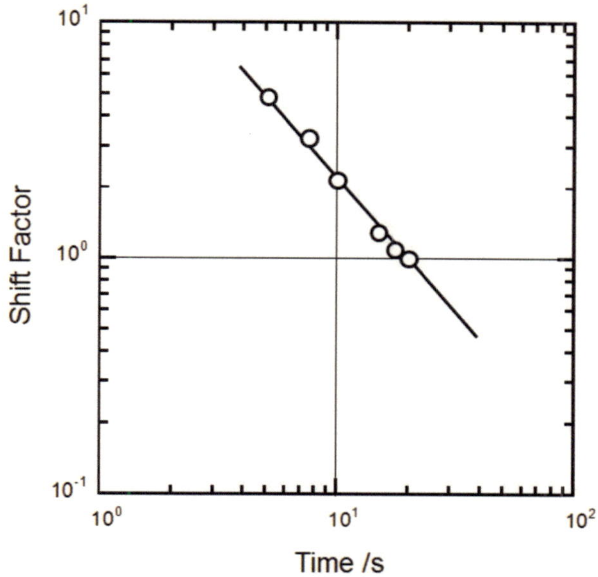

Fig. 7. Shift factor of Poisson's ratio of iPP for reference elongation speed of 20 mm/min.

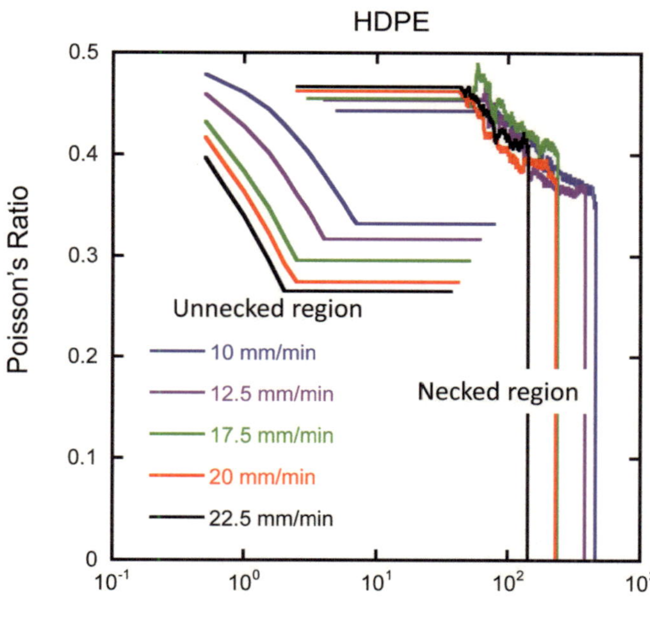

Fig. 8. Elongation speed dependence of Poisson's ratio of HDPE in the unnecked and necked regions plotted against elongation time.

In the case of LDPE, the necking phenomenon is unclear; no clear necking shoulders in the deformed specimens were observed over the entire experimental range of elongation speeds. Therefore, the Poisson's ratios in the center of the specimens are plotted against the elongation time. As shown in Figure 9, the minimum point of the Poisson's ratio corresponds to the initial point of necking, and the maximum point is taken as the natural drawing point. The overall values of Poisson's ratio are relatively higher than those of HDPE and iPP (which have higher crystallinity), but the Poisson's ratio markedly decreases with elongation time in the post necking or strain-hardening region similar to HDPE and iPP. It is interesting to note that the maximum point shifts to shorter times as the elongation speed increases and the Poisson's ratio at break was 0.33-0.36, independent of the elongation speed. Consequently, we can construct a master curve of Poisson's ratio vs elongation time. Similar to iPP, we have a linear relationship between the shift factor and the elongation time.

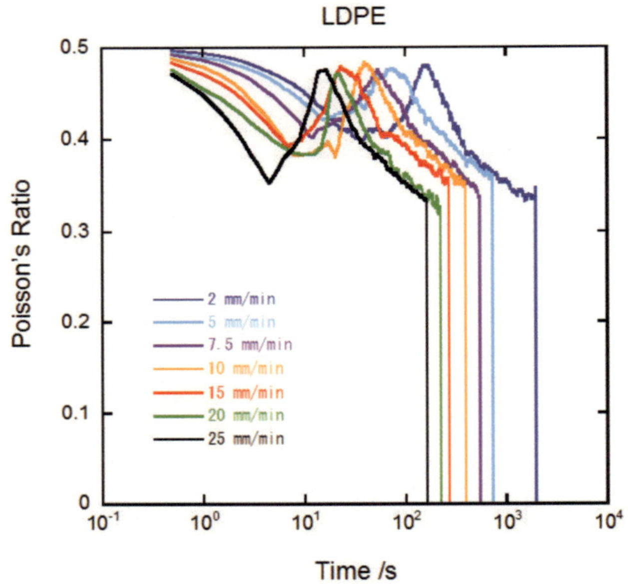

Fig. 9. Elongation speed dependence of Poisson's ratio of LDPE plotted against elongation time.

3.2 Temperature dependence

To investigate the temperature dependence, we equipped the tensile tester with an environmental chamber where the temperature was controlled within ±0.1 °C. Double-edge-notched specimens with a gauge length of 2mm were employed because of the limited size of the chamber.

Figure 10 shows the temperature dependence of the Poisson's ratio of iPP measured at a constant elongation speed of 20 mm/min (nominal strain rate of 10 /min). The onset times of necking and strain-hardening were almost independent of temperature and the Poisson's ratio increases with increasing temperature. The break time is also independent of temperature. Consequently, the Poisson's ratio-time curves longitudinally shift to higher values with increasing temperature. This indicates that the effect of temperature is not intrinsically coupled with that of elongation speed or nominal strain rate.

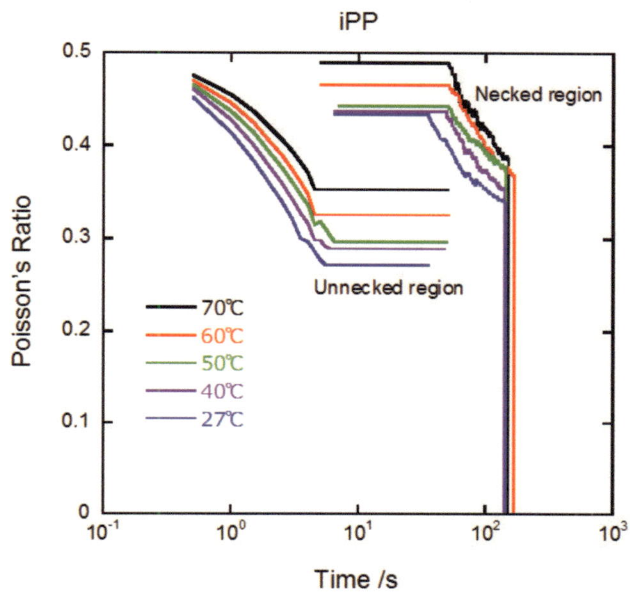

Fig. 10. Temperature dependence of Poisson's ratio of iPP in the unnecked and necked regions plotted against elongation time.

HDPE and LDPE did not reach breaking point owing to their high drawability. As shown in Figure 11, the temperature dependence of Poisson's ratio in HDPE is very similar to that in iPP. The overall level of the Poisson's ratio shifts to higher with increasing temperature, although the Poisson's ratio around the breaking point could not be identified. For LDPE, the minimum (neck initiation) and maximum (neck termination) points of Poisson's ratio appear to be insensitive to the temperature and the overall values approach 0.5 as the temperature increases (see Figure 12).

3.3 Crystallinity dependence

Figure 13 shows the crystallinity dependence of Poisson's ratio of iPP at a constant elongation speed of 20 mm/min, initial gauge length of 2 mm (nominal strain rate of 10/min) and 25 °C. In a similar way to the temperature dependence, the onset times of the necking and the strain-hardening as well as the break time were independent of the degree of crystallinity. The Poisson's ratio increases with decreasing crystallinity. This tendency is the same as that shown in the temperature dependence (Figure 10). It is interesting to note that the effects of softening of the materials due to lowering crystallinity are coupled with the effects of softening due to increasing temperature. In the tensile behavior, changing the crystallinity of the materials at a fixed temperature has an equivalent effect on the tensile deformation as changing the temperature for an iPP material with fixed crystallinity. Nitta et al. showed similar effects in the creep behavior of HDPE under fixed true stresses (Nitta &

Maeda, 2010). They showed that time-temperature superposition in creep behavior and crystallinity-time superposition are equivalent for HDPE samples having a wide range of crystallinities. The increase in amorphous phase fraction enhances the overall molecular mobility or extends the experimental timescale, corresponding to the enhancement of molecular mobility due to a rise in temperature.

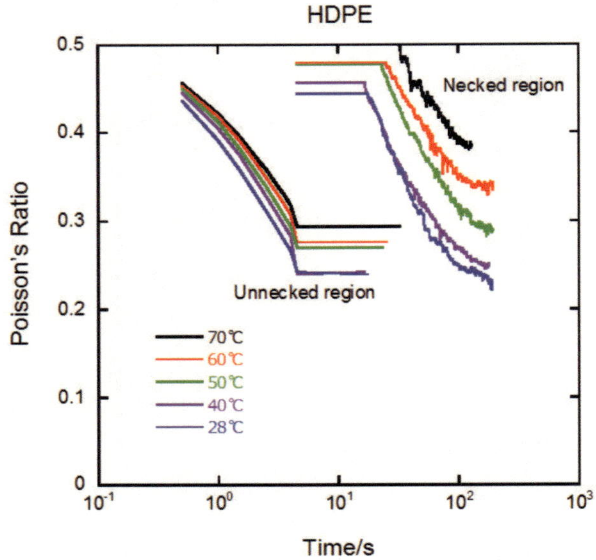

Fig. 11. Temperature dependence of Poisson's ratio of HDPE at the unnecked and necked regions plotted against elongation time.

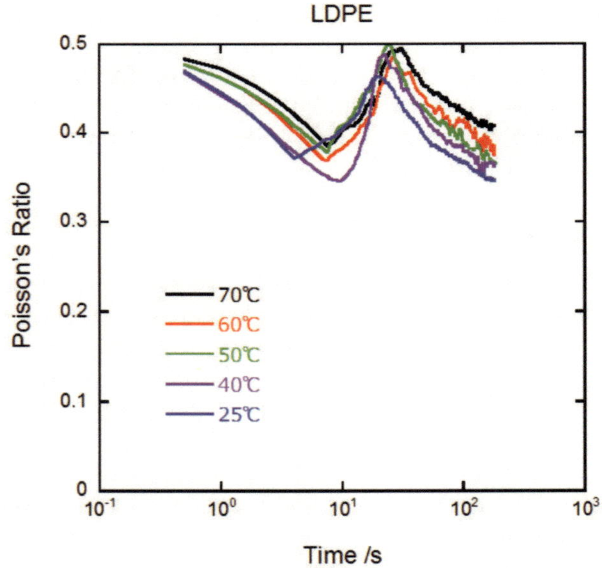

Fig. 12. Temperature dependence of Poisson's ratio of LDPE plotted against elongation time.

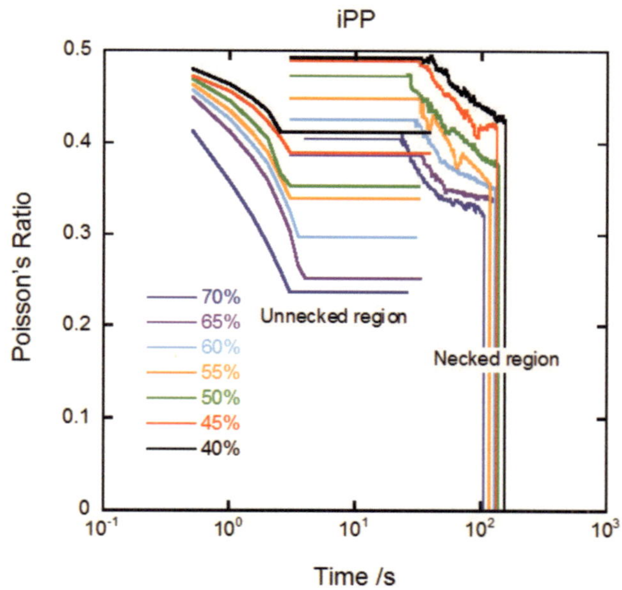

Fig. 13. Crystallinity dependence of Poisson's ratio of iPP in the unnecked and necked regions plotted against elongation time.

4. Constitutive relation

4.1 Stress-strain relation

Using the Poisson's ratio data, we can convert the nominal load-elongation time data into the true stress-elongation time and true stress-natural (true) strain curves. The true stress values in the necked and unnecked regions were estimated separately from the thickness data of each region. The Hencky strain or true strain values in the necked and unnecked regions were determined separately from the dimension of each region because of the inhomogeneous deformation.

Figures 14 and 15 show the elongation speed dependence of true stress-elongation time and true stress-natural strain curves in the necked region and unnecked region, respectively for iPP and HDPE at different elongational speeds (5-25 mm/min). It is important to note that two stressed states exist in the specimen during necking and the strains in both regions do not proceed during necking. It was found that the stress shows significant nonlinearity; for example, there is a maximum point in the unnecked region, whilst the true stress monotonically increases in a concave upward manner in the necked region. The upsweep in true stress in the necked parts becomes stronger as the elongation speed increases, and the break point was reduced at higher elongation speeds. There are no clear differences in true stress behavior between iPP and HDPE.

The temperature dependences of true stress-elongation time curves of iPP and HDPE are shown in Figures 16 and 17. Again, there are no clear differences in true stress behavior between iPP and HDPE. It is interesting to note that the slope of the initial time region is not sensitive to the temperature and the rise in temperature lowers the yield stress levels at the maximum. For LDPE, the elongation speed dependence of true stress-time behavior is similar to its temperature dependence, and the overall stress value becomes lower at higher temperatures. It is noted here again that HDPE did not reach a break point in the temperature dependence data.

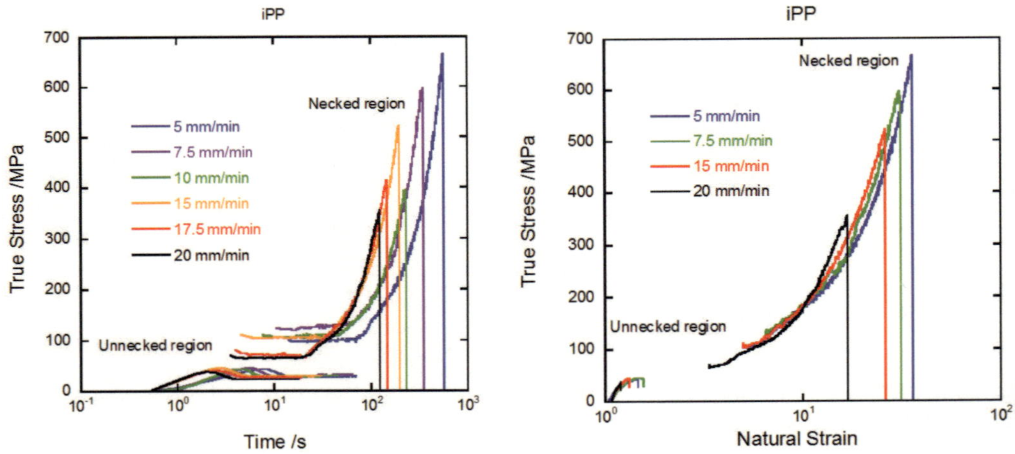

Fig. 14. Elongation speed dependence of true stress-elongation time and true stress- natural strain curves for iPP.

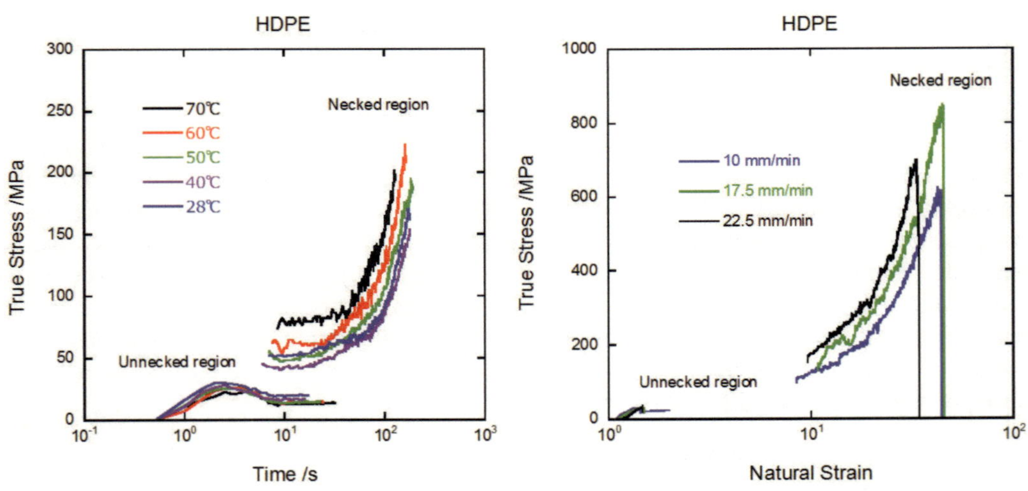

Fig. 15. Elongation speed dependence of true stress-elongation time curves for HDPE.

Figures 18 shows the elongation speed dependence of true stress-elongation time curves for LDPE. Because LDPE shows no clear necking process, the true stress-elongation time curves are not clearly divided into unnecked and necked regions. In the broad necking region, a stepwise increase in the true stress is seen in the curves. The true stress during necking propagation is an average of the true stress values at the maximum Poisson's ratio and the minimum Poisson's ratio.

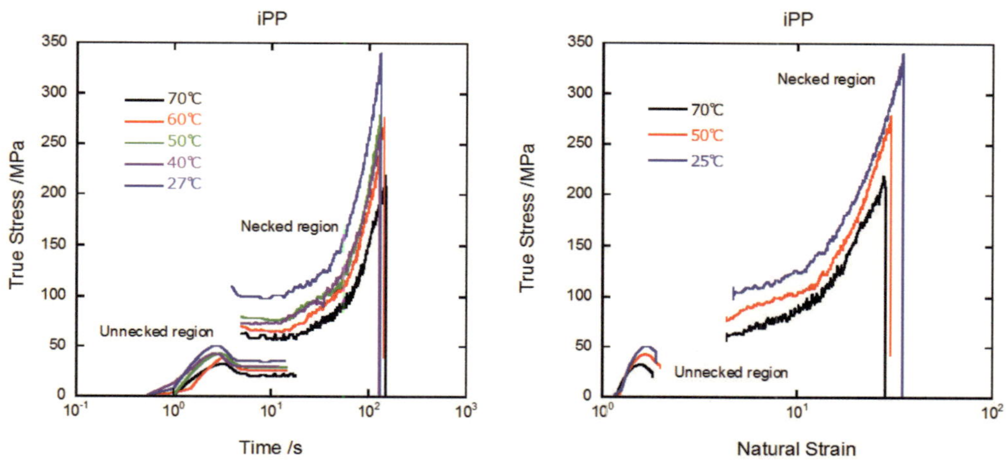

Fig. 16. Temperature dependence of true stress-elongation time and true stress- natural strain curves for iPP.

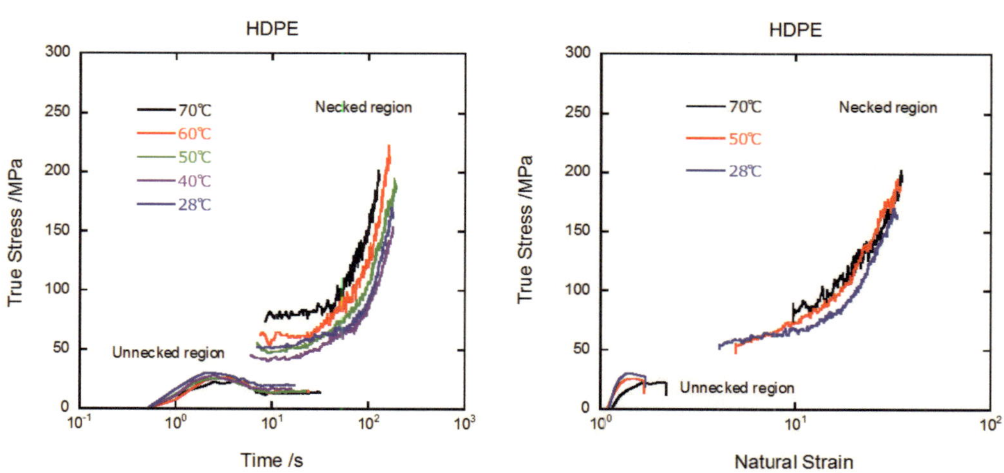

Fig. 17. Temperature dependence of true stress-elongation time and true stress- natural strain curves for HDPE.

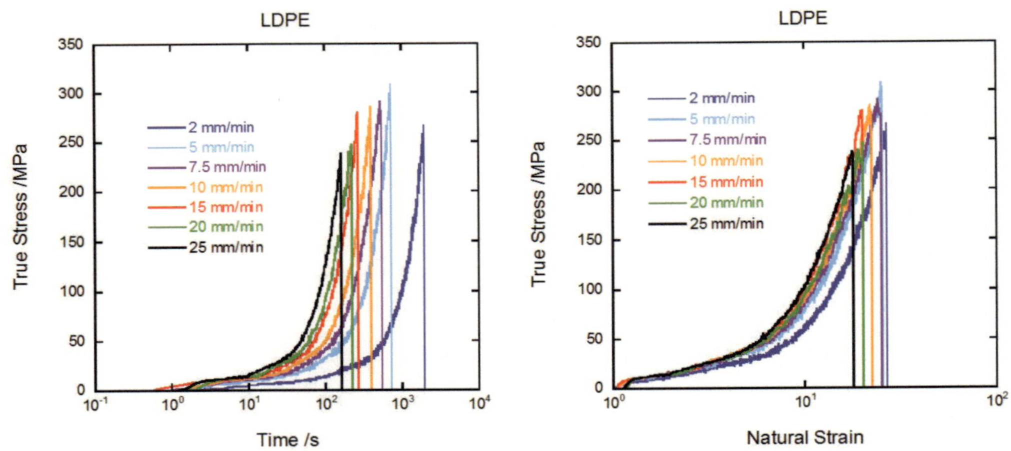

Fig. 18. Elongation speed dependences of true stress-elongation time and true stress-natural strain curves for LDPE.

4.2 Mechanical characteristics

Here we consider the yield stress as a measure of nonlinearity in the unnecked region, i.e. in the pre-necking region, and the strain-hardening coefficient as a measure of nonlinearity in the post-necking region. The yield stress was estimated from the maximum point of each true stress-elongation time curve. The strain-hardening coefficient, defined by Haward et al. (Haward & Thackray, 1968), is given by the modulus based on the true stress reported by the deformation following a neo-Hookean description.

According to Eyring's kinetic rate theory (Halsey et al., 1945) an equal number of plastic flow units move in the direction of the applied stress σ_p over the potential barrier shifted by the applied mechanical energy $V^*\sigma_p$, where V^* is termed the activation volume but its physical meaning is unclear. When the backward flow is negligible, the frequency ν_p of the flow in the forward direction is given by

$$\nu_p = \nu_0 \exp\left[-\frac{\Delta H - V^*\sigma_p}{RT}\right] \tag{1}$$

where R is the gas constant and ΔH is the original activation energy for plastic flow. Assuming that the frequency that a flow unit can surmount the potential barrier is related to the rate of change of strain $\dot{\varepsilon}$, then we have

$$\sigma_p = \frac{\Delta H}{V^*} - \frac{RT}{V^*}\ln\frac{\dot{\varepsilon}_0}{2\dot{\varepsilon}} \tag{2}$$

where $\dot{\varepsilon}_0$ is the constant pre-exponential factor which is larger than the conventional $\dot{\varepsilon}$. Rearranging gives the following relation:

$$\sigma_p = \frac{\Delta H}{V^*}\left(1 - \frac{T}{T_p}\right) = \gamma_p\left(T_p - T\right) \tag{3}$$

where $T_p = \Delta H / R(\ln \dot{\varepsilon}_0 / \dot{\varepsilon})^{-1}$ and $\gamma_p = \Delta H / V^* T_p$. The yield stress approaches zero as the temperature approaches T_p, which is in the vicinity of the α-relaxation temperature, where the crystalline phases start to melt or are in a quasi-molten state (Nitta & Tanaka, 2001). The wide applicability of Eyring model has been demonstrated for a number of polymers (Ward & Sweeney, 2004). Assuming that the plastic stress σ_p corresponds to the true yield stress, the temperature dependence of the yield stress follows Equation (3) as shown in Figure 19.

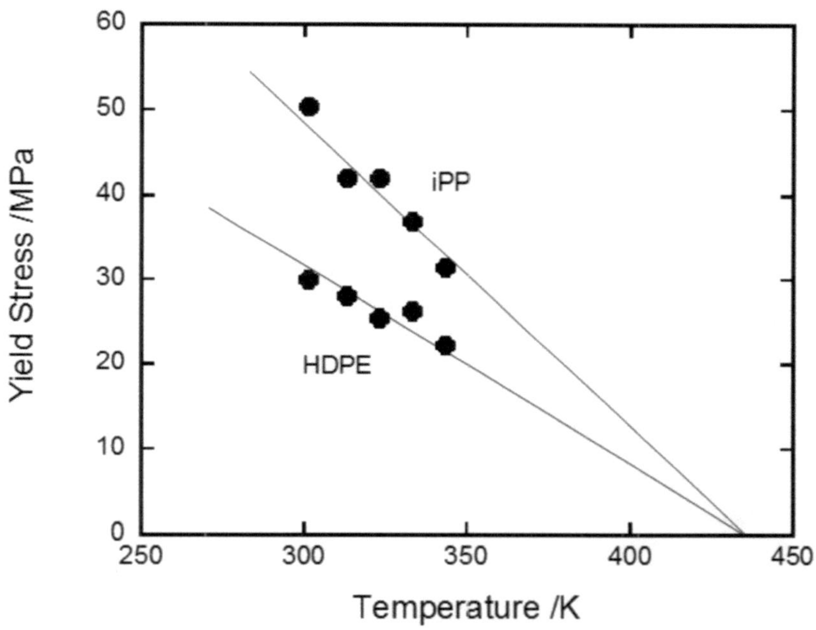

Fig. 19. Temperature dependence of the yield stress of iPP and HDPE.

In addition, Gent et al. presented an interesting concept that the mechanical energy for plastic deformation, yielding and necking corresponds to the thermodynamic melting work (Gent & Madan, 1989). If we accept their idea, Equation (3) can be modified, and ΔH can be replaced by $\chi \Delta H_0$, where ΔH_0 may be assumed to be the energy of 100% crystalline material. Then, the yield stress of iPP becomes proportional to the crystallinity in weight fraction χ, which was experimentally confirmed as shown in Figure 20.

It has been demonstrated that the tensile stress in the post yielding region, measured over a wide range of elongation speeds and temperatures is given by an additive equation of the plastic component and the network component (G'Sell & Jonas, 1981).

$$\sigma = \sigma_p(\dot{\varepsilon}, T) + \sigma_e(\lambda, T) \tag{4}$$

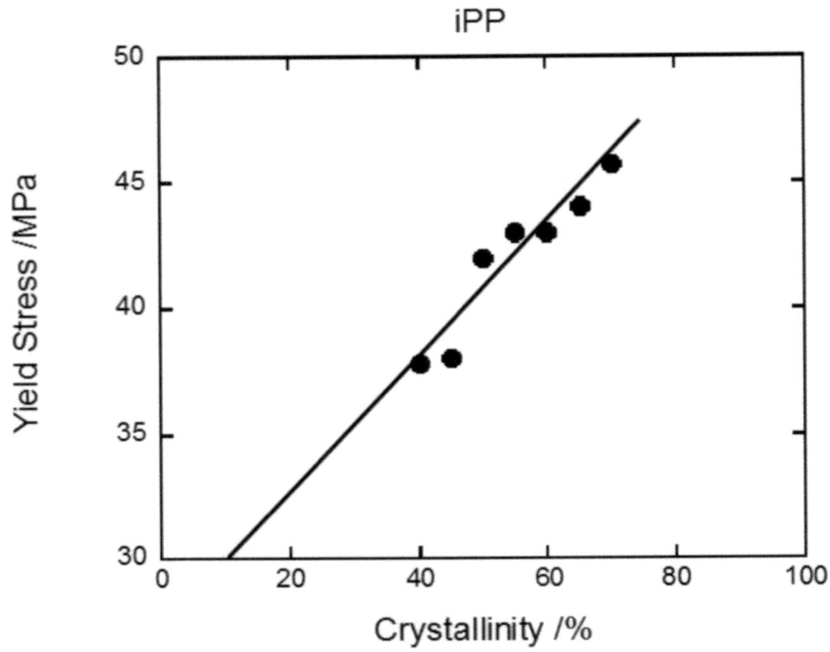

Fig. 20. Crystallinity dependence of the yield stress of iPP.

In the pre-necking region, the yield stress has a positive dependence on elongation speed and a negative dependence on temperature. These empirical characteristics of yielding flow can be expressed by an Eyring rate process.

According to Haward-Thackray theory, the true stress of iPP and HDPE is plotted as a function of the Gaussian strain modified by the transient Poisson's ratio data. The true stress σ and the modified Gaussian strain ε_G were defined using the following equations:

$$\sigma = \hat{\sigma}\lambda^{2\nu} \tag{5}$$

$$\varepsilon_G = \left(\frac{\lambda}{\lambda_{ND}}\right)^{1+2\nu} - 2\nu\frac{\lambda_{ND}}{\lambda} \tag{6}$$

where $\hat{\sigma}$ is the nominal stress and λ_{ND} is the natural draw ratio (the onset extensional ratio of the strain hardening). When λ_{ND}=1 and ν=0.5, we have a familiar relation: $\varepsilon_G = \lambda^2 - 1/\lambda$. A Gaussian plot for HDPE gives straight lines, and the slope of the line gives the strain hardening modulus G_p (see Figure 21). We have an empirical equation as follows:

$$\sigma - \sigma_N = G_p\varepsilon_G \tag{7}$$

where σ_N corresponds to the necking stress in the necked region. The results of iPP showed a deviation of the linearity at the initial Gaussian strains and the G_p values were estimated from the slope in the higher strain region.

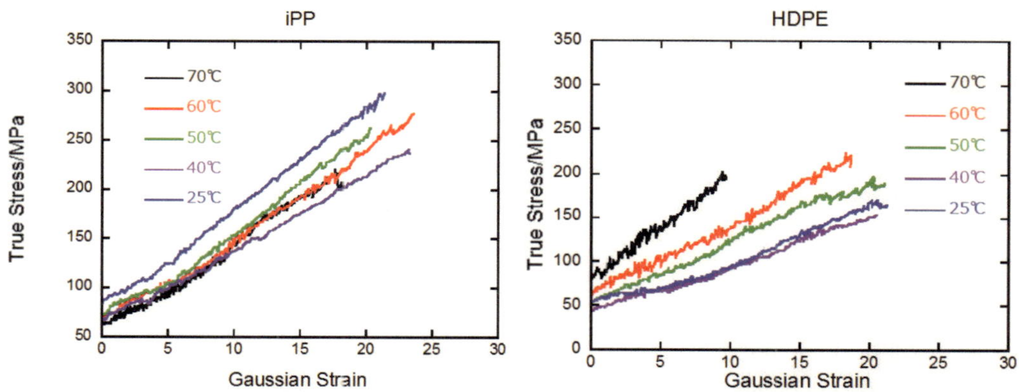

Fig. 21. Gaussian plots for iPP and HDPE

As shown in Figure 22, the G_p values increase linearly with temperature, empirically given by $G_p = \gamma_G(T - T_G)$ where γ_G is a positive constant and T_G is the temperature at G_p=0, which was almost 0 °C for HDPE. This indicates that entropic network deformation exists within the strain hardening process. It was found that the temperature sensitivity of HDPE is much greater than that of iPP. It should be noted here that the conventional G_p values, which are estimated from the replotting of the conventional Gaussian strain, $\lambda^2 - 1/\lambda.$, decrease with increasing temperature as demonstrated by Haward (Haward, 1993).

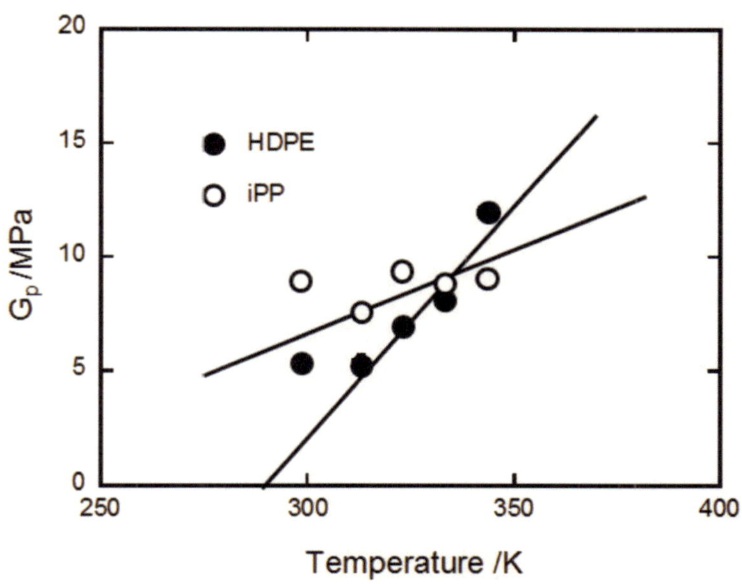

Fig. 22. Temperature dependence of the strain-hardening modulus of iPP and HDPE.

Considering that σ_N is also fitted by an Eyring equation (3) as demonstrated by Nitta et al. (Nitta & Takayanagi, 2006), we have the following simple constitutive equation:

$$\sigma = \gamma_p\left(T_p - T\right) + \gamma_G\left(T - T_G\right)\varepsilon_G \quad (8)$$

The first term is the plastic component with positive dependence on strain rate and negative dependence on temperature and the second term is associated with an entropic component with positive dependence on temperature.

5. Conclusion

Poisson's ratios in crystalline polymers were evaluated precisely from direct measurements of the sample dimensions during deformation over a wide range of extensional speeds, strain rates, temperatures and the degrees of crystallinities of the samples. True stress-strain curves with inhomogeneous deformation could be determined using the transient Poisson's ratio data in the necked portion and unnecked portions. Mechanical nonlinearity is closely related to the nonlinear behavior of Poisson's ratio in each region. We will now describe our conclusions for each region in more detail.

Elastic region: In the initial stage of deformation, Poisson's ratio cannot be precisely estimated using our experimental system. This is because the elongation is accelerated until reaching a preset elongation speed just after starting the elongation, and the acceleration causes large experimental errors when estimating precise values of lateral and transverse strains. The Poisson's ratio seems to start at zero, then reach a value consistent with the literature data and remain at that value during deformation up to the yield point.

Yielding and necking region: The Poisson's ratio decreases to a minimum value; the Poisson's ratio of the unnecked portion remains at this value during necking. This may be caused by volume expansion due to crazing, cracks, and voids. The Poisson's ratio of the necked part, (i.e. the natural drawn state), becomes nearly 0.5, and the total volume of deformation returns to the volume of the virgin specimen. The values of Poisson's ratio in the unnecked region depend on the extension ratio. As the elongation speed increases, the Poisson's ratio vs elongation curves completely shift to shorter times, resulting in an apparent decrease in Poisson's ratio with increasing elongation speed. On the other hand, a rise in temperature simply enhanced the Poisson's ratio to 0.5.

In the strain-hardening region: The true stress linearly increased with the Gaussian strain modified by the transient Poisson's ratio data and the slope of the line, corresponding to the strain-hardening coefficient, showed a positive dependence on temperature. In addition, the true stress-strain relation curves could be described by a linear relationship with the Gaussian strain modified by transient Poisson's ratio data.

6. Acknowledgment

This author (K.N) wishes to express his gratitude to the late Professor Motowo Takayanagi for his thoughts and suggestions.

7. References

Balta-Calleja, F.J., Kilian, H.-G., (1985). A novel concept in describing elastic and plastic properties of semicrystalline polymers: polyehtylene. *J. Colloid & Pollym. Sci.*, 263, 9, 697-707.

Carothers, W.H., Hill, W.J. J. Am. Chem. Soc, 54, 1579 (1932).

Ericksen, J. J. (1975). Equilibrium of bars. *J. Elasticity*, 5, 191-201

G'Sell, C., Jonas, J.J. (1981). Yield and transitent effects during the plastic deforamtion of solid polymers, *J. Mater. Sci.*, 16, 1956-1974

Halsey, G., White, H.J, Eyring, H.(1945). Mechanical properties of textiles I. *Tex. Res. J.*, 15, 9, 295-311

Haward, R.N., (1993). Strain hardening of thermoplastics, *Macromolecules*, 26, 22, 5860-5869.

Heise, B., Kilian, H.-G., Wulff, W. (1980). Deformation and microstructure in uniaxially stretched PE. *Prog. Collod & Polym. Sci.*, 67, 143-148, ISSN 0340-255X.

Huchingson, J. W., Neale, K.W. (1983). Neck propagation. *J. Mech. Phys. Solid*, 31, 405-426

Kilian, H.-G., Pietralla, M. (1978). Anisotorpy of thermal diffusivity of uniaxial stretched polyethylenes. *Polymer*, 19, 664-672.

Kilian, H.-G. (1981). Equation of state of real networks. *Polymer*, 22, 209-217.

Kilian, H.-G. (1981). A molecular interpretation of the parameters of the van der Waals equation of state for real networks. *Polymer Bull.*, 3, 151-158.

Kuriyagawa, M., Nitta, K.-H. (2011) Structural explanation on natural draw ratio of metallocene-catalyzed high density polyethylene, *Polymer*, 52, 3469-3477

Mark, J.E. (1993). *Physical Properties of Polymers* (3rd), Cambridge Univ., ISBN 0 521 53018 0, Cambridge, UK.

Nitta, K.-H., and Tanaka, A. Dynamic mechanical properties of metallocene catalyzed linear polyethylenes, 42 (2001),1219-1226.

Nitta, K.-H., Takayanagi, M. (2006). Application of catastrophe theory to the neck-initiation of semicrustalline polymers induced by the intercluster links. *Polymer J.* 38, 8, 757-766.

Nitta, K.-H., Takayanagi, M. (2003). Novel proposal of lLamellar clustering process for elucidation of tensile yield behavior of linear polyethylenes. *J. Macromol. Sci.-Phys.* 42, 1, 109-128

Nitta, K.-H., Maeda, H. (2010). Creep behavior of high density polyethylene under a constant true stress. *Polymer Testing*, 29, 60-65.

Peterlin A, Meinel G. (1971). Small-angle x-ray diffraction studies of plasticity deformed polyethylene. III Samll draw ratios. *Makromol Chem.* 142,1, 227-240

Peterlin A. (1971). Molecular model of drawing polyethylene and polypropylene. *J. Mater.Sci.*, 6, 490-508

Peterlin A. (1975). Plastic deformation of polymers with fibrous structure. Colloid Polym Sci. 253, 10, 809-823

Strobl, G.R. (1996). *The Physics of Polymers*, Springer, ISBN 3-540-60768-4, Berlin

Séguéla, R., Darras, O. (1994). Phenomenological aspects of the double yield of polyethylene and related copolymers under tensile loading, *J. Mater. Sci.*, 29, 5342-5352

Séguéla, R. (2007). On the natural draw ratio of semi-crystalline polymers: Review of the mechanical, physical and molecular aspects. *Macromol. Mater. Eng*. 292, 235-244

Tagawa, T.; Ogura, K. (1980), Piled-lamellae structure in polyethylene film and its deformation. J. Polym. Sci. Polym. Phys., 18,5,971-979

Ward, I.M., Sweeney, J. (2004) *An Introduction to The Mechanical Properties of Solid Polymers.* Wiley, IBSN: 047-149626X, New York.

6

Batch Foaming of Amorphous Poly (DL-Lactic Acid) and Poly (Lactic Acid-co-Glycolic Acid) with Supercritical Carbon Dioxide: CO_2 Solubility, Intermolecular Interaction, Rheology and Morphology

Hongyun Tai
School of Chemistry, Bangor University, Bangor, United Kingdom

1. Introduction

Poly (lactic acid) (PLA) is a biobased aliphatic polyester prepared by condensation polymerization of lactic acid (2-hydroxy propionic acid) or ring opening polymerization of lactide in the presence of suitable catalysts (Garlotta, 2001; Lim et al., 2008). Lactic acid and lactide are chiral molecules (Figure 1). Lactic acid has two optical isomers, L-(+)-lactic acid or (S)-lactic acid and its mirror image D-(−)-lactic acid or (R)-lactic acid. Lactide is the cyclic di-ester of lactic acid and has three stereoisomers, i.e. meso-lactide, D-lactide and L-lactide. The majority of the commercially produced lactic acid is made by bacterial fermentation of carbohydrates. PLA has great potentials as an alternative to petroleum-based synthetic polymers and has been exploited as biodegradable thermoplastics for packaging and as FDA approved biomaterials for tissue engineering and drug delivery (Dorgan et al., 2001; Langer &Peppas, 2003).

PLA with the different ratio of L and D stereoisomers shows significantly different physical properties, for example PLA with high content of L-lactic acid (P_LLA) (e.g. greater than 90 %) is semicrystalline and PLA with more than 20% D-lactic acid content is amorphous (Tsuji, 2005; Lim et al., 2008). The crystallinity level of PLA decreases with D-lactic acid content and molecular weight. The copolymerization of lactic acid and glycolic acid has been used to tailor the biodegradability of the polymer. The rate of degradation of the polymers increases with the level of glycolic acid in the copolymer and decreases with molecular weight and crystallinity. Semicrystalline PLA can be converted into end-use products using conventional melt processing methods, including extrusion, injection molding, blow molding, casting, blown film, thermoforming, foaming, blending and fibre spinning (Lim et al., 2008). These processing methods require an elevated temperature of 10 to 20 °C above the melting point (ca. 175 °C) of the polymer. Therefore, they are not applicable when biologically active guest species are required to be incorporated into polymer host in situ to fabricate drug delivery and tissue engineering devices. Common techniques for the fabrication of 3-D porous tissue engineering scaffolds include solvent casting/salt leaching (Lu et al., 2000; Murphy et al., 2002) and moulding/salt leaching (Hou et al., 2003; Sosnowski et al., 2006). In addition, for tissue engineering and drug delivery, double emulsion solvent evaporation method has been used to prepare microparticles/microspheres (Freitas et al., 2005; Jiang et al., 2005; Kim et al., 2006) and electrospinning has been used to prepare micro/nanofibres (Venugopal &Ramakrishna, 2005; You et al., 2006; Kim et al., 2010). However, organic solvents are often required in these fabrication approaches.

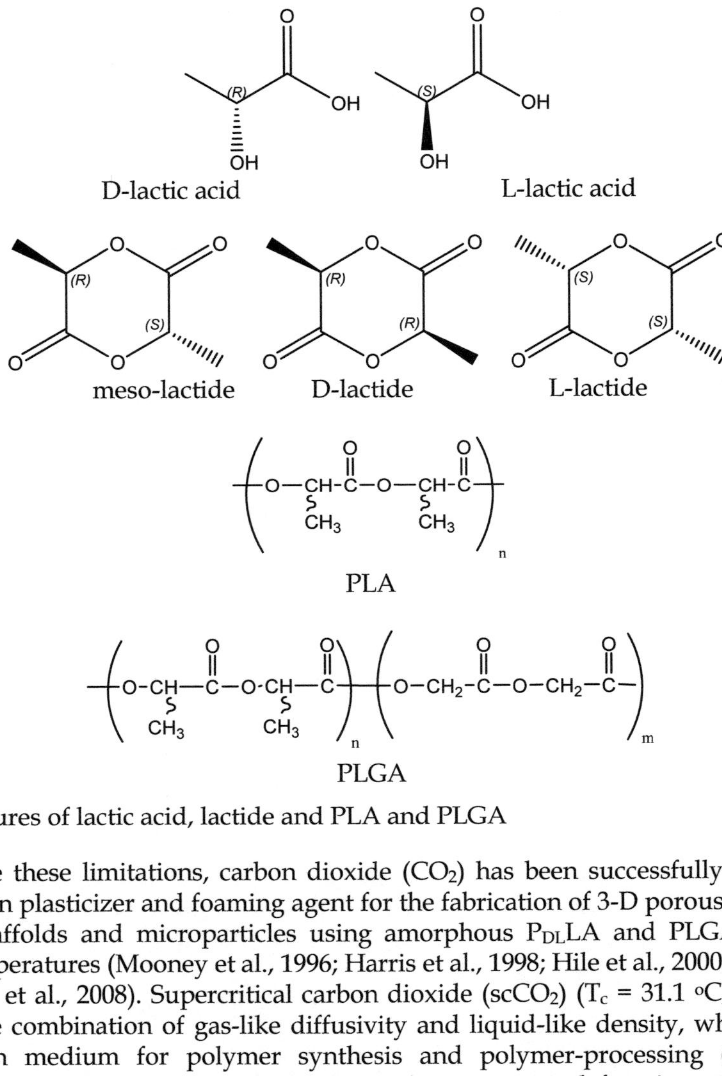

Fig. 1. Structures of lactic acid, lactide and PLA and PLGA

To overcome these limitations, carbon dioxide (CO_2) has been successfully utilized as an efficient green plasticizer and foaming agent for the fabrication of 3-D porous biodegradable monolith scaffolds and microparticles using amorphous $P_{DL}LA$ and PLGA polymers at ambient temperatures (Mooney et al., 1996; Harris et al., 1998; Hile et al., 2000; Howdle et al., 2001; Davies et al., 2008). Supercritical carbon dioxide ($scCO_2$) (T_c = 31.1 °C, P_c = 73.8 bar) has a unique combination of gas-like diffusivity and liquid-like density, which makes it a unique clean medium for polymer synthesis and polymer-processing (Cooper, 2001; Tomasko et al., 2003; Woods et al., 2004). In the supercritical foaming process, a small amount of CO_2 dissolved into the polymers can dramatically reduce the glass transition temperature (T_g) and the viscosity of the polymers, which make it possible to incorporate bioactive compounds into the polymer in situ at a low temperature thus to limit the loss of their activities. After the release of CO_2 from the polymer, a porous structure can be formed.

In supercritical foaming process, control of pore structure is crucial because pore size, porosity and interconnectivity of scaffolds strongly influences cell growth behavior and drug release profile. The key for this is to control pore nucleation and growth which are mainly influenced by the amount of CO_2 dissolved in the polymer and the rate of CO_2 escaping from the polymer. Therefore, CO_2 solubility and polymer viscosity are two important aspects to study in order to control pore structure of scaffolds in supercritical foaming.

In this chapter, the recent studies on controlling the amorphous $P_{DL}LA$ and PLGA supercritical foaming process will be reviewed, particularly on the studies of the solubility of CO_2 in PLA and PLGA (Pini et al., 2007; Pini et al., 2008), the intermolecular interaction of CO_2 with PLA and PLGA (Tai et al., 2010), the rheological property of PLA and PLGA/CO_2 mixture (Tai et al., 2010), and the effects of processing conditions on the porous structures of the PLA and PLGA foams (Tai et al., 2007; Tai et al., 2007; White et al., 2011).

2. CO_2 solubility in PLA and PLGA

To control and optimise the supercritical foaming of amorphous PLA and PLGA polymers, it is important to study the solubility and diffusivity of the CO_2 in these biodegradable polymers. The solubility represents the amount of the CO_2 that can be dissolved in them at equilibrium conditions. CO_2 solubility in a wide range of polymers has been studied using gravimetric (Zhang et al., 2003; Oliveira et al., 2006) and spectroscopy (Duarte et al., 2005) methods. The sorption and swelling of PLA and PLGA polymers have been studied using different measurement techniques and also simulated using different mathematical modeling. These include a quartz crystal microbalance for the measurement and a dual-mode sorption model and Flory-Huggins equation for modelling used by Oliveira and coworkers (Oliveira et al., 2006; Oliveira et al., 2006); an external balance for the measurement and perturbed-hard-sphere-chain equation of state for modeling used by Elvassore and coworkers (Elvassore et al., 2005); and a pressure decay technique for the measurement and Sanchez-Lacombe equation of state for modeling used by Liu and coworkers (Liu &Tomasko, 2007). Oliveira *et al.* studied the solubility of CO_2 in PLA with two different L:D content, 80:20 (amorphous) and 98:2 (20 % crystallinity) (Oliveira et al., 2006). The experiments were performed between 30 and 50 °C and up to 50 bar. It was found that CO_2 is slightly more soluble in PLA 80:20 than in PLA 98:2. The solubility of CO_2 in both polymers shows the same trend, increasing with pressure and decreasing with temperature. Three models were used to correlate the experimental results, dual-mode sorption model, Flory-Huggins equation and an extended Flory-Huggins equation. Elvassore et al. applied a gravimetric method to obtain CO_2 absorption isotherms on three different polymers (PLGA 4852, PLGA 5446, and PLGA 5347) at 40 °C, which were then correlated using a perturbed-hard-sphere-chain equation of state (Elvassore et al., 2005). Liu and Tomasko used a pressure decay technique to measure CO_2 sorption isotherms of three polymers (PLGA 7525, PLGA 5050 and $P_{DL}LA$ homopolymer) at temperatures of 30, 40 and 60 °C and pressures up to 100 bar (Liu &Tomasko, 2007).

Pini et al. studied the solubility of amorphous $P_{DL}LA$ and PLGA polymers (Table 1) at 35 °C and up to 200 bar by measuring their sorption and swelling in CO_2 using a magnetic suspension microbalance and visualization method, and the Sanchez-Lacombe equation of state was used for correlating the experimental data (Pini et al., 2007; Pini et al., 2008). The $P_{DL}LA$ homopolymer shows the largest values of sorption and swelling (sorption and fractional swelling up to 0.5 g CO_2/g polymer and 0.68, respectively, at 200 bar), whereas the PLGA5050 has the lowest affinity to CO_2 (Figure 2). The sorption and swelling decrease with increasing the content of glycolic acid in PLGA copolymers. The Sanchez-Lacombe equation of state is the most widely used model to describe the solubility of CO_2 in polymers due to its simplicity, well-defined physical meaning, and the ability to extend available data to high temperature and pressures (Tomasko et al., 2003). Therefore, all data have been correlated by the Sanchez-Lacombe equation of state, demonstrating that this model is able to represent the actual behavior with reasonable accuracy using literature values for the pure component parameters and by fitting a single binary interaction parameter (Pini et al., 2008). CO_2 solubility and diffusivity are influenced by both the molecular structure (the interaction between CO_2 and polymer molecular chains) and the morphology (crystalline or amorphous, related with free volume) of polymers interested. Shieh and Lin suggested that the sorption process at or below Pc was mainly driven by carbonyl groups and above Pc by the degree of crystallinity such that the higher the degree of crystallinity, the lower CO_2 solubility in the polymer (Shieh &Lin, 2002). It is easy to understand that a strong interaction between CO_2 and polymer, and a low crystallinity (large free volume) of a polymer can lead to high gas solubility.

Polymers	Resources	Composition (LA:GA)[a]	M_w[b] (KD)	PDI[c]	T_g[d] (°C)
P_{DL}LA 15K	Resomer	100:0	15	2.34	41.8
P_{DL}LA 52K	Purac	100:0	52	1.87	46.9
PLGA 85:15	Lakeshore	85:15	77	1.70	48.6
PLGA 75:25	Resomer	75:25	72	1.75	50.4
PLGA 65:35	Lakeshore	65:35	52	1.69	49.1
PLGA 50:50	Resomer	50:50	53	1.59	47.0

[a] Copolymer composition: the mole ratio of lactic acid (LA) and glycolic acid (GA) in the copolymer; [b] Weight average molecular weight, determined by GPC; [c] Polydispersity, determined by GPC; [d] Glass transition temperature, determined by DSC.

Table 1. PLA and PLGA Polymers (Tai et al., 2010)

3. Intermolecular interaction of CO_2 with PLA and PLGA

Despite numerous studies of polymers with gases or supercritical fluids, there is a little molecular level information on interactions within these polymer systems. There was the general perception that polymer swelling or gas sorption is a purely physical phenomena till Kazarian *et al.* first studied the specific intermolecular interactions between CO_2 and polymers by FTIR and ATR-IR spectroscopy (Kazarian et al., 1996). It is suggested that the reduction of T_g is a thermodynamic effect due to intermolecular interactions between CO_2 and the polymer and not simply a hydrostatic pressure effect (Tomasko et al., 2003). Nalawade *et al.* characterized polyesters, poly(ethylene glycol) and polyphenylene oxide by using a modified Fourier transform-infrared spectroscopy set-up under sub- and supercritical CO_2 conditions (Nalawade et al., 2006). Analysis of the corresponding spectra shows evidences of weak interaction (Lewis acid-base) between CO_2 and polymers. In particular, shifts to higher wavelengths of the maximum absorption of chain groups of the polymer and the modification of the absorption band of CO_2 represent a qualitative evidence of such interactions. In general, polymers with ether group display higher interaction strength than polyesters. The chain flexibility aids dissolution of CO_2 in polymers and carbonyl or ether groups that are accessible in the backbone or on side chains can specifically interact with CO_2 (Kazarian et al., 1996; Tomasko et al., 2003; Nalawade et al., 2006).

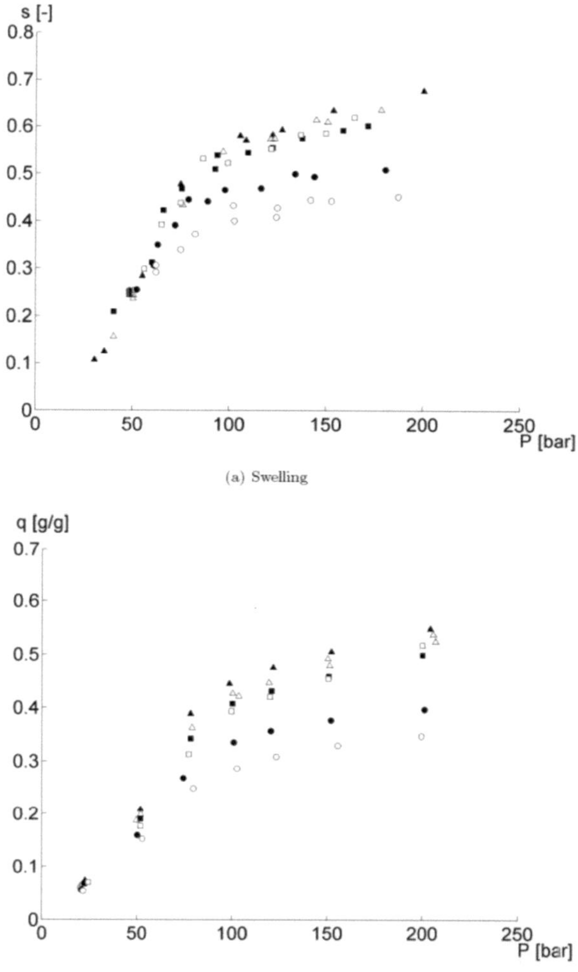

(a) Swelling

(b) Sorption

Fig. 2. Swelling and sorption isotherms at 35.0 °C as a function of pressure P for PLA15k (▲), PLA52k (Δ), PLGA8515 (■), PLGA7525 (□), PLGA6535 (●) and PLGA5050 (○). (a) Swelling s and (b) Sorption q. Reproduced with permission from (Pini et al., 2008)

Tai et al. adopted a high pressure attenuated total reflection Fourier transform infrared (ATR-IR) to investigate the interactions of CO_2 with PLA and PLGA polymers with the glycolic acid (GA) content in the copolymers as 15, 25, 35 and 50 % respectively (Table 1). Shifts and intensity changes of IR absorption bands of the polymers in the carbonyl region (~1750 cm^{-1}) are indicative of the interaction on a qualitative level. The spectra for $P_{DL}LA$

and PLGA polymers were recorded at 35 °C and 80 bar after soaking for 1 hour. The maximum absorptions in carbonyl region bands for all these polymers located at the similar position at ~1753.5 cm^{-1} (Figure 3, b). In comparison with the spectra for the pure polymers (Figure 3, a), P$_{DL}$LA had the largest shift (5.6 cm^{-1}) and PLGA 5050 had the smallest shift (1.3 cm^{-1}). This indicated that P$_{DL}$LA had the greatest interaction with CO_2 and the interaction decreased with the increase of GA content in the polymers. This is in good agreement with the results obtained from the sorption and swelling studies, in which the solubility of CO_2 in PLGA copolymers was found to decreases with the increase in the glycolic acid content (Pini et al., 2007; Pini et al., 2008).

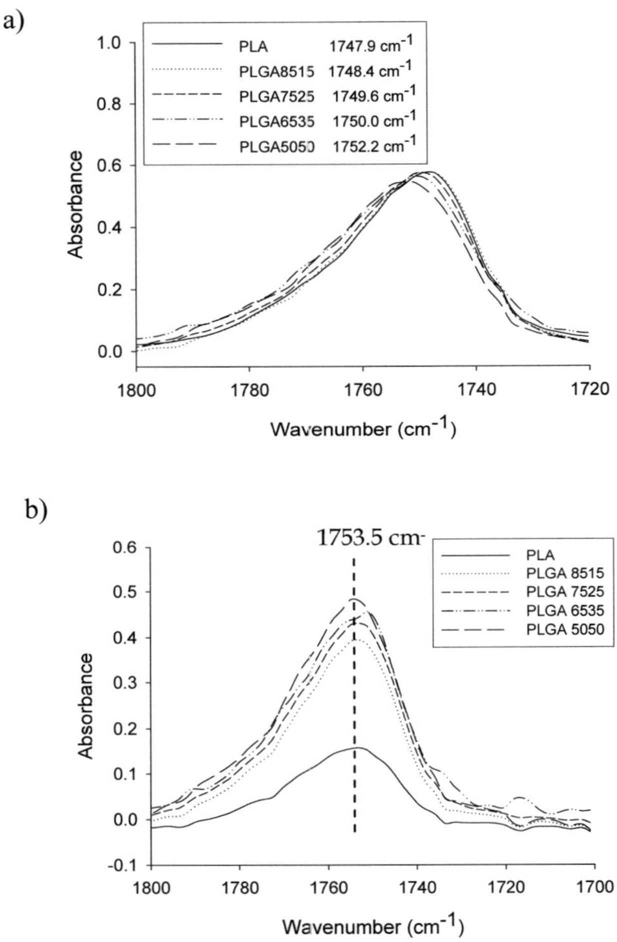

Fig. 3. FTIR spectra (carbonyl region bands) of P$_{DL}$LA and PLGAs. (a) Sepctra for the pure polymers obtained in the absence of CO_2, the absorbance bands in carbonyl region shift to high wavenumber with increasing GA content in the PLGA copolymers. (b) Spectra obtained for the polymers in the presence of CO_2 at 35 °C and 80 bar after soaking for 1 hour. Note, the maximum absorption in carbonyl region bands (picked by Omnic software) located at the position of ~1753.5 cm^{-1}. Reproduced with permission from (Tai et al., 2010)

The interaction of polymers with CO_2 is dependent upon the chemical structure. The affinity of CO_2 with polyesters is largely due to the interaction of CO_2 molecules with the carbonyl group on the polymer chains. Lactic acid (LA) possesses an extra methyl group which could lead to at least two-opposing consequences. One is to increase the steric hindrance and then lower the interaction between the carbonyl group and the CO_2 molecules, whereas the other one is to increase the available free volume in the matrix due to the steric effect. The latter factor could have played a dominant role in determining the CO_2 behavior in PLGA polymers, leading to a higher solubility for PLGA with a high LA content.

4. Rheology of PLA and PLGA

To process PLA polymers with conventional techniques, such as extrusion, injection molding, blow molding, casting, blown film, thermoforming and foaming, it is extremely important to understand the melt rheological behaviour of PLA which is highly dependent on temperature, molecular weight and shear rate (Lim et al., 2008). The melt rheology of semicrystalline P_LLA has been studied at elevated temperatures above its melting point (Dorgan et al., 1999; Dorgan et al., 2000; Lehermeier &Dorgan, 2001; Palade et al., 2001; Dorgan et al., 2005). Cooper-White *et al.* studied the dynamic viscoelastic behavior of P_LLA with molecular weights ranging from 2 to 360 kg/mol, over a broad range of reduced frequencies (approximately 1 X 10^{-3} s^{-1} to 1 X 10^3 s^{-1}), using time-temperature superposition principle (Cooper-White &Mackay, 1999). The temperature range used was between 170 to 220 °C. Melts are shown to have a critical molecular weight of approximately 16 kg/mol, and an entanglement density of 0.16 mmol/cm^3 (at 25 °C). Fang *et al.* studied viscosities of two types of PLA resins (amorphous and semicrystalline) at 150 and 170 °C and at various shear rates (30, 50, 70, 90, 110, 130, and 150 rpm screw speeds) with a tube rheometer on an extruder (Fang &Hanna, 1999). The viscosity data was calculated from the pressure profiles and the volume flow rate. The effects of resin type, temperature and shear rate on melt viscosity were determined. Under the same processing conditions, semicrystalline PLA had a higher shear viscosity than amorphous PLA. As the temperature increased, the shear viscosity decreased for both types of PLA. The PLA melt was characterized as a pseudoplastic, non-Newtonian fluid. Dorgan *et al.* also studied the melt rheology of PLA by measuring master curves of polymers. The effects of variable L-content (Palade et al., 2001; Dorgan et al., 2005), entanglement and chain architecture (Dorgan et al., 1999), linear and star polymer chain architecture (Dorgan et al., 2000) and blending chain architectures (Lehermeier &Dorgan, 2001) on the rheological property of PLA were investigated. The dynamic frequency sweeps were performed at 180 °C to obtain zero shear viscosity data for a linear and a branched PLA (Dorgan et al., 2000). Longer relaxation times for the branched material, compared to the linear material, manifests itself as a higher zero shear rate viscosity. However, the branched material shear thins more strongly, resulting in a lower value of viscosity at high shear rate. Isothermal frequency sweeps were also performed at a wide range of temperatures from (56 to 180 °C) for semicrystalline P_LLA and amorphous $P_{DL}LA$ (Dorgan et al., 2005). For all compositions of PLA investigated, the weight average molecular weights were within the range of 10^5-10^6 g/mol.

CO_2 has been used in gas assisted extrusion process to reduce the glass transition temperature and the viscosity of the polymer, also to increase the miscibility of the different polymers in blending to form polymer composites (Garlotta, 2001; Lim et al., 2008).

However, there are very few studies about extrusion foaming of PLA (Reignier et al., 2007; Mihai et al., 2010). Mihai et al. studied the rheology and extrusion foaming of chain-branched PLA at a temperature between 180 °C to 200 °C (Mihai et al., 2010). Two PLA grades, an amorphous and a semicrystalline one, were branched using a multifunctional styrene-acrylic-epoxy copolymer. The branching of PLA and its foaming were achieved in one-step extrusion process. CO_2 up to 9% was used to obtain foams from the two PLA branched using chain-extender contents up to 2%. The foams were investigated with respect to their shear and elongational behavior, crystallinity, morphology, and density. The addition of the chain-extender led to an increase in complex viscosity, elasticity, elongational viscosity, and in the manifestation of the strain-hardening phenomena. Differences in foaming behavior were attributed to crystallites formation during the foaming process. The rheological and structural changes associated with PLA chain-extension lowered the achieved crystallinity but slightly improved the foamability at low CO_2 content. Corre et al studied the batch foaming of modified PLA with $scCO_2$ (Corre et al., 2011). Improvement of the melt viscosity and elasticity was achieved by the use of an epoxy additive during a reactive extrusion process. Rheological characterizations confirmed an increase of the melt strength due to this chain extension process. Foaming was then performed on the neat and modified PLAs using a batch process with $scCO_2$ as blowing agent using the saturation temperature of 165 °C and pressure ranging from 96 to 142 bar. Depending on the foaming parameters, foams with a cellular structure ranging from macro scale to micro scale have been obtained.

The viscosity of polymer/CO_2 mixture plays an important role for the control of porous structures in CO_2 foaming process. The dissolved CO_2 in the polymers reduces the viscosity and T_g of the polymers dramatically (Tomasko et al., 2003; Woods et al., 2004). These reductions are closely related to the solubility of CO_2 in the polymers. The higher the solubility of CO_2 in a polymer, the higher is the reduction in the viscosity and the glass transition temperature (Nalawade et al., 2006). To study the rheological property of PLA in the presence of CO_2, high pressure rheometer or high pressure viscometer are required. There are several studies on the viscosity reduction of polymer melts or fluid samples under CO_2 using high pressure concentric cylinder rotational viscometer (Flichy et al., 2003), high pressure slit die (Royer et al., 2001) and capillary viscometer(Qin et al., 2005), and high pressure magnetically levitated sphere rheometer (Royer et al., 2002). Tai et al. studied the rheological properties of amorphous $P_{DL}LA$ and PLGA polymers in the presence and absence of CO_2 using a parallel plate rheometer (Tai et al., 2010). The viscosity curves of $P_{DL}LA$, PLGA 8515 and PLGA 6535 were recorded at 35 °C and 100 bar after CO_2 saturated the polymers (soaking for 24 hours). It was found that the viscosity curves (i.e., shear viscosity versus shear rate) of polymer/CO_2 mixtures were similar in shape to those of pure polymers (Figure 4, a and b). The shear thinning phenomenon was observed for all the polymers, which is important information for extrusion and injection processing of the polymers. The comparison of zero viscosity data for the polymers indicated that the dissolved CO_2 (*ca.* 25-30 wt %) lowered the viscosity of $P_{DL}LA$ and PLGA at 35 °C to the similar level for the pure polymers at a high temperature of 140 °C (Figure 4, c). Moreover, a greater viscosity reduction for $P_{DL}LA$ was observed comparing to PLGAs. These rheological data also demonstrated that the interactions of CO_2 with amorphous $P_{DL}LA$ and PLGA polymers decrease with increasing GA content in the polymers.

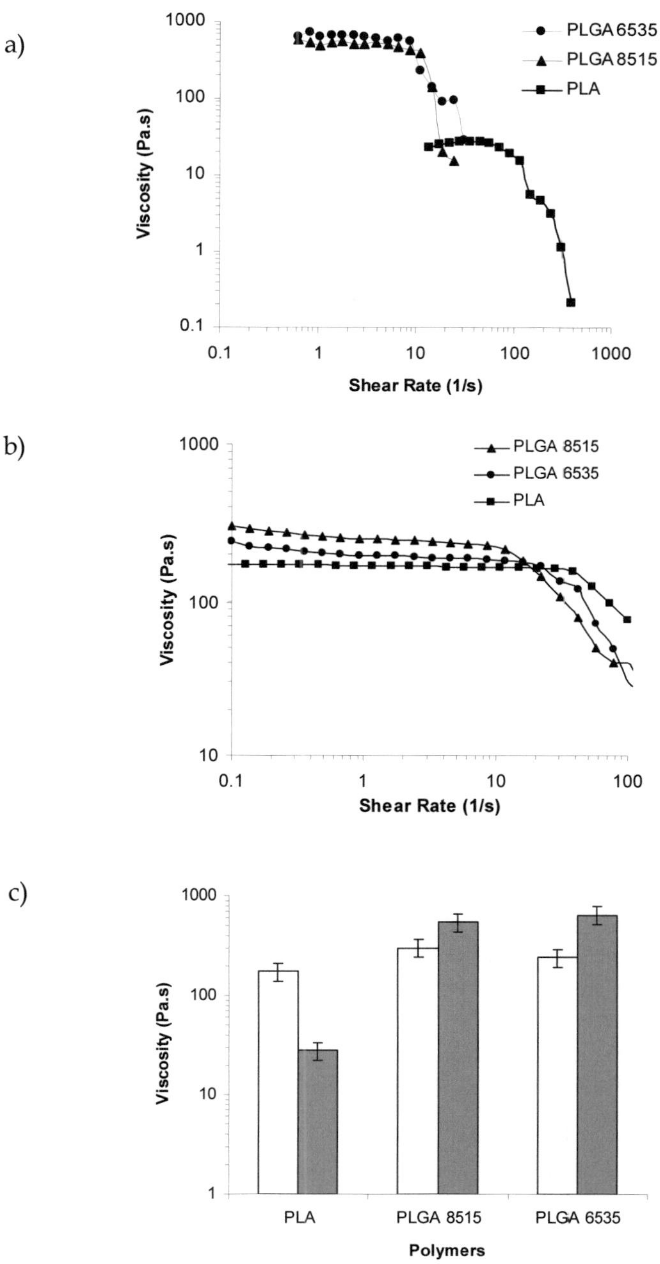

Fig. 4. Shear viscosity of $P_{DL}LA$, PLGA 8515 and PLGA 6535 with and without CO_2. (a) Flow curves for CO_2-plastisized polymers at 35 °C and 100 bar after soaking for 24 hours; (b) Flow curves for polymer melts at 140 °C and atmosphere pressure; (c) Comparison of the zero viscosities of the polymers at 140 °C and atmosphere (white bars) with those at 35 °C and 100 bar CO_2 pressure (grey bars). Reproduced with permission from (Tai et al., 2010)

5. Morphology of supercritical foamed PLA and PLGA scaffolds

Tissue engineering scaffolds require a controlled pore size and structure to host tissue formation and drug release. The chemical composition of the polymers and the morphology (pore size, porosity and interconnectivity) of the porous scaffolds are crucial because these parameters influence cell filtration, migration, nutrient exchange, degradation and drug release rate. CO_2 batch foaming process was used to fabricate foamed scaffolds in which the escape of CO_2 from a plasticized polymer melt generates gas bubbles that shape the developing pores. Mooney et al formed PLGA scaffolds (a mole ratio of lactic acid and glycolic acid (L/G) was 50:50) by CO_2 pressure quenching method at 20-23 °C and 55 bar (Mooney et al., 1996). Pishko et al. produced PLGA (L/G ratio as 80:20 and 65:35) scaffolds *via* CO_2 pressure quenching method using a water-in-solvent emulsion (aqueous protein phase and organic polymer solution phase) at conditions in the supercritical region (35 °C, 80 bar) (Hile et al., 2000; Hile &Pishko, 2004). Howdle and Shakesheff developed a single step $scCO_2$ foaming process using polymer powder samples to generate porous scaffolds at high pressure (170-230 bar) and short soaking time (0.5-2 hours) with a controlled venting rate (venting time between 2 minutes and 2 hours) at 35 °C (Howdle et al., 2001). The produced scaffolds with an interconnected porous structure have been used for growth factor and gene delivery. For example, bone morphogenetic protein 2 (BMP-2) has been encapsulated into $P_{DL}LA$ scaffolds for bone tissue engineering by this supercritical fluid mixing and foaming (Yang et al., 2004). Bone formation was observed due to the release of the osteoinductive protein BMP-2 from $P_{DL}LA$ scaffolds both in vitro and in vivo (Yang et al., 2004; Yang et al., 2006). These scaffolds have also been used to study adenoviral gene transfer into primary human bone marrow osteoprogenitor cells (Howard et al., 2002; Partridge et al., 2002). Polyamidoamine polymer (PAA)/DNA complexes has been incorporated into supercritical $P_{DL}LA$ scaffold, exhibiting a slow release and extended gene expression profile (Heyde et al., 2007).

The process of CO_2 batch foaming involves a simultaneous change in phase in the CO_2 and the polymer, resulting in rapid expansion of a surface area and changes in polymer rheological properties. Hence, the process is difficult to control with respect to the desired final pore size and structure. Tai et al. performed a detailed study of the effects of polymer chemical composition, molecular weight and batch foaming conditions on final scaffold characteristics (Tai et al., 2007; White et al., 2011). A series of amorphous $P_{DL}LA$ and PLGA polymers with various molecular weights (from 13 KD to 96 KD) and/or chemical compositions (the mole percentage of glycolic acid in the polymers was 0, 15, 25, 35 and 50 respectively) were employed. Processing parameters under investigation were temperature (from 5 to 55 °C), pressure (from 60 to 230 bar), soaking time and venting rate. The results demonstrated that the pore size and structure of the supercritical $P_{DL}LA$ and PLGA scaffolds can be tailored by careful control of the processing conditions. A higher pressure and a longer soaking time allowed more CO_2 molecules to diffuse into the polymer matrix, leading to a higher nucleation density and hence the production of smaller pores. Higher temperatures produced foams with larger pores because increased diffusion rates facilitated pore growth. In addition, reducing the rate of depressurization allowed a longer period for pore growth and therefore larger pores were formed than with rapid depressurization. The pore size of scaffolds also decreased with increasing glycolic acid content in the PLGA copolymers (Figure 5).

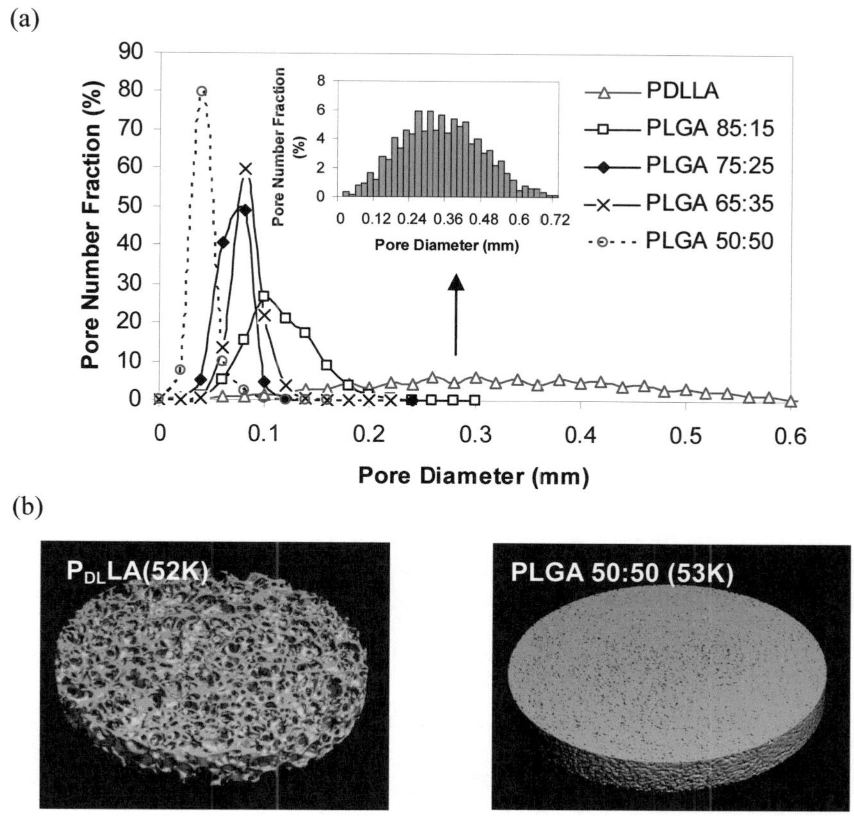

Fig. 5. Scaffolds fabricated with $P_{DL}LA$(52K), PLGA 85:15(77K), PLGA 75:25(72K), PLGA 65:35(52K), and PLGA 50:50(53K) polymers. (a) pore size distribution determined by Micro CT; (b) Micro CT 3-D images. Processing conditions: T = 35 °C, P = 230 bar, FT = 20 minutes, ST = 60 minutes, VT = 60 minutes. Reproduced with permission from (Tai et al., 2007)

6. Conclusion

Batch foaming of amorphous poly (DL-lactic acid) and poly (lactic acid-co-glycolic acid) with supercritical carbon dioxide is a solvent-free approach and without the need of an elevated temperature for the fabrication of 3-D porous scaffolds. CO_2 sorption and swelling isotherms at 35 °C and up to 200 bar on a variety of homo- and copolymers of lactic and glycolic acids (glycolic acid content as 0, 15, 25, 35 and 50 % respectively) have been studied. A high pressure attenuated total reflection Fourier transform infrared (ATR-IR) have been successfully used to study the interactions of CO_2 with these polymers. Shifts of the maximum absorption of carbonyl groups (C=O) of the polymers and the absorption intensity changes of both carbonyl and CO_2 bands represent such interactions. The viscosity of CO_2-plasticised polymers has been measured directly using high pressure parallel plate rheometer at 35 °C and 100 bar and the results were compared with the viscosity of the polymer melts at 140 °C. All data demonstrated that the interactions of CO_2 with PLGA polymers decreased with the increase of GA contents in the copolymers. These investigations provided fundamental understandings on the control of $scCO_2$ foaming process for the fabrication of $P_{DL}LA$ and PLGA porous scaffolds. Moreover, the experiments have been performed to tailor the pore size and structure of the $P_{DL}LA$ and PLGA porous scaffolds by altering the processing conditions. The results demonstrated that a higher pressure and a longer soaking time allowed more CO_2 molecules to diffuse into the polymer matrix, leading to a higher nucleation density and hence the production of smaller pores. Higher temperatures produced foams with larger pores because increased diffusion rates facilitated pore growth. In addition, reducing the rate of depressurization allowed a longer period for pore growth and therefore larger pores were formed than with rapid depressurization. The pore size of scaffolds also decreased with increasing glycolic acid content in the PLGA copolymers. These findings empower the definition of $scCO_2$ batch foaming conditions to tailor the pore size and structure of scaffolds for potential application as controlled release devices for growth factor delivery in Tissue Engineering.

7. References

Cooper-White, J. J. &Mackay, M. E. (1999). Rheological properties of poly(lactides). Effect of molecular weight and temperature on the viscoelasticity of poly(l-lactic acid). *Journal of Polymer Science Part B-Polymer Physics* 37(15): 1803-1814.

Cooper, A. I. (2001). Recent developments in materials synthesis and processing using supercritical CO_2. *Advanced Materials* 13(14): 1111-1114.

Corre, Y. M., Maazouz, A., Duchet, J. &Reignier, J. (2011). Batch foaming of chain extended PLA with supercritical CO_2: Influence of the rheological properties and the process parameters on the cellular structure. *Journal of Supercritical Fluids* 58(1): 177-188.

Davies, O. R., Lewis, A. L., Whitaker, M. J., Tai, H. Y., Shakesheff, K. M. &Howdle, S. M. (2008). Applications of supercritical CO_2 in the fabrication of polymer systems for drug delivery and tissue engineering. *Advanced Drug Delivery Reviews* 60(3): 373-387.

Dorgan, J. R., Janzen, J., Clayton, M. P., Hait, S. B. &Knauss, D. M. (2005). Melt rheology of variable L-content poly(lactic acid). *Journal of Rheology* 49(3): 607-619.

Dorgan, J. R., Lehermeier, H. &Mang, M. (2000). Thermal and rheological properties of commercial-grade poly(lactic acid)s. *Journal of Polymers and the Environment* 8(1): 1-9.

Dorgan, J. R., Lehermeier, H. J., Palade, L. I. &Cicero, J. (2001). Polylactides: Properties and prospects of an environmentally benign plastic from renewable resources. *Macromolecular Symposia* 175: 55-66.

Dorgan, J. R., Williams, J. S. &Lewis, D. N. (1999). Melt rheology of poly(lactic acid): Entanglement and chain architecture effects. *Journal of Rheology* 43(5): 1141-1155.

Duarte, A. R. C., Anderson, L. E., Duarte, C. M. M. &Kazarian, S. G. (2005). A comparison between gravimetric and in situ spectroscopic methods to measure the sorption of CO2 in a biocompatible polymer. *Journal of Supercritical Fluids* 36(2): 160-165.

Elvassore, N., Vezzu, K. &Bertucco, A. (2005). Measurement and modeling of CO_2 absorption in poly (lactic-co-glycolic acid). *Journal of Supercritical Fluids* 33(1): 1-5.

Fang, Q. &Hanna, M. A. (1999). Rheological properties of amorphous and semicrystalline polylactic acid polymers. *Industrial Crops and Products* 10(1): 47-53.

Flichy, N. M. B., Lawrence, C. J &Kazarian, S. G. (2003). Rheology of poly(propylene glycol) and suspensions of fumed silica in poly(propylene glycol) under high-pressure CO_2. *Industrial & Engineering Chemistry Research* 42(25): 6310-6319.

Freitas, S., Merkle, H. P. &Gander, B. (2005). Microencapsulation by solvent extraction/evaporation: reviewing the state of the art of microsphere preparation process technology. *Journal of Controlled Release* 102(2): 313-332.

Garlotta, D. (2001). A literature review of poly(lactic acid). *Journal of Polymers and the Environment* 9(2): 63-84.

Harris, L. D., Kim, B. S. &Mooney, D. J. (1998). Open pore biodegradable matrices formed with gas foaming. *Journal of Biomedical Materials Research* 42(3): 396-402.

Heyde, M., Partridge, K. A., Howdle, S. M., Oreffo, R. O. C., Garnett, M. C. &Shakesheff, K. M. (2007). Development of a slow non-viral DNA release system from P(DL)LA scaffolds fabricated using a supercritical CO_2 technique. *Biotechnology and Bioengineering* 98(3): 679-693.

Hile, D. D., Amirpour, M. L., Akgerman, A. &Pishko, M. V. (2000). Active growth factor delivery from poly(D,L-lactide-co- glycolide) foams prepared in supercritical CO_2. *Journal of Controlled Release* 66(2-3): 177-185.

Hile, D. D. &Pishko, M. V. (2004). Solvent-free protein encapsulation within biodegradable polymer foams. *Drug Delivery* 11(5): 287-293.

Hou, Q. P., Grijpma, D. W. &Feijen, J. (2003). Porous polymeric structures for tissue engineering prepared by a coagulation, compression moulding and salt leaching technique. *Biomaterials* 24(11): 1937-1947.

Howard, D., Partridge, K., Yang, X. B., Clarke, N. M. P., Okubo, Y., Bessho, K., Howdle, S. M., Shakesheff, K. M. &Oreffo, R. O. C. (2002). Immunoselection and adenoviral genetic modulation of human osteoprogenitors: in vivo bone formation on PLA scaffold. *Biochemical and Biophysical Research Communications* 299(2): 208-215.

Howdle, S. M., Watson, M. S., Whitaker, M. J., Popov, V. K., Davies, M. C., Mandel, F. S., Wang, J. D. &Shakesheff, K. M. (2001). Supercritical fluid mixing: preparation of thermally sensitive polymer composites containing bioactive materials. *Chemical Communications*(01): 109-110.

Jiang, W. L., Gupta, R. K., Deshpande, M. C. &Schwendeman, S. P. (2005). Biodegradable poly(lactic-co-glycolic acid) microparticles for injectable delivery of vaccine antigens. *Advanced Drug Delivery Reviews* 57(3): 391-410.

Kazarian, S. G., Vincent, M. F., Bright, F. V., Liotta, C. L. &Eckert, C. A. (1996). Specific intermolecular interaction of carbon dioxide with polymers. *Journal of the American Chemical Society* 118(7): 1729-1736.

Kim, S. J., Jang, D. H., Park, W. H. &Min, B. M. (2010). Fabrication and characterization of 3-dimensional PLGA nanofiber/microfiber composite scaffolds. *Polymer* 51(6): 1320-1327.

Kim, T. K., Yoon, J. J., Lee, D. S. &Park, T. G. (2006). Gas foamed open porous biodegradable polymeric microspheres. *Biomaterials* 27(2): 152-159.

Langer, R. &Peppas, N. A. (2003). Advances in biomaterials, drug delivery, and bionanotechnology. *Aiche Journal* 49(12): 2990-3006.

Lehermeier, H. J. &Dorgan, J. R. (2001). Melt rheology of poly(lactic acid): Consequences of blending chain architectures. *Polymer Engineering and Science* 41(12): 2172-2184.

Lim, L. T., Auras, R. &Rubino, M. (2008). Processing technologies for poly(lactic acid). *Progress in Polymer Science* 33(8): 820-852.

Liu, D. H. &Tomasko, D. L. (2007). Carbon dioxide sorption and dilation of poly(lactide-co-glycolide). *Journal of Supercritical Fluids* 39(3): 416-425.

Lu, L., Peter, S. J., Lyman, M. D., Lai, H. L., Leite, S. M., Tamada, J. A., Uyama, S., Vacanti, J. P., Langer, R. &Mikos, A. G. (2000). In vitro and in vivo degradation of porous poly(DL-lactic-co-glycolic acid) foams. *Biomaterials* 21(18): 1837-1845.

Mihai, M., Huneault, M. A. &Favis, B. D. (2010). Rheology and Extrusion Foaming of Chain-Branched Poly(lactic acid). *Polymer Engineering and Science* 50(3): 629-642.

Mooney, D. J., Baldwin, D. F., Suh, N. P., Vacanti, L. P. &Langer, R. (1996). Novel approach to fabricate porous sponges of poly(D,L-lactic- co-glycolic acid) without the use of organic solvents. *Biomaterials* 17(14): 1417-1422.

Murphy, W. L., Dennis, R. G., Kileny, J. L. &Mooney, D. J. (2002). Salt fusion: An approach to improve pore interconnectivity within tissue engineering scaffolds. *Tissue Engineering* 8(1): 43-52.

Nalawade, S. P., Picchioni, F., Janssen, L., Patil, V. E., Keurentjes, J. T. F. &Staudt, R. (2006). Solubilities of sub- and supercritical carbon dioxide in polyester resins. *Polymer Engineering and Science* 46(5): 643-649.

Nalawade, S. P., Picchioni, F., Marsman, J. H. &Janssen, L. (2006). The FT-IR studies of the interactions of CO_2 and polymers having different chain groups. *Journal of Supercritical Fluids* 36(3): 236-244.

Oliveira, N. S., Dorgan, J., Coutinho, J. A. P., Ferreira, A., Daridon, J. L. &Marrucho, I. M. (2006). Gas solubility of carbon dioxide in poly(lactic acid) at high pressures. *Journal of Polymer Science Part B-Polymer Physics* 44(6): 1010-1019.

Oliveira, N. S., Goncalves, C. M., Coutinho, J. A. P., Ferreira, A., Dorgan, J. &Marrucho, I. M. (2006). Carbon dioxide, ethylene and water vapor sorption in poly(lactic acid). *Fluid Phase Equilibria* 250(1-2): 116-124.

Palade, L. I., Lehermeier, H. J. &Dorgan, J. R. (2001). Melt rheology of high L-content poly(lactic acid). *Macromolecules* 34(5): 1384-1390.

Partridge, K., Yang, X. B., Clarke, N. M. P., Okubo, Y., Bessho, K., Sebald, W., Howdle, S. M., Shakesheff, K. M. &Oreffo, R. O. C. (2002). Adenoviral BMP-2 gene transfer in mesenchymal stem cells: In vitro and in vivo bone formation on biodegradable polymer scaffolds. *Biochemical and Biophysical Research Communications* 292(1): 144-152.

Pini, R., Storti, G., Mazzotti, M., Tai, H. Y., Shakesheff, K. M. &Howdle, S. M. (2007). Sorption and swelling of poly(D,L-lactic acid) and poly(lactic-co-glycolic acid) in supercritical CO2. *Macromolecular Symposia* 259: 197-202.

Pini, R., Storti, G., Mazzotti, M., Tai, H. Y., Shakesheff, K. M. &Howdle, S. M. (2008). Sorption and swelling of poly(DL-lactic acid) and poly(lactic-co-glycolic acid) in supercritical CO2: An experimental and modeling study. *Journal of Polymer Science Part B-Polymer Physics* 46(5): 483-496.

Qin, X., Thompson, M. R., Hrymak, A. N. &Torres, A. (2005). Rheology studies of polyethylene/chemical blowing agent solutions within an injection molding machine. *Polymer Engineering and Science* 45(8): 1108-1118.

Reignier, J., Gendron, R. &Champagne, M. F. (2007). Extrusion foaming of poly(lactic acid) blown with Co-2: Toward 100% green material. *Cellular Polymers* 26(2): 83-115.

Royer, J. R., DeSimone, J. M. &Khan, S. A. (2001). High-pressure rheology and viscoelastic scaling predictions of polymer melts containing liquid and supercritical carbon dioxide. *Journal of Polymer Science Part B-Polymer Physics* 39(23): 3055-3066.

Royer, J. R., Gay, Y. J., Adam, M., DeSimone, J. M. &Khan, S. A. (2002). Polymer melt rheology with high-pressure CO_2 using a novel magnetically levitated sphere rheometer. *Polymer* 43(8): 2375-2383.

Shieh, Y. T. &Lin, Y. G. (2002). Equilibrium solubility of CO_2 in rubbery EVA over a wide pressure range: effects of carbonyl group content and crystallinity. *Polymer* 43(6): 1849-1856.

Sosnowski, S., Wozniak, P. &Lewandowska-Szumiel, M. (2006). Polyester scaffolds with bimodal pore size distribution for tissue engineering. *Macromolecular Bioscience* 6(6): 425-434.

Tai, H. Y., Mather, M. L., Howard, D., Wang, W. X., White, L. J., Crowe, J. A., Morgan, S. P., Chandra, A., Williams, D. J., Howdle, S. M. &Shakesheff, K. M. (2007). Control of pore size and structure of tissue engineering scaffolds produced by supercritical fluid processing. *European Cells & Materials* 14: 64-76.

Tai, H. Y., Popov, V. K., Shakesheff, K. M. &Howdle, S. M. (2007). Putting the fizz into chemistry: applications of supercritical carbon dioxide in tissue engineering, drug delivery and synthesis of novel block copolymers. *Biochemical Society Transactions* 35: 516-521.

Tai, H. Y., Upton, C. E., White, L. J., Pini, R., Storti, G., Mazzotti, M., Shakesheff, K. M. &Howdle, S. M. (2010). Studies on the interactions of CO2 with biodegradable poly(DL-lactic acid) and poly(lactic acid-co-glycolic acid) copolymers using high pressure ATR-IR and high pressure rheology. *Polymer* 51(6): 1425-1431.

Tomasko, D. L., Li, H. B., Liu, D. H., Han, X. M., Wingert, M. J., Lee, L. J. &Koelling, K. W. (2003). A review of CO_2 applications in the processing of polymers. *Industrial & Engineering Chemistry Research* 42(25): 6431-6456.

Tsuji, H. (2005). Poly(lactide) stereocomplexes: Formation, structure, properties, degradation, and applications. *Macromolecular Bioscience* 5(7): 569-597.

Venugopal, J. &Ramakrishna, S. (2005). Applications of polymer nanofibers in biomedicine and biotechnology. *Applied Biochemistry and Biotechnology* 125(3): 147-157.

White, L. J., Hutter, V., Tai, H. Y., Howdle, S. M. &Shakesheff, K. M. (2011). The effect of processing variables on morphological and mechanical properties of supercritical CO2 foamed scaffolds for tissue engineering. *Acta Biomaterialia* 2012, 8(1): 61-71.

Woods, H. M., Silva, M., Nouvel, C., Shakesheff, K. M. &Howdle, S. M. (2004). Materials processing in supercritical carbon dioxide: surfactants, polymers and biomaterials. *Journal of Materials Chemistry* 14(11): 1663-1678.

Yang, X. B., Green, D., Roach, H. I., Anderson, H. C., Howdle, S. M., Shakesheff, K. M. &Oreffo, R. O. C. (2006). The effect of an admix of bone morphogenetic proteins on human osteoprogenitor activity in vitro and in vivo. *Tissue Engineering* 12(4): 1002-1003.

Yang, X. B., Whitaker, M. J., Sebald, W., Clarke, N., Howdle, S. M., Shakesheff, K. M. &Oreffo, R. O. C. (2004). Human osteoprogenitor bone formation using encapsulated bone morphogenetic protein 2 in porous polymer scaffolds. *Tissue Engineering* 10(7-8): 1037-1045.

You, Y., Lee, S. J., Min, B. M. &Park, W. H. (2006). Effect of solution properties on nanofibrous structure of electrospun poly(lactic-co-glycolic acid). *Journal of Applied Polymer Science* 99(3): 1214-1221.

Zhang, C., Cappleman, B. P., Defibaugh-Chavez, M. &Weinkauf, D. H. (2003). Glassy polymer-sorption phenomena measured with a quartz crystal microbalance technique. *Journal of Polymer Science Part B-Polymer Physics* 41(18): 2109-2118.

Part 2

Pipe Flow and Porous Media

7

Resistance Coefficients for Non-Newtonian Flows in Pipe Fittings

Veruscha Fester[1], Paul Slatter[2] and Neil Alderman[3]
[1]*Cape Peninsula University of Technology,*
[2]*Royal Melbourne Institute of Technology,*
[3]*BHR Group,*
[1]*South Africa*
[2]*Australia*
[3]*United Kingdom*

1. Introduction

The focus of this chapter is to provide a review of the loss coefficient data for laminar flow of non-Newtonian fluids in pipe fittings. Since the total pressure change in a piping system generally consists of three components: (i) the frictional pressure loss in the pipe, (ii) the frictional pressure loss arising from flow through fittings and (iii) the pressure loss or gain resulting from elevation changes, this review will also deal with laminar and turbulent pipe flow of non-Newtonian fluids and the application of viscometry for flow in pipes and fittings. The rheological models relevant to industrial fluids such as mine tailings and sewage sludges are introduced, with particular emphasis on yield stress, or viscoplastic, fluids.

Hooper (1981) presented a two-K method for determining the loss coefficient for laminar and turbulent flow through various fittings and valves. This method consists of two factors, one for laminar flow, K_1 and the other for turbulent flow, K_{turb}. Unlike that for K_{turb}, there is little data available for K_1. Experimental data over the full range of laminar and turbulent flow are presented for flow of Newtonian and non-Newtonian fluids in various fittings. The experimental procedures for the accurate determination of loss coefficients are described.

Current practice for laminar flow through various fittings is to present the loss coefficient as a function of an appropriate Reynolds number. Different Reynolds numbers developed for non-Newtonian fluids have been evaluated to determine their ability to establish the necessary requirement of dynamic similarity for flow of viscoplastic fluids in various fittings.

The laminar to turbulent transition in pipe fittings are also discussed. The experimental work done to date on contractions, expansions, valves and orifices is reviewed in addition to similar work published in literature.

The magnitude of errors that can be obtained using the incorrect loss coefficient is demonstrated by means of a worked example. This chapter will provide the pipeline design engineer dealing with non-Newtonian fluids with the necessary information critical for energy efficient design.

2. Rheological models

The main types of flow behaviour exhibited by fluids under steady-state shear include:

- Newtonian
- non-Newtonian
  - shear-thinning and dilatant (shear-thickening)
  - viscoplastic

Over a limited shear rate range (of one or two decades), a fluid can exhibit a single class of behaviour characterised by the flow curves of Sections 2.1, 2.2.1 or 2.2.2. Over a wider shear rate range (within 10^{-6} to $10^{6}\,s^{-1}$), most fluids exhibit more than one class of flow behaviour.

It is difficult to predict the type of flow behaviour that a fluid will exhibit under given flow conditions (for example, at a given temperature, pressure and concentration). Nevertheless, there are textbooks such as those of Laba (1993) and Steffe (1996) that give examples that can serve as illustrations only. The user should be aware that a rheological test is the only sure method of ascertaining the rheological behaviour of a fluid (Alderman, 1997).

2.1 Newtonian behaviour

These are fluids for which an infinitesimal shear stress will initiate flow and for which the shear stress is directly proportional to the shear rate. The flow curve, at a given temperature and pressure, is therefore linear and passes through the origin as is shown in Figure 1(a). The slope of the flow curve, which is constant, is the viscosity. Re-plotting the flow curve in the form of a viscosity curve as shown in Figure 1(b) clearly depicts a constant viscosity with respect to shear rate.

$$\eta = \frac{\tau}{\dot{\gamma}} \tag{1}$$

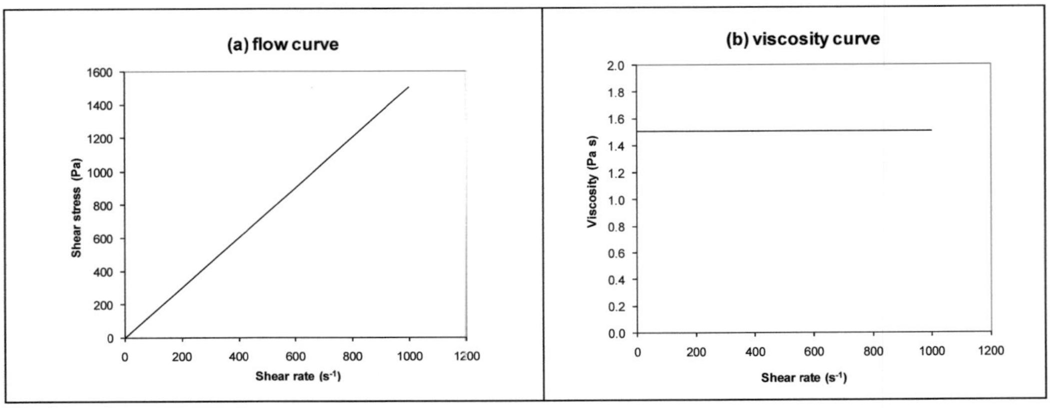

Fig. 1. Flow and viscosity curves for fluids exhibiting Newtonian behaviour

Any deviation from Newtonian behaviour is said to be non-Newtonian.

2.2 Non-Newtonian behaviour

2.2.1 Shear-thinning and dilatant (shear-thickening) behaviour

Two departures from Newtonian behaviour, namely **shear-thinning** and **dilatant** (shear-thickening) behaviour, are depicted in Figure 2.

Shear-thinning behaviour is observed when an infinitesimal shear stress will initiate flow, that is, the flow curve passes through the origin, Figure 2(a) and the viscosity decreases with increasing shear rate as shown in Figure 2(b). This behaviour is sometimes incorrectly termed thixotropy because the equilibrium flow curve of a thixotropic material is often shear-thinning. However, unlike shear-thinning behaviour, thixotropy is a time-dependent property. Dilatant (shear-thickening) behaviour is observed when an infinitesimal shear stress will initiate flow, that is, the flow curve passes through the origin, Figure 2(a) and the viscosity increases with increasing shear rate as shown in Figure 2(b).

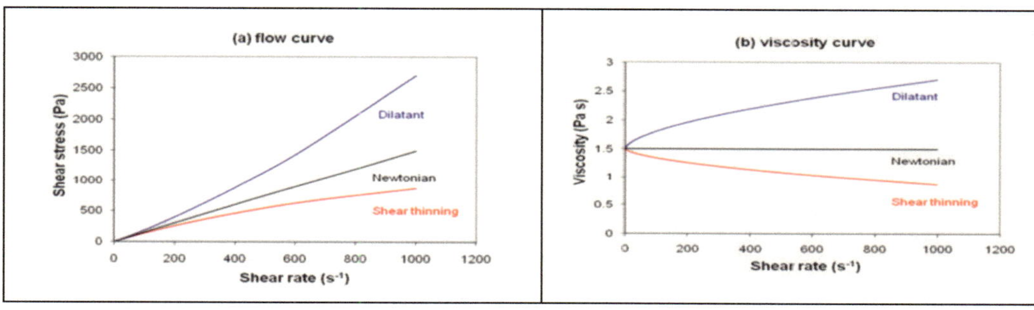

Fig. 2. Idealised shear-thinning and dilatant (shear–thickening) behaviour

For polymeric systems of low concentrations, low molecular weights or at temperatures well above the glass transition temperature, the variation of viscosity with shear rate as a function of concentration, molecular weight or temperature is shown typically in Figure 3.

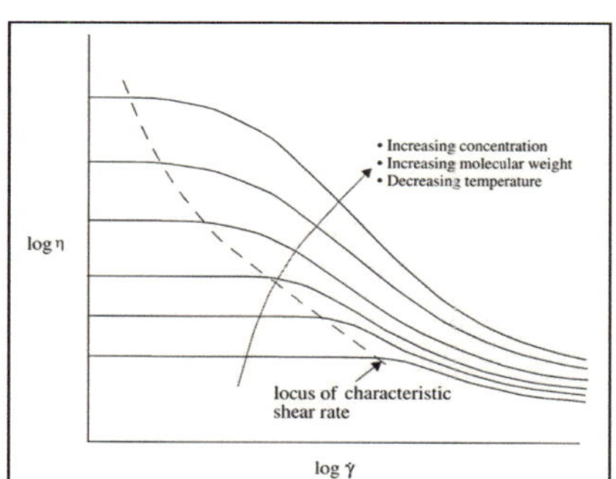

Fig. 3. Non-Newtonian viscosity of polymeric systems of low concentrations, low molecular weight or at temperatures well above the glass transition temperature (Alderman, 1997)

It can be seen that as the concentration or molecular weight is increased or the temperature is decreased, both the zero and infinite shear viscosities, η_0 and η_∞ increases with the change in the zero shear viscosity being much greater than that for infinite shear viscosity. The difference between η_0 and η_∞ can be very large, as much as two or three orders of magnitude even, if the molecular weight or concentration is sufficiently high. The characteristic shear rate, where the curve starts to deviate from the η_0 line, decreases with increasing concentration or molecular weight or with decreasing temperature. The characteristic shear rate, where the curve starts to deviate from the η_0 line, decreases with increasing concentration or molecular weight or with decreasing temperature.

For particulate systems containing a dispersed phase having negligible inter-particle attraction, the variation of viscosity with shear rate as a function of solids content, shown typically in Figure 4, is complex. At low solids concentrations, the curve depicts shear-thinning behaviour with well-defined zero and infinite shear viscosities. The lower boundary of the shear-thinning region can be identified as the curve depicted in Figure 4. At higher solids concentrations, dilatant behaviour is observed at high shear rates. This becomes more pronounced with increasing solids concentration. The characteristic shear rate decreases with increasing concentration until at very high concentrations it falls outside the lowest measurable shear rate. The magnitude of the slope at the characteristic shear rate also increases until at some concentration, it attains a value of -1. If replotted as a flow curve, this would show a yield stress. Here, the flow behaviour is viscoplastic.

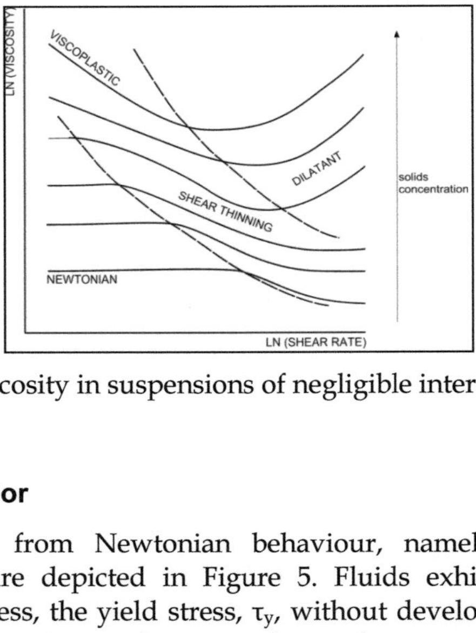

Fig. 4. Non-Newtonian viscosity in suspensions of negligible inter-particle attraction (Alderman, 1997)

2.2.2 Viscoplastic behavior

Two further departures from Newtonian behaviour, namely **Bingham plastic** and **viscoplastic** behaviour, are depicted in Figure 5. Fluids exhibiting viscoplasticity will sustain a certain shear stress, the yield stress, τ_y, without developing continuous flow. For stresses below τ_y, the shear rate remains zero whereas for stresses above τ_y, the fluid flows with a shear rate dependent on the excess stress (τ - τ_y). Bingham plastic behaviour is observed when there is a linear relationship between the shear stress in excess of the yield stress, τ_y and the resulting shear rate. The flow curve for a Bingham plastic fluid is linear but does not pass through the origin as shown in Figure 5(a). Viscoplastic behaviour is observed when the rate of increase in shear stress with shear rate in excess of the yield stress, τ_y decreases with increasing shear rate. The flow curve, shown in Figure 5(a), has the same characteristic shape as that for a fluid exhibiting shear-thinning behaviour but does not pass through the origin. The flow curve for both fluids cuts the shear stress-axis above the origin at τ_y on the linear plot, Figure 5(a). The viscosity for both fluids decreases with shear rate similar to that for a shear-thinning fluid, Figure 5(b). However, unlike the shear-thinning viscosity which is finite at zero shear rate, the Bingham plastic or viscoplastic viscosity tends to infinity as shear rate is reduced to zero.

Both polymeric and particulate systems can possess a yield stress. This occurs particularly for cross-linked polymers, polymeric gels and filled systems, and for high solids content suspensions and high dispersed phase emulsions and foams. In the case of suspensions, the yield stress arises because of particle-particle frictional interaction. In these materials, it is found that the Bingham plastic behaviour is obtained at high shear rates, as an asymptotic behaviour, Figure 5(a). The intercept of the extrapolated Bingham plastic asymptote is commonly called the Bingham yield stress, τ_{yB}, to distinguish it from the yield stress, τ_y.

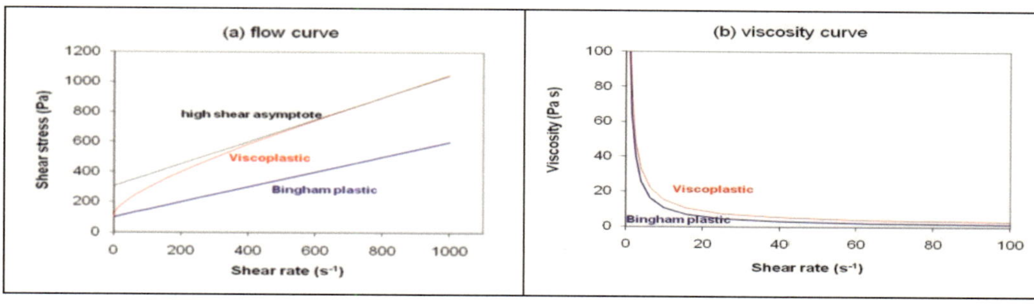

Fig. 5. Idealised Bingham plastic and viscoplastic behaviour

2.3 Viscosity and flow curve models

Fluids exhibiting Newtonian behaviour have constant viscosities as defined by Eqn. (1), and single point measurement at a convenient shear rate is sufficient to describe such a fluid. However, non-Newtonian fluids have viscosities that are shear rate-dependent, and a single point measurement is inadequate to describe the flow behaviour. Here, the relevant shear rate range in the engineering application must be assessed and used in determining the measurement conditions for the viscometer/rheometer. Shear rate-dependent viscosity is often referred to as apparent viscosity, η_a, but there is no need to make this distinction if it is accepted that viscosity, as defined in Eqn. (1), can be variable. However, it is essential that values of η are quoted with their corresponding values of shear rate (or shear stress).

The flow curve or viscosity curve data obtained from viscometric measurements under steady-state shear can be described mathematically in terms of rheological models (constitutive equations) and are amenable to curve fitting. The general form of a rheological model is fitted over a relevant shear rate range to the flow curve or the viscosity curve by least-squares regression analysis so that a specific rheological model can be obtained. This specific equation is then used for material characterization, engineering design applications or product formulation.

The rheological model for Newtonian fluids contains just one constant, η_N. Many models have been proposed to describe the non-Newtonian flow behaviour of fluids, although the majority of these are of little value for engineering design applications and serve more as theoretical analyses.

In general, the simpler models (for example, Newtonian, Bingham plastic and power law) are used for fitting narrow shear rate ranges (say, one decade), giving straight lines on linear or logarithmic plots. Other models such as Herschel-Bulkley (1926) and Casson (1959) for two or three decades and Sisko (1958) for four decades or more are better for wider shear rate ranges. For the widest shear rate range achievable in practice (that is, 10^{-6} to 10^{6} s^{-1}), it is necessary to use a more complex model such as the Cross (1965) model or the Carreau (1968) model.

Of the numerous rheological models available in the literature, the most commonly-used for engineering applications are described here.

2.3.1 Newtonian model

This is the simplest of all flow curve models and is given by

$$\tau = \eta_N \dot{\gamma} \tag{2}$$

with η_N being the Newtonian viscosity.

2.3.2 Power law model

This model originally proposed by de Waele (1923) and Ostwald (1925) is described by the two-parameter equation:

$$\tau = K \dot{\gamma}^n \tag{3}$$

where K is the consistency coefficient in units of Pas^n and n is the power law exponent. This equation can be used to describe Newtonian behaviour when n = 1, shear-thinning behaviour when n < 1 or dilatant behaviour when n > 1. On a log-log plot, the model is a straight line with a slope of n. Values of n typically ranges from 0.2 to about 1.4. The further the value of n is from unity, the more non-Newtonian is the fluid.

2.3.3 Bingham plastic model

This two parameter model (Bingham, 1922) is described by

$$\tau = \tau_{yB} + \eta_B \dot{\gamma} \tag{4}$$

where τ_{yB} is the Bingham yield stress and η_B is the Bingham plastic viscosity.

2.3.4 Herschel-Bulkley model

This model also known as the generalised Bingham plastic model, is a three parameter yield/power law model (Herschel-Bulkley, 1926), given by

$$\tau = \tau_{yHB} + K \dot{\gamma}^n \tag{5}$$

where τ_{yHB} is the Herschel-Bulkley yield stress. This equation describes viscoplastic behaviour when n < 1. Because power law (τ_{yHB} =0; shear-thinning when n < 1 or dilatant when n > 1), Newtonian (τ_{yHB} = 0 and n = 1) and Bingham plastic behaviour (n= 1) can be regarded as special cases, the model represents the flow behaviour of a wide range of fluids without being too difficult to handle mathematically.

2.3.5 Casson model

A popular alternative to the Herschel-Bulkley model is the theoretical two-parameter model of Casson (1959). This is given by

$$\sqrt{\tau} = \sqrt{\tau_{yC}} + \sqrt{\eta_C \dot{\gamma}} \tag{6}$$

where τ_{yC} is the Casson yield stress and τ_C is the Casson viscosity.

2.4 Viscosity and flow curve measurement

Commercial viscometers and rheometers employ a wide range of geometries for viscosity and flow curve measurement. These can be grouped into two main types: rotational viscometers and tube viscometers.

2.4.1 Rotational viscometers

Rotational viscometers, which rely on rotational motion to achieve simple shear flow, can be operated either in the controlled rate or controlled stress mode. In controlled-rate instruments, there are two methods of applying the rotation and measuring the resultant torque. The first method is to rotate one member and measure the torque exerted on the other member by the test sample, whilst the second method involves the rotation of one member and measuring the resultant torque on the same member, The rotating member is either at constant speed which can be sequentially stepped or with a steadily-changing speed ramp. The resultant torque is measured by a torsion spring. In controlled-stress instruments, either a constant torque (which can be sequentially changed) or a torque ramp is applied to the member, and the resultant speed is measured. The more common geometries used in rotational viscometry are shown in Figure 6.

2.4.2 Tube viscometers

Tube viscometers are generally once-through batch devices consisting of either a horizontal or vertical length of precision-bored, straight tube through which the test fluid is passed at varying rates from a reservoir. The diameters of the tube can typically range from 1 to 5 mm. Essentially, there are two types of tube viscometer, the controlled flow rate and the controlled pressure, as shown in Figure 7.

In the controlled flow rate tube viscometer, a piston forces the fluid through a horizontal or vertical tube at a constant flow rate and the resultant pressure drop is measured. In the controlled pressure viscometer, compressed air (or nitrogen) is applied to drive the fluid through a horizontal or vertical tube and the resultant volumetric flow rate is measured.

2.4.3 Practical considerations

Flow curve measurements can be made using all of these geometries but there are several drawbacks for each geometry which need to be considered for each fluid under test and each specific application. Guidance on the use of these geometries for obtaining relevant viscometric data required for a particular application is given by Alderman and Heywood (2004a, 2004b).

Coaxial cylinder 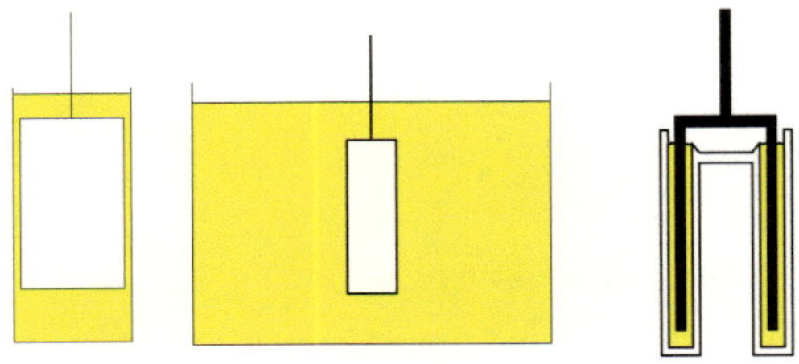 (a) Bob-in-cup (b) single bob (c) Moore-Davis	The simplest geometry (a) consists of a bob (inner cylinder) located in a cup (outer cylinder) with the test sample contained in the narrow annular gap between the bob and the cup. Other variations of the coaxial cylinder viscometer that are commonly used include (b) a bob rotating in a large container that approximates to the "infinite sea" situation (that is, the container to bob radius ratio is at least 10 and (c) the Moore-Davis double cylinder viscometer used for low viscosity fluids.
Rotating disc 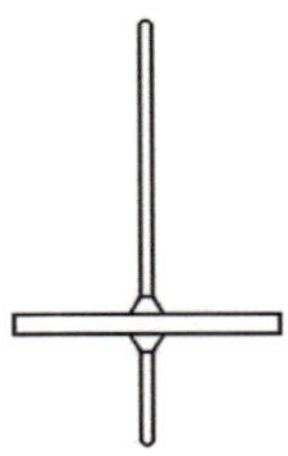	This consists of a disc rotating in a large container (normally a 600 ml plastic beaker with an internal diameter of 97mm) of test material. The torque exerted by the test fluid on the rotating disc is measured as a function of rotational speed.
Cone-and-plate 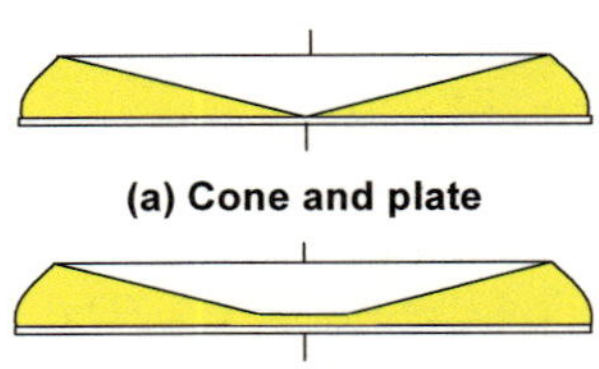 (b) Truncated cone and plate	Usually, the sample under test is contained between the exterior angle of a cone and a flat plate, (a). The axis of the cone is set normal to the plate with the cone apex touching the plate. Cone angles typically range from 0.25 to 1°. However, for solid/liquid suspensions and emulsions, a cone with a truncated apex (b) is often used to minimise problems due to particle jamming. In controlled-rate instruments, either the cone or the plate is rotated at a fixed speed and the resultant torque via the cone is measured. In controlled-stress instruments, a fixed torque or a torque ramp is applied to the cone and the resultant speed is measured.
Parallel plate 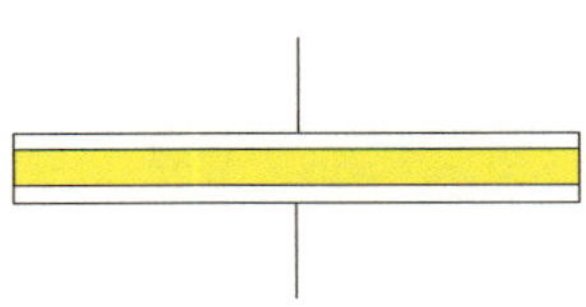	The sample under test is held in the gap between two identical circular flat plates. The gap between the two plates can be varied, typically up to 5 mm for plates of about 25 mm radius. In controlled-rate instruments, either the top or bottom plate is rotated at a fixed speed and the resultant torque is measured. In controlled-stress instruments, a constant torque or a torque ramp is applied to the top plate and the resultant speed is measured.

Fig. 6. Types of rotational viscometer

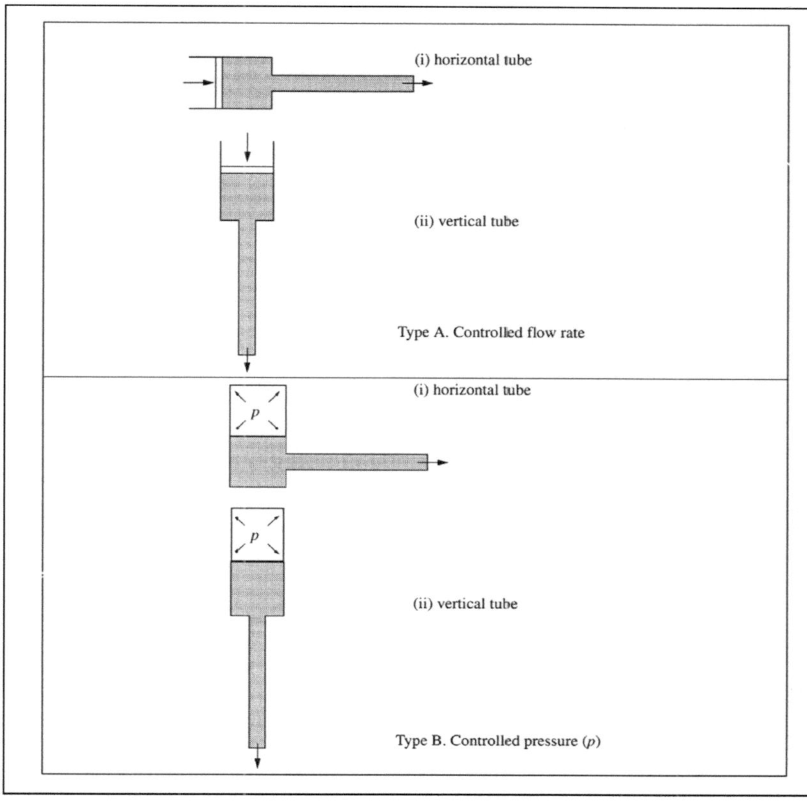

Fig. 7. Types of tube viscometer

For the correct end-use of viscometric data in any processing application, the flow curve must be measured under appropriate conditions of temperature and shear. Hence, the shear rate (or shear stress) to be covered by the viscometer should be matched to that applicable to the particular flow process. With the range of viscometers commercially available, it is technically possible to measure viscosity over 12 orders of magnitude of shear rate from 10^{-6} to 10^{6} s^{-1}. However, most processing applications require viscosity data over no more than two or three orders of magnitude of shear rate. Shear rates typical of some flow processes can be found in Alderman and Heywood (2004a). There are quick and simple methods available for defining the shear rate range for any processing application, Alderman and Heywood (2004a). If a method for shear rate estimation in some processing applications is not available, a useful approach is to define the flow region of interest, determine differences in fluid velocity at two points across the region (this will often be the surface of some moving element such as a pump impeller or mixer agitator) and divide this velocity difference by the distance of separation between the two points. These typical shear rates can be used as a basis for choosing a viscometer.

As the equations for calculating the shear stress, shear rate and viscosity assume the flow in the viscometer is laminar, a check must be made to ensure the validity of the viscometric data. Details on how this is done are outlined in Alderman and Heywood (2004b).

2.5 Viscosity and flow curve data generation and interpretation

Figure 8 summarises the key steps to ensuring the measurements of the flow curve are both accurate and relevant (Alderman and Heywood, 2004a; 2004b).The need for accuracy may sometimes introduce additional laboratory experiments, particularly if end effect and/or wall slip errors are incurred. However, ensuring that only the viscometric data that are relevant to the application of interest are measured will minimise the overall effort. Having selected the most appropriate viscometer for the sample under test and the application of interest, the next step is to generate the flow curve.

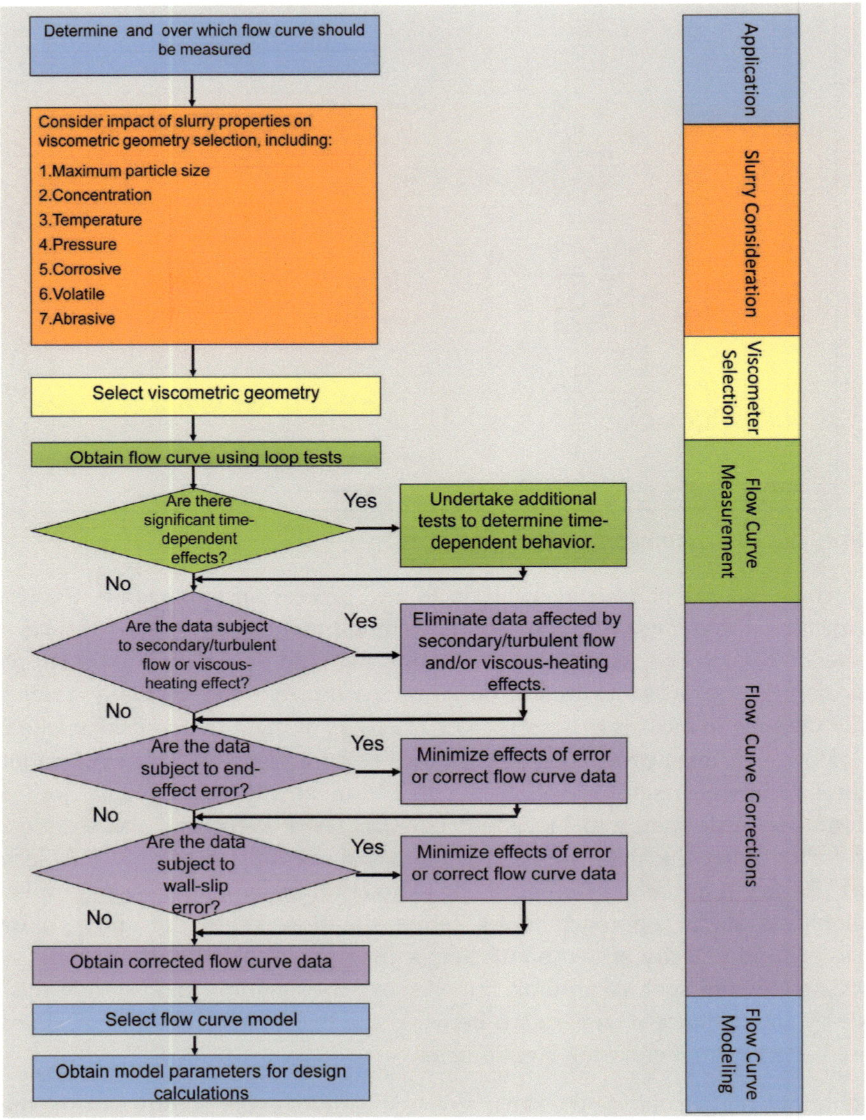

Fig. 8. Flowchart for making accurate and relevant flow curve measurements (Alderman and Heywood, 2004b)

2.5.1 Rotational viscometers

With controlled rate rotational instruments, the flow curve measurement is usually done by carrying out at least three cycles of a fixed time where the speed is varied, either by a sequence of step speed changes or a steadily-changing speed ramp, between minimum and maximum values whilst measuring the resultant torque. With controlled stress rotational instruments, the torque is varied in a similar manner whilst the resultant speed is measured. The time over which the speed or torque is varied is left to the operator to decide. However, a good starting point would be the time that gives 30 s per step.

Repeated shear cycles on the test sample will enable one to determine whether the sample exhibits time-dependent flow behaviour such as thixotropy. If the up and down curves for the first and successive cycles coincide, the sample is undergoing steady-state shear. However, if hysteresis loops between the up and down curves are observed for each successive cycle, the sample is exhibiting time-dependent flow behaviour. In such cases, it is advisable to repeat the experiment with the speed (or torque) held constant until the torque (or speed) attains a steady value before changing the speed (or torque) to the next value. This will yield an equilibrium flow curve in which the up and down curves coincide.

The average of the torque versus speed data at which the up and down curves coincide is the first step of the calculation procedure for obtaining the corrected flow curve. Care must be taken to ensure that the flow curve is not affected by the four error sources: secondary/turbulent flow, viscous heating, end effect and wall slip. Further details of this procedure can be found in Alderman and Heywood (2004b) for the appropriate viscometric geometry used.

2.5.2 Tube viscometers

With controlled-rate tube viscometers (Type A in Figure 7), the flow curve is obtained by carrying out pressure drop measurements at different constant flow rates. For each measurement, a new sample is often required due to the small capacity of the sample reservoir. With controlled-stress tube viscometers (Type B in Figure 7), the flow curve is generated by measuring the flow rate as a function of pressure drop on the sample. The pressure drop versus flow rate data is the first step of the calculation procedure for obtaining the corrected flow curve. Care must be taken to ensure that the flow curve is not affected by the four error sources: secondary/turbulent flow, viscous heating, end effect and wall slip. Further details of this procedure can be found in Alderman and Heywood (2004b).

2.6 Viscosity and flow curve interpretation for engineering design

Having completed the calculation procedure for the corrected flow curve (and hence the corrected viscosity curve), the data may be amenable to a single curve fit (Brown & Heywood, 1991). Sometimes, because of considerable scatter in the data, it may be more appropriate to construct at least two curves: a mean curve obtained from regression analysis using all the data and an upper bound curve obtained from regression analysis using $(\tau, \dot{\gamma})$ data selected from the curve that was initially drawn by eye. The upper bound curve would normally represent the worst case for many engineering applications and would lead to a conservative design. Further factors can cause difficulties in attempting to draw a single flow curve through the data. These factors include the use of two or more different viscometric geometries which may give differing degrees of phase separation during shear, sample variability taken from the same batch, and uncorrected errors associated with the use of any viscometric geometry.

For engineering design, the choice of flow curve model is limited by the design method to be used. For example, if the flow curve is to be used for pipeline design, the choice can be made from a number of models including Newtonian, power law, Bingham plastic, Cassonand Herschel Bulkley models. However if the flow curve is to be used for agitation in stirred tanks, the choice is either the Newtonian or the power law model. It is often not immediately obvious from the data which of the flow models should be selected. A decision will need to be made on which model to use. The following approach is suggested:

i. Plot all the $(\tau, \dot{\gamma})$ data on linear axes and separately, on double logarithmic axes. This is to assess the suitability of the Newtonian, Bingham plastic and power law models.
ii. If there is considerable scatter in the data, decide by eye or from the correlation coefficient obtained by linear regression analysis whether a straight line through the linear or the log-log plot gives the better representation. Similarly decide for the upper bound curve. If one of these alternatives is acceptable, the use of the Herschel-Bulkley model is probably not warranted.
iii. If neither of these alternatives appears satisfactory because there is significant curvature of the data on both linear and log-log plots, the following can arise:
    a. If there is data curvature on the log-log plot with the slope of the curve increasing with shear rate axis and if the linear plot does not produce a straight line then the Herschel-Bulkley model should adequately describe the data.
    b. If there is data curvature on the log-log plot and the slope of the curve decreasing with shear rate axis, then the use of the Herschel-Bulkley model is inappropriate as this implies a negative yield stress parameter. However, a curve fit is possible and would result in a negative value for the yield stress parameter. Force-fit either a Bingham plastic or power law model to the data.

Estimates need to be made for the parameters defined in the flow models. As the Herschel-Bulkley model can be reduced to the Newtonian, power law and Bingham plastic models, a least squares regression analysis can first be performed on the $(\tau, \dot{\gamma})$ data to obtain τ_{yHB}, K and n. It may then be possible to simplify the model by setting the τ_{yHB} to zero if the estimate is close to zero and/or setting n to 1 if the estimate is close to unity.

Two methods are commonly used when carrying a regression analysis on the $(\tau, \dot{\gamma})$ data (Heywood & Cheng, 1984):

1. a non-linear least squares regression on unweighted data,
2. a non-linear least squares regression on weighted data.

In Method 1 it is assumed that the error e lies in τ:

$$\tau = \tau_{yHB} + K\dot{\gamma}^{n} + e \tag{7}$$

whereas in Method 2, the error is assumed to lie in ln $(\tau - \tau_{yHB})$:

$$\ln\left(\tau - \tau_{yHB}\right) = \ln K + n \ln \dot{\gamma} + e \tag{8}$$

Standard non-linear regression software packages can be used in either case. Alternatively, non-linear regression can be performed using Microsoft Excel via the 'Solver' tool (Roberts et al., 2001). Both methods will provide sets of τ_{yHB}, K and n estimates which give viscometric data predictions to ± 2% of the original data within the original shear rate range. Outside this shear rate range, agreement can be poor.

Examples of the extrapolated flow curves for the two sets of parameters obtained in the two regression methods are shown in Figure 9. Provided the relevant shear rate/shear stress window for the application is covered by the viscometer, either of the two regression methods can be used. However, the method that gives the most uniform data spread should be used.

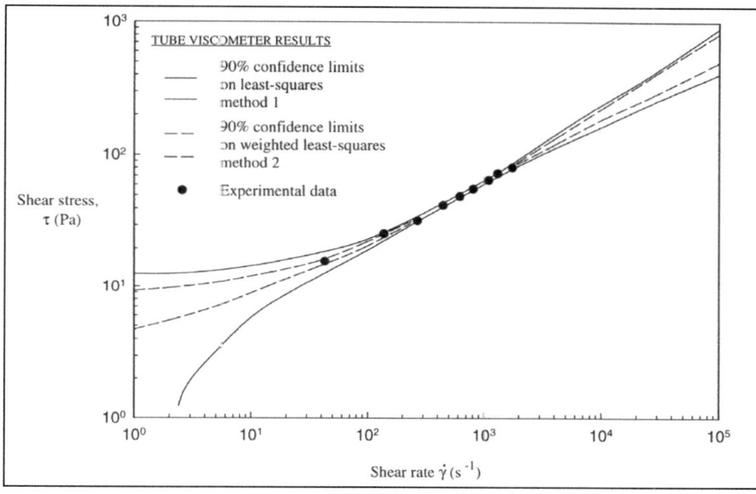

Fig. 9. Example of extrapolation of predicted ($\tau, \dot{\gamma}$) data outside the experimental range for an 8.0% digested sewage sludge (Heywood and Cheng, 1984)

3. Pipe flow fundamentals

The basic relationships for design in laminar, transitional and turbulent pipe flow are obtained by integration of the constitutive rheological relationship, over the cross-sectional area of the pipe.

3.1 Rheological approach

Industrial fluids exhibiting viscoplastic behaviour are often best modelled using the Herschel-Bulkley model (Govier & Aziz, 1972 and Hanks, 1979). The constitutive rheological equation is given by Eqn. (5).

3.2 Laminar flow

Equations for the design of laminar pipe flow can be derived by integrating Eqn. (5) over the circular pipe geometry (Govier & Aziz, 1972). Because of the yield stress, a central solid plug is formed where the point shear stress is less than the yield stress, as shown in Figure 10.

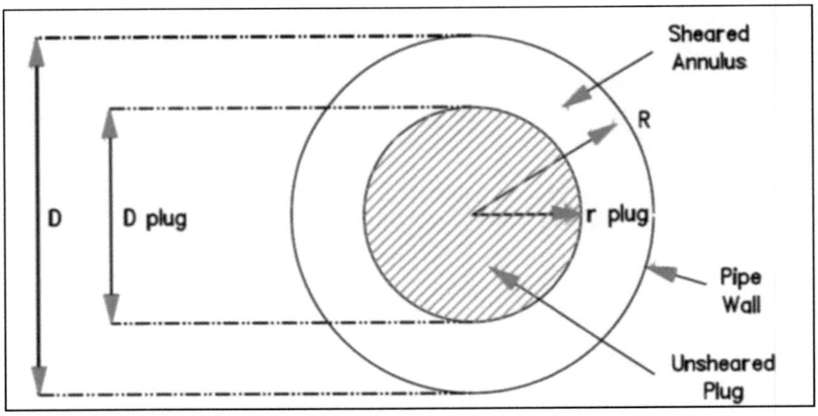

Fig. 10. Schematic showing the geometry of the unsheared plug

The radius of the plug is determined from the linear shear stress distribution and the yield stress as

$$r_{plug} = \frac{\tau_y}{\tau_0} R \tag{9}$$

In the annular region where the point shear stress exceeds the yield stress, the point velocity u is obtained by integration of the constitutive rheological relationship, and applying the no-slip assumption at the pipe wall

$$u = \frac{R}{K^{\frac{1}{n}} \tau_0} \frac{n}{n+1} \left[(\tau_0 - \tau_y)^{\frac{n+1}{n}} - (\tau - \tau_y)^{\frac{n+1}{n}} \right]. \tag{10}$$

The plug velocity u_{plug} is obtained at the point where the applied shear stress is equal to the yield stress as

$$u_{plug} = \frac{R}{K^{\frac{1}{n}} \tau_0} \frac{n}{n+1} (\tau_0 - \tau_y)^{\frac{n+1}{n}}. \tag{11}$$

These two velocity distributions are integrated over the pipe cross-sectional area appropriate to each, in order to yield the total volumetric flow rate Q as

$$\frac{32\,Q}{\pi D^3} = \frac{8\,V}{D} = \frac{4\,n}{K^{1/n} \tau_0^3} (\tau_0 - \tau_y)^{\frac{1+n}{n}} \left[\frac{(\tau_0 - \tau_y)^2}{1+3n} + \frac{2\tau_y(\tau_0 - \tau_y)}{1+2n} + \frac{\tau_y^2}{1+n} \right] \tag{12}$$

This relationship can be used for laminar pipe flow design.

3.3 Transitional flow

The approach used here for the prediction of the transition from laminar to turbulent flow is the modified Reynolds number Re_3 (Slatter, 1995; Slatter, 1999). This approach predicts a laminar to turbulent transition in the Reynolds number region of 2100. This approach was specifically developed to place emphasis on the viscoplastic nature of the material (Slatter, 1995). Using the fundamental definition that Re ∝ inertial/viscous forces, the final expression is

$$Re_3 = \frac{8\rho V_{ann}^{\ 2}}{\tau_y + K\left(\frac{8V_{ann}}{D_{shear}}\right)^n} \tag{13}$$

As shown in Figure 10, in the presence of a yield stress the central core of the fluid moves as a solid plug which fundamentally affects the stability of flow (Slatter, 1995, 1999). The unsheared plug is treated as a solid body in the centre of the pipe. The flow that the plug represents must be subtracted as it is no longer being treated as part of the fluid flow. The corrected mean velocity in the annulus V_{ann} is then obtained as follows:-

$$V_{ann} = \frac{Q_{ann}}{A_{ann}} = \frac{Q - Q_{plug}}{\pi\left(R^2 - r_{plug}^2\right)} \tag{14}$$

and

$$Q_{plug} = u_{plug} A_{plug} \tag{15}$$

The sheared diameter, D_{shear}, is taken as the characteristic dimension because this represents the zone in which shearing of the fluid actually takes place, and it is defined as

$$D_{shear} = D - D_{plug} \tag{16}$$

and

$$D_{plug} = 2r_{plug} \tag{17}$$

These relationships can be used for pipe flow design to determine the transition from laminar to turbulent flow.

3.4 Turbulent flow

The approach used here is the particle roughness turbulence approach (Slatter, 1996, 1999b, 2011). The point of departure of this approach is the classical logarithmic velocity distribution

$$\frac{u}{V^*} = A \ln\left(\frac{y}{d_x}\right) + B \tag{18}$$

The value of A is taken as the inverse of the von Karman universal constant, $A = 1/\chi = 2.5$. B is the classical roughness function (Schlichting, 1960). Integrating and rearranging we get the mean velocity V as

$$\frac{V}{V^*} = \frac{A}{\chi} \ln\left(\frac{R}{d_{85}}\right) + B - 3.75 \tag{19}$$

For smooth wall turbulent flow, this reduces to

$$\frac{V}{V^*} = 2.5 \ln\left(\frac{R}{d_{85}}\right) + 2.5 \ln Re_r + 1.75 \tag{20}$$

For fully developed rough wall turbulent flow, this reduces to

$$\frac{V}{V^*} = 2.5 \ln\left(\frac{R}{d_{85}}\right) + 4.75 \tag{21}$$

which will yield a constant value for the Fanning friction factor, f

$$\frac{1}{\sqrt{f}} = 4.07 \log\left(\frac{3.34D}{d_{85}}\right) \tag{22}$$

4. Minor losses in pipe systems

Head losses, in addition to those due to straight pipe friction, are always incurred at pipe bends, junctions, contractions, expansions and valves. These additional losses are due to eddy formation generated in the fluid at the fitting. In the case of long pipelines of several kilometres, these local losses may be negligible, but for short pipelines they may be greater than the straight pipe frictional losses (Chadwick & Morfett, 1993). A general theoretical treatment for local head losses is not available, but it is usual to assume rough turbulence (where the friction factor is independent of the Reynolds number) since it leads to a simple equation (Chadwick & Morfett, 1993). The prediction of losses in pipe fittings is either based on (King, 2002):

a. The fitting will contribute to the energy dissipation an amount equivalent to an additional length of pipe that is calculated as a multiple of the pipe diameter.
b. The kinetic energy is dissipated as the fluid flows through the fitting and the loss is calculated in terms of the number of velocity heads that are lost.

Here, the losses in pipe fittings will be expressed as the number of velocity heads lost.

On dimensional grounds, the head loss in a fitting will depend upon the fluid velocity, fluid properties and the geometry of the fitting as follows (Edwards et al., 1985):

$$\frac{2gh_{fitt}}{V^2} = fn(\text{Reynolds number, geometry}) \tag{23}$$

This can be expressed as a function of the velocity energy head given by

$$h_{fitt} = k_{fitt}\frac{V^2}{2g} \tag{24}$$

where h_{fitt} is the local head loss and k_{fitt} is the fitting loss coefficient.

Since the pressure drop across the fitting, Δp_{fitt} is given by $\Delta p_{fitt}=\rho g h_{fitt}$, Eqn. (24) can be rewritten as

$$k_{fitt} = \frac{\Delta p_{fitt}}{\frac{1}{2}\rho V^2} \tag{25}$$

The loss coefficient, k_{fitt}, is the non-dimensionalised difference in overall pressure between the ends of two long straight pipes when there is no fitting and when the real fitting is installed (Miller, 1978). The flow lengths over which pressure losses occur start from a few diameters upstream to several pipe diameters downstream of the actual length of the fitting. This is known as the region of influence or interference (see Figure 11). The pressure loss across the fitting Δp_{fitt} should be measured across this region. It can be the measured static pressure drop (Δp_s) or the total pressure that is $\Delta p_{tot} = \Delta p_s + \frac{1}{2}\rho V^2$. It is therefore important to state whether k_{fitt} is based on the static or total pressure (Miller, 1978). If there is a change in pipe diameter, the convention is to use the higher mean flow velocity (V) of either the upstream or the downstream pipe.

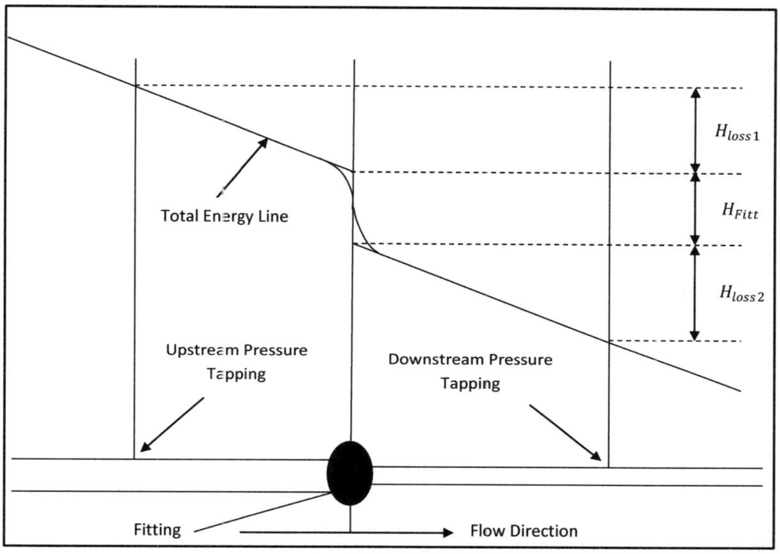

Fig. 11. Definition of energy head loss

With the exception of abrupt contractions and expansions, all other fittings have a physical length. There are three distinct conventions for estimating the length of the straight pipe in the test section (Perry & Chilton, 1973):

i. the actual length of the centreline of the entire system is taken;
ii. the lengths of the individual pieces of pipe that are actually straight are summed up;
iii. the distances between the intersections of the extended centrelines of the successive straight pipes are added.

4.1 Turbulent flow loss coefficient

For turbulent flow through fittings, with the exception of bends, the loss coefficient is independent of the Reynolds number because inertia forces dominate. Experimental work showed that this is true for Newtonian fluids (Miller, 1978, Crane, 1999) and non-Newtonian fluids (Edwards et al., 1985, Ma 1987, Turian et al., 1998).

4.2 Laminar flow loss coefficient

For laminar flow, the loss coefficient is inversely proportional to the Reynolds number (Hooper, 1981; Edwards et al., 1985; Ma, 1987; Pienaar, 1998) and the data loci is presented as a straight line of slope x = –1 in most cases when tests are conducted to sufficiently low Reynolds numbers, i.e. Re < 10.

$$k_{fitt} = \frac{K_1}{Re^x} \tag{26}$$

where K_1 is the laminar flow loss coefficient constant and is a characteristic of a specific fitting involving its dimensions.

4.3 Generalised correlation for loss coefficient

Hooper (1981) provided a generalised correlation for the determination of the loss coefficient spanning Reynolds number from laminar to turbulent flow

$$k_{fitt} = \frac{K_1}{Re} + k_{turb}\left(1 + \frac{1}{D_i}\right) \tag{27}$$

where K_1 and k_{turb} are determined experimentally and D_i is the pipe diameter in inches (King, 2002).

4.4 Determination of loss coefficients

The energy losses across a fluid-conveying conduit in a fluid are normally accounted for using the mechanical energy balance:

$$z_1 + \frac{\alpha_1 V_1^2}{2g} + \frac{p_1}{\rho g} = z_2 + \frac{\alpha_2 V_2^2}{2g} + \frac{p_2}{\rho g} + \sum_{i=1}^{N} h_{loss} \tag{28}$$

where subscripts 1 and 2 refer to the upstream and downstream conditions respectively, V is the mean flow velocity, z is the elevation from the datum, α is the kinetic energy correction factor and p is the static pressure and where there are N sources of energy loss (Edwards et al., 1985). Each term in the expression represents energy per unit weight of fluid, known as energy head or head loss, and is a statement of the law of conservation of energy as applied to fluid flow. The head loss is in units of metres. For the case of a head loss in a fitting, the energy equation may be rewritten as:

$$z_1 + \frac{\alpha_1 V_1^2}{2g} + \frac{p_1}{\rho g} = z_2 + \frac{\alpha_2 V_2^2}{2g} + \frac{p_2}{\rho g} + h_1 + h_{fitt} + h_2 \tag{29}$$

where h_1 and h_2 refer to the friction head loss in the straight pipe upstream and downstream of the fitting and h_{fitt} refers to the head loss in the fitting. The head loss, h, in the straight pipe can be calculated from

$$\Delta h = \frac{4fL}{D}\left[\frac{V^2}{2g}\right] \tag{30}$$

In practice, it is difficult to measure the pressure drop across the fitting only (Ward Smith, 1976) to distinguish between the incompletely developed and the fully-developed flow region prior to and after the fitting. One method is to extrapolate the fully developed pressure gradient to the fitting centreline if the pressure is measured along the length of the two pipes. The second method is to measure the total pressure drop across the system by using two pressure taps only. The latter is experimentally a cheaper method, as only one pressure transducer is required. There are difficulties associated with both methods that could influence the results obtained to some extent. For the pressure gradient to be determined, it is required to select the points that are in the fully developed friction gradient region, but this is not always an easy task, as the distance of interference changes with Reynolds number (Pal & Hwang, 1999). For the total pressure drop method on the other hand, one needs to ensure that the additional losses are accounted for by ensuring that there is significant length of straight pipe. The difficulty is often encountered that to determine the loss in the fitting, two large numbers are being subtracted to obtain a very small number, resulting in significant errors or negative results (Sisavath, 2002). The following was found when data obtained for Newtonian lubrication oil flowing through sudden contractions was analysed using the pressure grade line approach as well as the total pressure drop approach to obtain values of k_{turb} and K_{1con} and compared to results found in literature as shown in Figure 12.

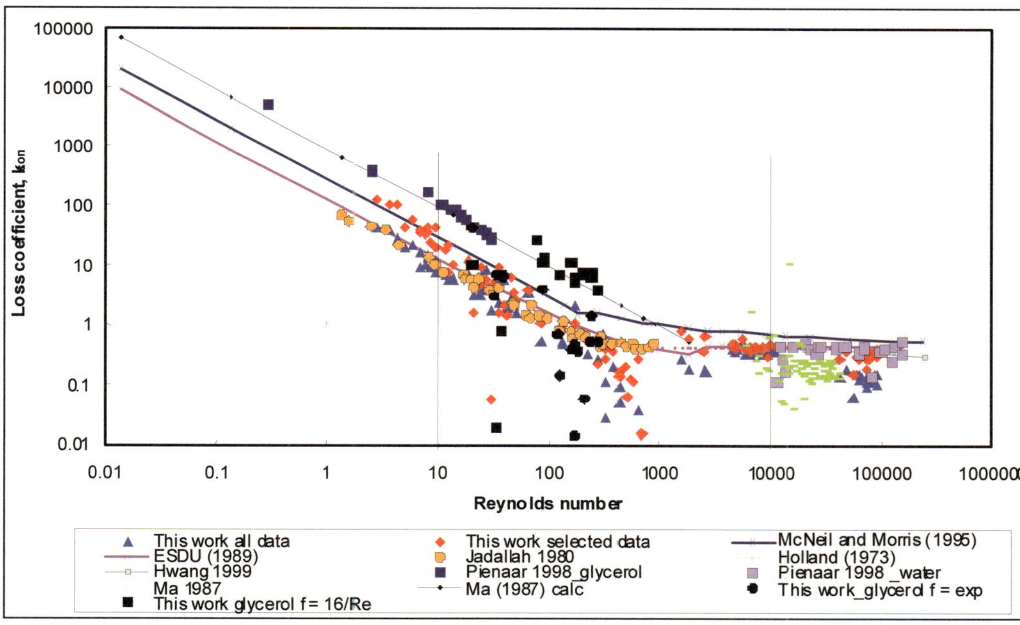

Fig. 12. Comparison of different analysis method to determine loss coefficient (Pienaar & Slatter, 2004)

Between Re = 10 - 10 000, the fact that all the data points are used for extrapolation or that some data are selected gives little difference in the loss coefficient calculated. It is clear that the range of data selected for analysis of results has a significant effect at Reynolds numbers less than 10 and greater than 10 000. Good agreement was found between this work using all data and that of Jadallah (1980) and the prediction of ESDU (1989).

- At Re > 10 000, the results were significantly lower than predicted results and it was obvious that in this range the data closed to the contraction plane had to be excluded from the pressure gradient analysis.
- At Re < 10, good agreement was found between this work using selected data and the prediction of McNeil and Morris (1995).
- Discrepancies between K_1 values are often due to the fact that results are not obtained at Reynolds numbers less than 10 in the purely viscous range before onset of turbulence.

It is evident that it is not only the physical experimental measurements that contribute to the large scatter of results, but also the analytical approach applied to the experimental results to obtain loss coefficients. It is important to analyse experimental results in various ways by assessing experimental data very carefully.

4.5 Dynamic similarity

Complete analytical solutions for engineering problems involving the flow of real fluids is seldom attainable and experiments on models of different physical size are often a necessary part of the design process. This investigation is typical of this approach. In order to correctly interpret the qualitative and quantitative data obtained from such experiments, it is necessary to understand the relationship between models of different size. The concepts of dimensional analysis and physical, geometric, kinematic and dynamic similarity are introduced as they apply to the specific problem of modelling and data interpretation of fittings of different size. Dimensional analysis enables the magnitudes of individual quantities relevant to a physical problem to be assembled into dimensionless groups, often referred to by name. The dimensionless group of specific interest here is the Reynolds number. These groups assist in the interpretation of model studies by ensuring that the conditions under which tests and observations take place at one size fitting are the same as those at the other size fitting.

Physical similarity, like dimensional analysis, helps to ensure that the conditions under which tests and observations take place at one scale are the same as those on another scale. The models at different scale are said to be physically similar in respect of specified physical quantities (eg, velocity), when the ratio of corresponding magnitudes of these quantities between the two systems are everywhere the same. For any comparison between models, the sets of conditions associated with each must be physically similar.

Geometric similarity is similarity of shape. The requirement is that any ratio of length in one model to the corresponding length in another model is everywhere the same. This ratio is referred to as the scale factor. Geometric similarity is the first requirement of physical similarity. Kinematic similarity is similarity of motion and requires similarity of both length and time interval.

Dynamic similarity is similarity of forces. Since there may be several kinds of forces acting on a fluid particle, it is usually impossible to satisfy dynamic similarity for all of them simultaneously. The justification for comparing observations from one model flow system to another is that the fluid behaviour in both systems is similar thus implying kinematic similarity. Geometric similarity alone does not imply dynamic similarity. The requirement for kinematic similarity is to have both geometric and dynamic similarity. This produces geometric similarity of flow patterns and it is this which is of prime importance in this study.

It was shown by Edwards et al. (1985) and Fester and Slatter (2009) that if geometric similarity is maintained, the Reynolds number can be used to establish dynamic similarity in globe and gate valves. Although Edwards et al. (1985) initially indicated that for turbulent flow in globe valves, the turbulent loss coefficient is geometry dependent, this was later found to be due to the globe valves not being geometrically similar. Fester and Slatter (2009) found that the turbulent loss coefficient is in fact independent of the valve size for the geometrically similar globe valves tested. This was found to be in agreement with Edwards et al.'s (1985) findings for gate valves. It will only depend on the valve opening for carefully machined valves. However, results for diaphragm valves agreed with Edwards et al.'s statement that for significant changes in cross-sectional areas, the result becomes size dependent. The viscous force dominated range at Re < 10 is not affected by valve size. In fact, the K_1 value obtained for non-Newtonian fluids is in excellent agreement with that provided by Hooper (1981). As the Reynolds number increases beyond the critical Reynolds number as inertia forces begin to dominate, the loss coefficient is sensitive to geometry and valve size. Results for valves from different manufacturers are remarkably different in this region and this is directly related to the flow path. Furthermore, the valves were rubber-lined, and this influenced the actual final diameter of the valve.

4.6 Transition from laminar to turbulent flow

Miller (1978) presented general idealized curves of the laminar to turbulent transition region as shown in Figure 13. The transition is not always sharp since it can be the entire region from where the loss coefficient starts to deviate from the trend of being inversely proportional to the Reynolds number to where it becomes independent of the Reynolds number. Sometimes a minimum is reached before the loss coefficient increases and then takes a constant value in turbulent flows. The transition region starts where the flow upstream of the fitting is laminar, but downstream turbulence is introduced (Jameson & Villemonte, 1971). Since the Stokes flow region is between 1 < Re < 10, it is important that tests must be conducted in this range to ascertain the onset of transitional flow. Various

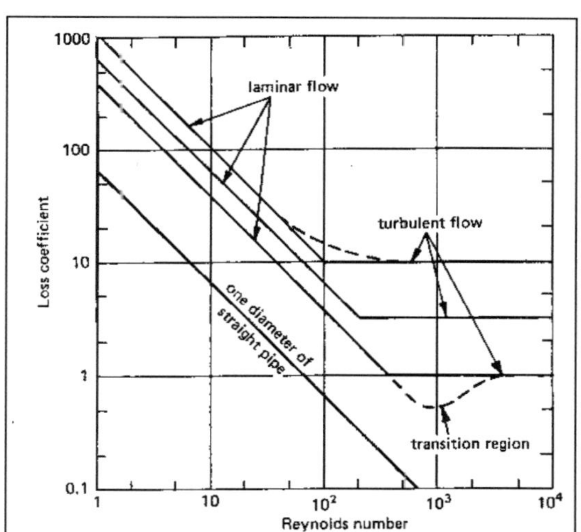

Fig. 13. Trends in loss coefficients in the laminar to turbulent transition region (Miller, 1978)

conflicting transition Reynolds numbers that have been reported is mainly due to the different definitions used for onset of transitional flow. It has been defined as either the intersection between the straight lines for laminar flow and turbulent flow (Ma, 1987) or as the point where the data starts to deviate from the laminar flow line (Fester & Slatter, 2009). The latter always occur at Reynolds numbers lower than that predicted by the intersection method. For short orifices, the ratio of the Reynolds number from the onset of deviation from laminar flow to that of fully developed turbulent flow was found to be ≈0.008 for β ratios ranging from 0.2 to 0.7. For sudden contractions and long orifices, this minimum value is not as pronounced as for short orifices as shown in Figures 14 to 16.

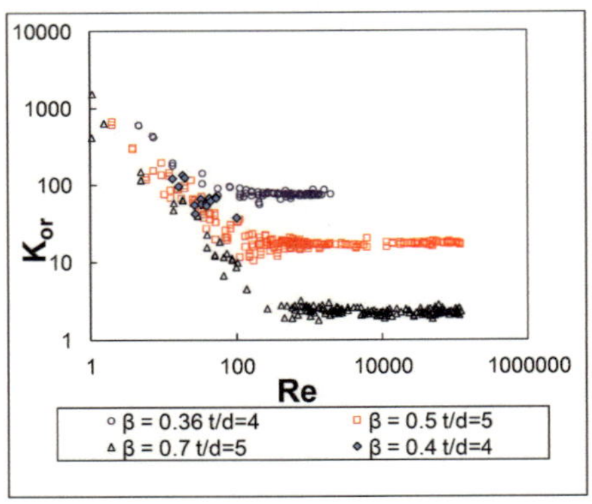

Fig. 14. Data showing transition region for a long orifice (Fester et al., 2010)

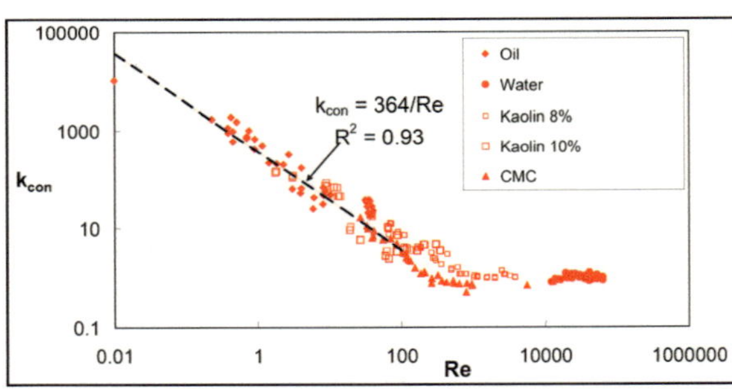

Fig. 15. Data showing transition region for a sudden contraction (Fester et al., 2008)

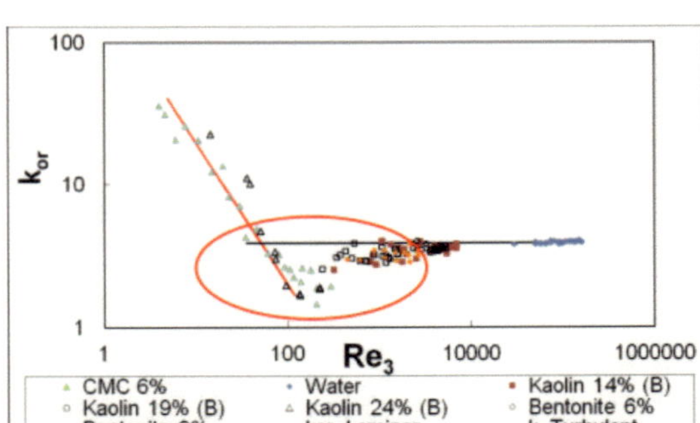

Fig. 16. Data showing transition region for a short orifice (Ntamba, 2011)

4.7 Loss coefficient data

Loss coefficients for both laminar and turbulent flow of Newtonian and non-Newtonian fluids through various types of fittings are provided for engineering design calculations.

4.7.1 Contractions and expansions

For sudden contractions and expansions where β is the ratio of the downstream (d_d) to upstream (d_u) pipe diameter, K_1 values for non-Newtonian fluids were found to be similar to those found for Newtonian fluids. Here, the non-Newtonian behaviour of the fluid was accounted for by using the appropriate Reynolds number for the fluid. This was the conclusion drawn by Edwards et al.(1985), Ma (1987), Pienaar (1998), Pal and Hwang (1999) and Fester et al. (2008). Mika (2011) was, however, unable to establish dynamic similarity for laminar flow of ice slurries of various concentrations in sudden contractions. Tables 1 and 2 show a comparison of the results obtained by various workers for sudden contractions and expansions respectively. In general, there was much better agreement between values in turbulent flow than those in laminar flow for which no quantitative agreement was found amongst the work of different researchers or correlations to predict losses through sharp-edged contractions and expansions.

STUDY	Fluid	β_{con}	K_{1con}	k_{con}
Hooper, 1981	-	-	160	-
Edwards et al., 1985	50% glycerol/water Lubricating oil	0.445	110	0.45
Edwards et al., 1985	50% glycerol/water Lubricating oil, CMC, China clay	0.660	59	0.33
Ma, 1987	Laterite & Gypsum slurries	0.5	900	0.23
Pienaar, 1998	100% Glycerol, Kaolin, CMC	0.463	640	0.414
Pienaar, 1998	100% Glycerol, Kaolin, CMC	0.204	1300	0.44
Pal & Hwang, 1999	Oil-in-water Emulsions	0.49	-	0.43

Table 1. Loss coefficient data for sudden contractions

Reference	Fluid	β_{exp}	K_{1exp}	k_{exp}
Idelchik, 1966	Newtonian fluids	-	30	-
Edwards et al., 1985	Glycerol, CMC Solutions, China clay	1.97	139	0.55
Edwards et al., 1985	Glycerol, CMC Solutions, China clay	1.52	87.7	0.32
Edwards et al., 1985	Glycerol, CMC Solutions, China clay	2.18	150	0.62
Ma, 1987	Laterite, Gypsum	2	115	0.551
Pienaar, 1998	Water, Glycerol, Kaolin, CMC	2.16	959	0.954
Pienaar, 1998	Water,Glycerol,Kaolin, CMC	4.9	1408	0.918
Turian et al., 1998	Laterite, Gypsum	2	-	0.551
Pal & Hwang, 1999	Oil-in-water emulsions	0.49	-	0.49

Table 2. Loss coefficient data for sudden expansions

4.7.2 Valves

Valves form an integral part of slurry pipework systems. These fall within the category of fittings that influence slurry flow in pipework by opening, closing, diverting, mixing or partially obstructing the flow passage (Whitehouse, 1993). An isolation (on-off) valve is a valve designed for use in either the fully open or closed position. Those suitable for slurry service include knife gate, parallel gate, diaphragm, pinch, plug, ball, butterfly and rotating disc (Alderman & Heywood, 1996).

A regulating (throttling) valve is a valve designed for use in all positions between fully open and fully closed. Those suitable for slurry service include globe, diaphragm, pinch and segmented ball (Alderman & Heywood, 1996). Work on frictional pressure losses arising from flow through valves has been carried out with Newtonian slurries using both isolation and regulating valves. Initially, work with non-settling, non-Newtonian slurries appears to have been restricted to gate and globe valves (Edwards et al., 1985; Turian et al., 1998; Pal & Hwang, 1999). This work was extended to diaphragm valves by Mbiya et al. (2009) and Kabwe et al. (2010).

4.7.2.1 Globe and gate valves

A summary of loss coefficient data for gate and globe valves are provided in Table 3 ranging from 12.5 mm to 40 mm. Figure 17 shows experimental data obtained for 12.5, 25, 40 and 65 mm globe valves from two different manufacturers. A range of Newtonian and non-Newtonian fluids were tested for Reynolds numbers covering from 0.05 to 1000000 obtained from three different test rigs (Fester & Slatter, 2009). In laminar flow, good agreement was obtained with those obtained by Hooper for a standard globe valve, but in turbulent flow, the predicted values for a 1 inch globe valve approximated those for the 65 mm valve. The laminar to turbulent transition for the 12.5, 25 and 40 mm valves was found to be smooth whereas a minimum was obtained for the 65 mm valve.

Type	Fluid	Setting	K_{1valve}	x	k_{valve}
Gate: Hooper, 1981	Newtonian	Full open	300	1	0.1
Globe Standard: Hooper, 1981	Newtonian	Full open	1500	1	4
Globe Angle/Y: Hooper, 1981	Newtonian	Full open	1000	1	2
Gate: 1 inch Turian et al., 1998	Laterite and Gypsum slurries	Full open	320	1	0.80
Gate: 2 inch Turian et al., 1998	Laterite and Gypsum slurries	Full open	320	1	0.17
Gate: 1 inch Edwards et al., 1985	CMC solutions China clay slurries	Full open	273	1	-
Gate: 2 inch Edwards et al., 1985	CMC solutions China clay slurries	Full open	273	1	-
Globe: 1 inch Edwards et al., 1985	CMC solutions China clay slurries	Full open	1460	1	122

Type	Fluid	Setting	K_{1valve}	x	k_{valve}
Globe: 2 inch Edwards et al., 1985	CMC Solutions China clay slurries	Full open	384	1	25.4
Globe: 1 inch Pal & Hwang, 1999	Oil-in-water emulsions	Full open	62	0.53	-
Globe: 1 inch Pal & Hwang, 1999	Oil-in-water emulsions	Half open	169	0.53	-
Globe: 1 inch Turian et al., 1998	Laterite and Gypsum slurries	Full open	-	-	10.0
Globe: 25, 15, 40 mm Fester et al, 2009	Water, CMC, Koalin slurries	Full open	700	1	12
Globe:, 25, 15, 40 mm Fester et al, 2009	Water, CMC, Koalin slurries	Half open	1200	1	23

Table 3. Loss coefficient data for globe and gate valves

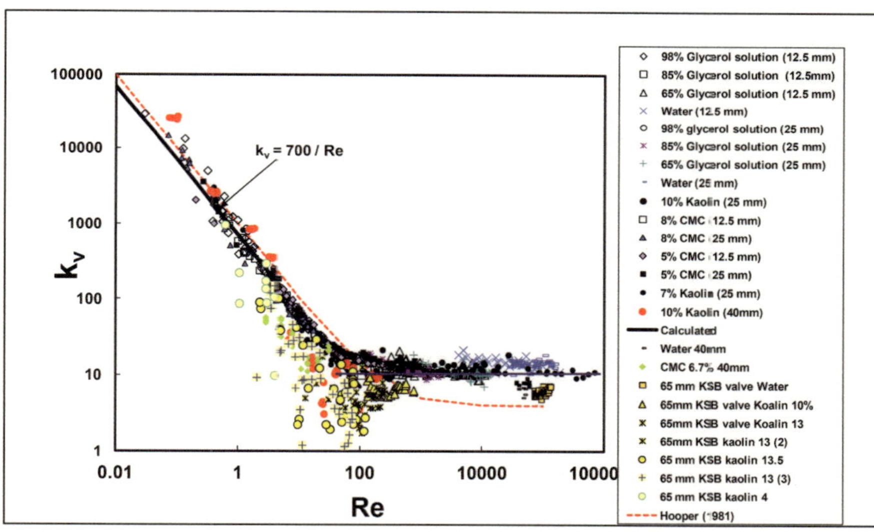

Fig. 17. Comparison of experimental pressure loss coefficient with various correlations (Fester & Slatter, 2009)

4.7.3 Diaphragm valves

Extensive experimental work was conducted on straight-through diaphragm valves from two manufacturers (Kabwe et al., 2010). The nominal valve sizes were 40, 50, 65, 80 and 100 mm. The fluids tested were water, carboxymethyl cellulose solutions and kaolin suspensions. It was observed that at Reynolds numbers below 10, the loss coefficient is independent of the valve opening for valves from both manufacturers. This purely laminar flow regime can be modelled using a K_{1valve} constant of 1000. This is in agreement with the value provided by Hooper (1981). However, in the fully turbulent regime this loss coefficient for the Natco valves were found to be higher than those for Saunders valves. This may be attributed to the more tortuous flow path of the Natco valve compared with that of the Saunders valve. A new correlation was developed to account for the prediction of head losses through diaphragm valves at various opening positions that would be useful for design purposes. This is given by

$$k_{valve} = \frac{1000}{Re} + \frac{\lambda_{\Omega}}{\theta^{2.5}} \tag{31}$$

where λ_{Ω} is k_{valve} at 100% opening and θ is the valve opening. The results obtained for the valves are given in Table 4.

4.7.4 Orifices

Loss coefficient data for orifices are provided in Table 5 and 6. Johansen, in 1930, initiated the first detailed studies on sharp edged concentric orifices but as recent as 2007 the lack of pressure loss coefficient data for β ratios of 0.2, 0.3, 0.4, 0.5 and 0.7 in literature for long and short orifices was identified (ESDU, 2007). Little work has been done with non-Newtonian fluid despite their importance in the field of polymer processing, flow of petroleum products, biomedical engineering, biochemical engineering, food processing, and mineral processing plants. In such applications, the flow remains laminar even at large flow rates (Bohra , 2004). Correlations found in literature are generally for turbulent flow regimes for which the loss coefficients mainly depends on the geometry of the orifice, practically independent of Reynolds number.

TYPE	size (mm)	k_{valve} at % opening			
		25	50	75	100
Saunders Diaphragm valve Kabwe et al., 2010	100	69	18	4.7	1.0
Saunders Diaphragm valve Kabwe et al., 2010	80	28	19	4.3	0.5
Saunders Diaphragm valve Kabwe et al., 2010	65	22	3.6	1.8	0.6
Saunders Diaphragm valve Kabwe et al., 2010	50	89	10	3.9	1.6
Saunders Diaphragm valve Kabwe et al., 2010	40	72	33	8.2	2.7
Natco Diaphragm valve Mbiya et al., 2009	100	100	29	10	1.4
Natco Diaphragm valve Mbiya et al., 2009	80	67	18	6.8	2.5
Natco Diaphragm valve Mbiya et al., 2009	65	63	16	2.8	1.2
Natco Diaphragm valve Mbiya et al., 2009	50	85	25	8.1	2.5
Natco Diaphragm valve Mbiya et al., 2009	40	211	35	18	8.1

Table 4. Loss coefficient data for diaphragm valves

4.7.4.1 Long orifices

In 1972, Lakshmana Rao and his co-workers obtained the pressure loss coefficients for laminar flow in five sharp square-edged long orifices with a constant β ratio of 0.2 whilst varying the thickness to diameter ratios from 0.48 to 10.11. Pressure loss coefficients in turbulent flow for orifices with β ratios of 0.36, 0.4, 0.5 and 0.7 with aspect ratios of 4, 4, 5 and 5 respectively are compared with those of Ward-Smith (1971) and Idel'chik et al. (1994). Excellent agreement was found between experimental work and turbulent flow correlations from Ward-Smith (1971) and Idel'chik et al. (1994), with maximum difference between the experimental data models were 9.87% and 12% respectively for the maximum beta ratio of 0.7. This difference was well within the experimental error obtained. The comparison shows clearly that correlations published by Ward-Smith (1971) and Idel'chik et al. (1994) can be used from Re > 1000, although the two models has an applicability range from Re > 10^4.

Due to lack of correlations to predict losses through long orifices for laminar flow, the correlations of Hasegawa et al. (1997) and Bohra (2004), although limited to β = 0.1 and 0.137 respectively, were evaluated at the higher orifice diameter ratios tested in this work. Hasegawa et al. (1997) shows good agreement in laminar flow for the Stokes flow region, Re < 10, for β = 0.5. It is however unable to predict turbulent flow. Bohra (2004) approximates turbulent flow data very well for all cases except for β = 0.7. However, there is good agreement in laminar flow for β = 0.7. The comparison (Figure 18) also revealed that an improvement of the models is required to predict losses accurately over a wide range of laminar and turbulent flow. No suitable correlation was found in the literature to predict pressure losses through long square edged orifices from laminar to turbulent flow regimes. The loss coefficient data obtained from the experimental work (Fester et al., 2010) is given in Table 5.

STUDY	FLUID	β_{Lor}	t/d	K_{1Lor}	k_{Lor}
Fester et al., 2010	CMC, Kaolin,	0.36	4	3500	76
Fester et al., 2010	CMC, Kaolin	0.40	4	2100	44
Fester et al., 2010	CMC, Kaolin	0.50	5	1500	17
Fester et al., 2010	CMC, Kaolin	0.70	5	860	2.3

Table 5. Loss coefficient data for long orifice

4.7.4.2 Short orifices

Although the work on orifices started in the early 1900s, it was mainly for determining discharge coefficients (Johansen, 1930; Medaugh & Johnson, 1940). A comparison of pressure loss characteristics of different geometries orifices and nozzles was done by Alvi et al. (1978). They found that the flow characteristics of orifices can be divided into three regimes: fully laminar region, re-laminarising region and turbulent region. Lakshmana Rao et al. (1977) investigated the critical Reynolds number for orifice and nozzle flows and found that the critical Reynolds number approached a constant value for low value of orifice or nozzle diameter to pipe diameter ratio. The work done on flow through orifice plate was carried out using Newtonian fluids. In such applications, the flow remains laminar even at large flow rates (Bohra et al., 2004). Edwards et al. (1985) used aqueous solutions of carboxymethylcellulose and suspensions of china clay in water through orifice plates.

Significant differences were found between the experimental results of Edwards et al. (1985) and those of Ntamba (2011) as shown in Table 6. However, the results of Ntamba (2011) shown in Figure 19 are in good agreement with that of Lakshmana Rao et al. (1977).

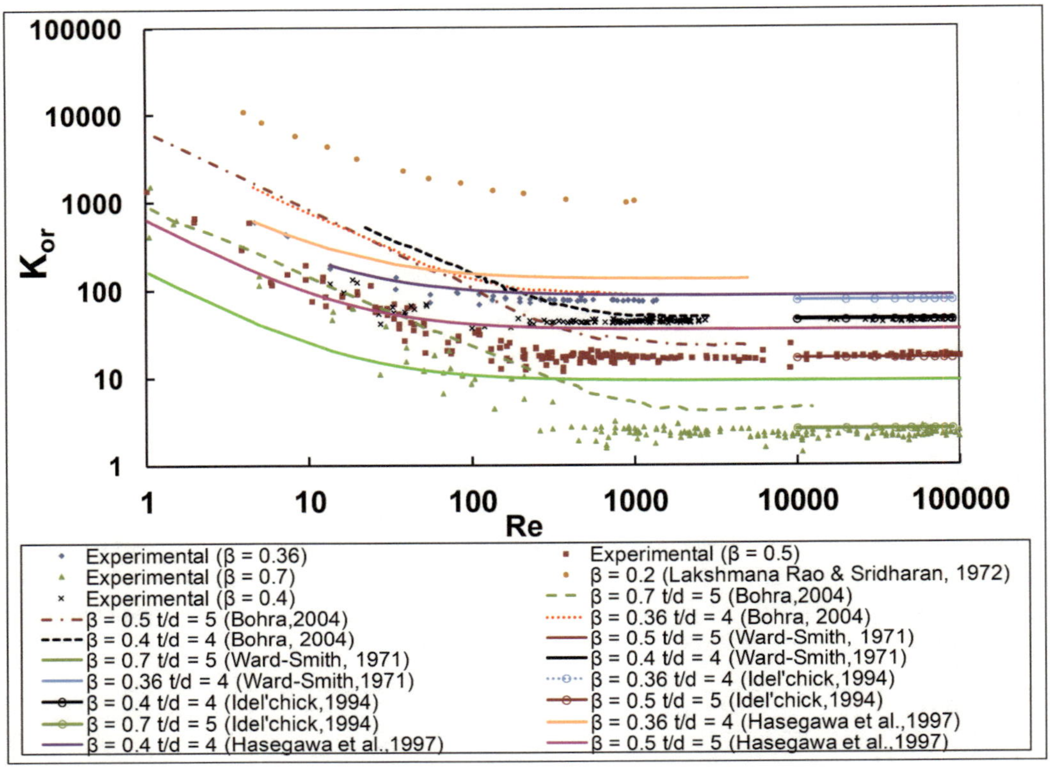

Fig. 18. Comparison of experimental pressure loss coefficient with correlations for long orifices

STUDY	FLUID	β_{or}	K_{1or}	k_{or}
Edwards et al., 1985	50% glycerol/water Lubricating oil	0.289	786	-
Edwards et al., 1985	50% glycerol/water Lubricating oil, CMC, China clay	0.577	154	-
Ntamba, 2011	Koalin, CMC, Bentonite	0.20	2250	1213
Ntamba, 2011	Koalin, CMC, Bentonite	0.30	1111	227
Ntamba, 2011	Koalin, CMC, Bentonite	0.57	340	14.2
Ntamba, 2011	Koalin, CMC, Bentonite	0.70	122	3.85

Table 6. Loss coefficient data for short orifices

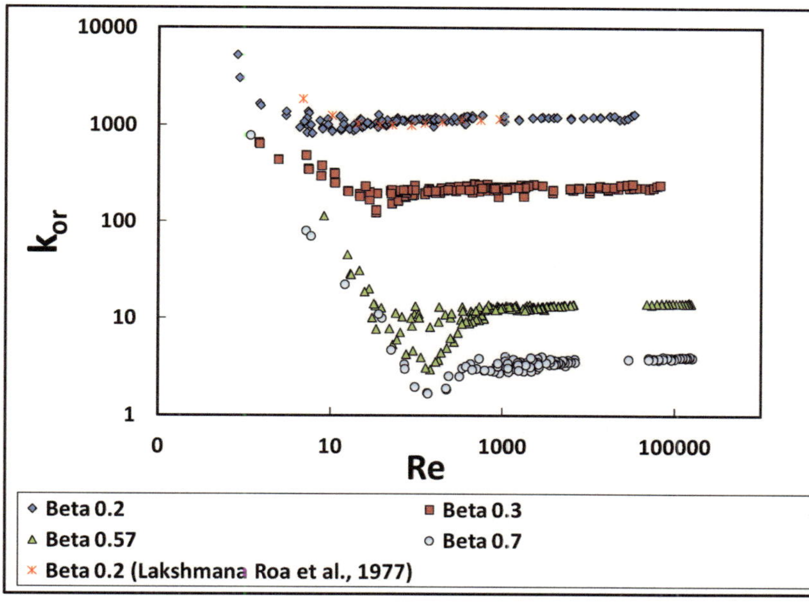

Fig. 19. Experimental pressure loss coefficient data for short orifices

5. Worked example[1]

In order to illustrate the effect that the fittings loss has in laminar viscoplastic flow, a simple system consisting of 10 m of straight 50 mm ID pipe and 5 fittings - the loss coefficient of the above diaphragm valve in laminar flow is $k_v=946/Re_3$, and for turbulent flow is constant at $k_v= 2.5$ - is set and analysed. The fluid used for the analysis is a viscoplastic paste ($\tau_y = 100$ Pa, K = 1 Pa.s, relative density = 1.5 and n = 1). These values were chosen so as to present a relatively simple viscoplastic rheology which would yield laminar flow in a 50 mm pipe at 3 m/s.

The objective of the analysis was to obtain the head loss as a function of volumetric flow rate. The operating flow rate considered was 0.006 m^3/s which corresponded to an operating average velocity V of 3 m/s in a 50 mm pipe.

Since the value of the laminar flow loss coefficient is not available to most designers, the effect of incorrectly using the turbulent - constant - value for design in laminar flow will be highlighted.

Figure 20 shows the fittings head losses in both laminar and turbulent flow. Figure 20 shows three principal differences between the fittings head losses in both laminar and turbulent flow:

1. The significant contribution of the laminar fittings loss to the start-up static head (at Q = 0).
2. The very different shape presentation between laminar and turbulent flow.
3. The significant difference in magnitude which arises.

[1] From Slatter & Fester (2010): reproduced with permission

5.1 Start-up static head

In this example, the contribution of the laminar fittings loss to the start-up static head (at Q = 0) is 6.5 m, as can be seen on the ordinate of Figure 20. Inspection of Eqn. (24) might lead one to expect a zero result at Q=0, as portrayed by the turbulent locus in Figure 20.

However, it must be understood that this non-zero value is a result of the combined effects of Eqns. (13), (24) and (26), and is a sign of the presence of a yield stress.

5.2 Shape presentation

The inherently parabolic shape of the turbulent locus in Figure 20 is a direct consequence of the quadratic form of Eqn. (24) and the constant value of the loss coefficient k_{fitt}=2.5 for this valve type in turbulent flow.

Equally, the inherently viscoplastic shape of the laminar locus in Figure 20 is a direct consequence of the hyperbolic form of Eqn. (26) combined with the emphasis of the role of the yield stress in Eqn. (13).

5.3 Difference in magnitude

Figure 20 shows that there are considerable differences in the magnitude of the fittings Head losses in laminar and turbulent flow. A direct comparison of these magnitudes is shown in Figure 21.

Figure 21 shows that the difference in magnitude is best expressed in orders of magnitude.

In this example, the fittings head losses in laminar flow shown in Figure 20 and Figure 21 exceed those in turbulent flow by several orders of magnitude.

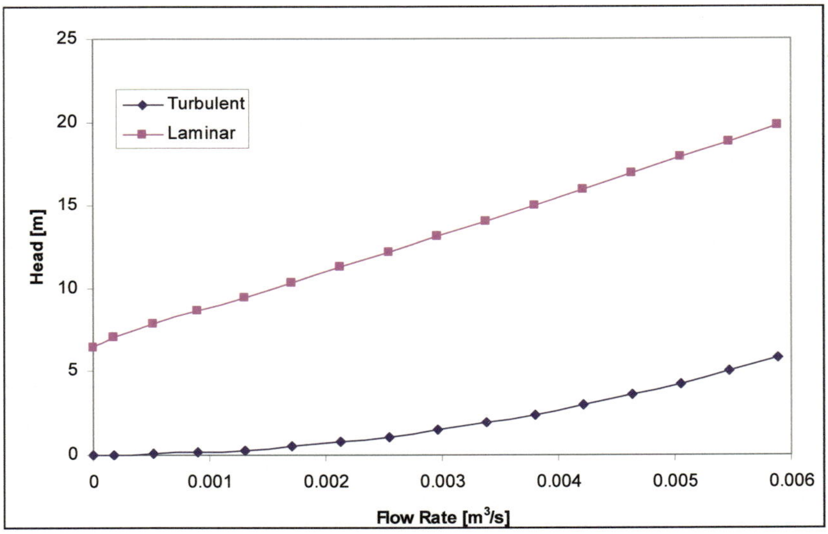

Fig. 20. Fittings Head Losses in Laminar and Turbulent flow

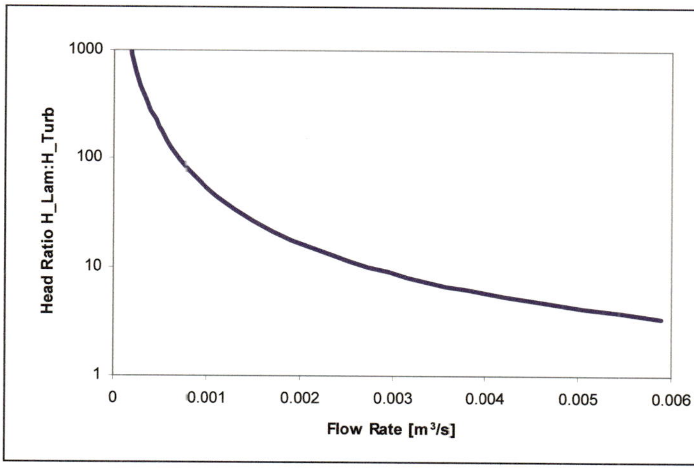

Fig. 21. Comparison of the Fittings Head Losses in Laminar and Turbulent flow shown in Figure 20

5.4 Pump power predictions

Whilst these three issues discussed above are of importance, the issue of primary practical interest for paste piping design is the error in power prediction, if the design is performed using the incorrect approach (Slatter & Fester, 2010). This situation is summarised for this example in Figure 22.

Figure 22 shows the pump operating points for both the laminar and turbulent design cases, as derived from the practical design case example presented above.

The three possible operating points arising from the different design approaches to the practical design case example are highlighted in Figure 22 and designated Point A, Point B and Point C. The coordinates and resulting motor power (brake power) requirements from each of these three operating points are presented in Table 7.

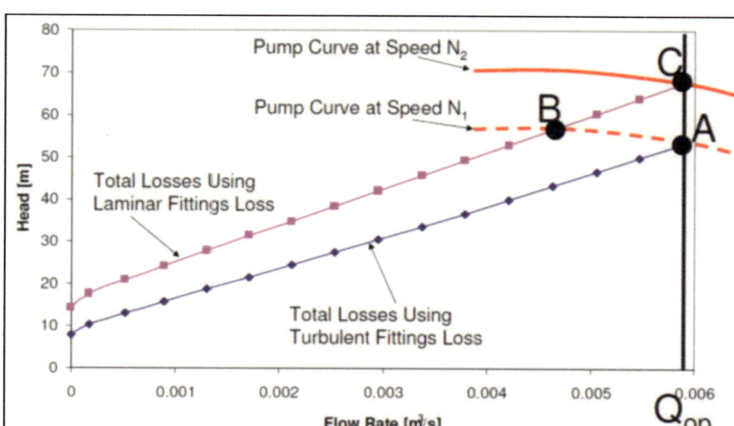

Fig. 22. Pump operating points for both the laminar and turbulent design cases

Point	A	B	C
Q [m3/s]	0.0059	0.0046	0.0059
H [m]	53	57	67
Fluid Power [kW]	4.6	3.9	5.8
Pump Efficiency	65%	50%	55%
Brake Power [kW]	7.1	7.7	10.6

Table 7. Pump operating point coordinates from Figure 22

If the design is performed using the incorrect approach (Slatter & Fester, 2010), the (incorrect) design operating point will be at Point A for this example, as shown in Figure 22 and Table 7.

If the designer follows the information published by the authors (Fester et al., 2007; Mbiya et al., 2009) , the (correct) design operating point will be at Point C for this example, as shown in Figure 21 and Table 7.

This presents at least two problems of profound practical importance:-

1. If the designer follows the first (incorrect) approach, then Figure 22 shows that - in reality - the pump will actually operate, not at Point A, but at Point B, at pump speed N_1. Point B presents an operating flow rate penalty exceeding 20% whilst being almost 10% under-powered, as shown in Table 7.
2. In order to achieve the desired flow rate, the operating point will need to move from Point B to Point C, i.e. the speed would need to be significantly increased to pump speed N_2. This increase in pump speed will require 50% more brake power, as shown in Figure 22 and Table 7.

The important practical message here is that the pump motor would be 50% undersized.

All of these issues are further exacerbated by the shallow intersection angle which the laminar flow system curve presents to a centrifugal pump curve, as shown Figure 22. This will result in unstable operation, as small changes in head (i.e. small changes in material consistency) will result in significant changes in operating point flow rate.

As indicated by other laminar flow investigations over several decades (e.g. Edwards et al., 1985; Fester et al., 2008), all fittings show the k_v values in laminar flow to be significantly greater than those in turbulent flow.

6. Conclusions

Minor losses are important in the efficient design of pipelines in laminar flow. The loss coefficients in laminar flow for all fittings are orders of magnitude greater than those in turbulent flow. The purely viscous driven flow regime is primarily at Re < 10, after which turbulence will be introduced that is dependent on the geometry of the fitting. For some fittings, the minima value obtained are very pronounced, especially for larger orifice ratios and larger valve sizes. For sudden contractions and long orifices, a smoother transition is observed. Dynamic similarity can be established for geometrically similar fittings using a Reynolds number that can accommodate the rheological parameters of the fluid. Pump sizing estimates for shorter lengths of pipelines with a number of fittings operating in laminar flow can be underpredicted by up to 50%. The significance of the impact on pump sizing would therefore appear to be representative of other fittings in general, but focussed research to substantiate this is required.

7. References

Alderman, N.J. & Heywood, N.I. (1996), The design, selection and application of valves for slurry pipeline service, *Proc. Hydrotransport 13*, Johannesburg, South Africa, 3-5 September.

Alderman, N.J. & Heywood, N.I. (2004a). Improving slurry viscosity and flow curve measurements, *Chemical Engineering Progress*, Vol.101 (4), pp. 38-44

Alderman, N.J. & Heywood, N.I. (2004b). Making accurate slurry flow curve measurements, *Chemical Engineering Progress*, Vol.101 (5), pp. 35-41

Alderman, N.J. (1997). Non-Newtonian Fluids: Guide to classification and characteristics, *ESDU 97034*, ESDU International plc, London

Banerjee T.K., Das M. & Das S.K. (1994). Non-Newtonian liquid flow through globe and gate valves. *Canadian Journal of Chemical Engineering*, Vol.72, (Apr), pp. 207–211

Barnes, H. A. (2002). *Viscosity*, Institute of Non-Newtonian Fluid Mechanics, University of Wales, Aberystwyth

Bohra, L. K. (2004). *Flow and Pressure Drop of Highly Viscous Fluids in Small Aperture Orifices.* MSc thesis. Georgia, U.S.A: Georgia Institute of Technology

Bingham, E.C. (1922). *Fluidity and Plasticity*, McGraw-Hill, New York

Brown, N.P. & Heywood, N.I. (1991). *Slurry Handling: Design of Solid-Liquid Systems.* Kluwer Publications, ISBN 1-85166-645-1, Dordrecht, Netherlands

Carreau P.J. (1968). PhD Thesis, Univ Wisconsin, Madison, 1968

Casson, N. (1959). A flow equation for pigmented-oil suspension for the printing ink type, In: Mill, C.C. (ed). *Rheology of Dispersed Systems.* pp. 84-104. Pergamon Press, New York

Chadwick, A.J. & Morfett, J.C. 1993. *Hydraulics in Civil and Environmental Engineering.* ISBN 0 4191 816 01. Taylor & Francis Books Ltd, London

Crane, Co. (1999). *Flow of Fluids Through Valves, Fittings and Pipe*, Technical paper No. 410. Crane Co., Chicago, IL, USA

Cross, M.M. (1965). Rheology of non-Newtonian flow: equation for pseudoplastic systems, *Journal of Colloid Science*, Vol.20, pp. 417-437

de Waele, A. (1923). Oil Color, *Chemical Association Journal*, Vol.6, pp. 23-88

Edwards, M.F., Jadallah, M.S.M. and Smith, R. (1985). Head losses in pipe fittings at low Reynolds numbers, *Chemical Engineering Research and Design*, Vol.63, (January 1985) pp. 43-50

Engineering Sciences Data Unit. 1989. *Pressure Losses in Flow Through a Sudden Contraction of Duct Area.* Data Item 89040.

ESDU. (2007). *Incompressible Flow Through Orifice Plates-A Review of the Data in the Literature.* London: Engineering science data unit. ISBN 978 1 86246 608 1, ISSN 0141-4011

Fester V.G. & Slatter P.T. (2009). Dynamic similarity for Non-Newtonian fluids in globe valves, *Trans IChemE, Part A, Chemical Engineering Research and Design*, Vol.87, pp. 291- 297, ISSN 0263-8762

Fester V.G., Chowdhury M.R. & Iudicello F. (2010). Pressure loss and discharge coefficients for non-Newtonian fluids in long orifices, *British Hydromechanics Research Group 18th International Conference on Slurry Handling and Pipeline Transport HYDROTRANSPORT 18*, Rio de Janeiro, Sept 22-24, 2010, pp. 309-323. ISBN: 978 1 85598 119 5.

Fester V.G., Mbiya B.M., & Slatter P.T. (2008), Energy losses of non-Newtonian fluids in sudden pipe contractions, *Chemical Engineering Journal*, Vol.145, No.1, pp. 57-63, ISSN 1385-8947

Fester, V.G., Kazadi, D.M., Mbiya, B.M. & Slatter P.T. (2007). Loss coefficients for flow of Newtonian and non-Newtonian fluids through diaphragm valves, *Trans IChemE, Part A, Chemical Engineering Research and Design*, Vol.85 (A9), pp. 1314–1324

Govier, G.W. & Aziz, K. (1972). *The Flow of Complex Mixtures in Pipes*, van Nostrand Reinhold Co.

Hanks, R. W.(1979).The axial laminar flow of yield-pseudoplastic fluids in a concentric annulus. *Ind. Eng. Chem. Proc. Des Dev*, 18, pp. 488-493

Hasegawa, T., Suganuma, M., & Watanabe, H. (1997). Anomaly of excess pressure drops of the flow through very small orifices. *Physics of Fluids*, Vol.9, pp. 1 - 3

Herschel, W.H. and Bulkley, R. (1926). Konsistenzmessungen von GummiBenzollosungen. *Kolloid-Z*, Vol.39, pp. 291-330

Heywood, N.I. & Cheng, D.C-H. (1984). Comparison of methods for predicting head loss in turbulent pipeflow of non-Newtonian fluids. T*rans. Institute for Measurement and . Control.*, Vol.6, pp. 33-45

Hooper, W.B. (1981). The two-K method predicts head losses in pipe fittings. *Chemical Engineering*, August, pp. 96–100

Idel'chik, I. E., Malyavskaya, G. R., Martynenko, O. G., & Fried, E. (1994). *Handbook of Hydraulic Resistance*. CRC press, London

Jadallah, M.S.M. (1980). *Flow in pipe fittings at low Reynolds numbers*. Unpublished PhD thesis, University of Bradford, UK.

Jamison, D.K. & Villemonte, J.R (1971). Junction losses in laminar and transitional flows. *Journal of the Hydraulics Division, Proceedings of the American Society of Civil Engineers*, HY 7, pp.1045-1063

Johansen, F.C. (1930). Flow through pipe orifices at low Reynolds numbers. *Proceedings of the Royal Society of London, Series A*, Vol.126, No.801, pp.231-245

Kabwe, A. M., Fester, V. G., & Slatter, P. T. (2010). Prediction of non-Newtonian head losses through diaphragm valves at different opening positions. *Chemical Engineering Research and Design*, Vol. 88, pp. 959-970

King, R.P. 2002. *Introduction to Practical Fluid Flow*. ISBN 0-7506-4885-6, Butterworth-Heinemann, Oxford

Laba, D. (1993). Rheological properties of Cosmetics and Toiletries, Marcel Dekker Inc., New York

Lakshmana Rao, N.S. & Shridharan, K. 1972. Orifice losses for laminar approach flow. *ASCE Journal of Hydraulics Division*, Vol.98, No.11, (November), pp. 2015-2034

Lakshmana Rao, N.S., Srhidharan, K. & Alvi, S.H. (1977). Critical Reynolds Number for orifice and nozzle flows in pipes. *Journal of Hydraulic Research, International Association for Hydraulic Research,* Vol. 15, No. 2, pp. 167-178

Ma, T.W. (1987). *Stability, rheology and flow in pipes, fittings and venturi meters of concentrated non-Newtonian suspensions,* Unpublished PhD thesis, University of Illinois, Chicago

Massey, B.S. (1970). *Mechanics of fluids*, Second edition, van Nostrand Reinhold Co

Mbiya, B.M., Fester, V.G. & Slatter P.T. (2009). Evaluating resistance coefficients of control diaphragm valves, *The Canadian Journal of Chemical Engineering*, Vol.87, pp. 704–714, ISSN 0008-4034

McNeil, D.A. & Morris, S.D. (1995). A mechanistic investigation of laminar flow through an abrupt enlargement and nozzle and its application to other pipe fittings. Report EUR 16348 EN

Medaugh, F.W. & Johnson, G.D. 1940. Investigation of the discharge and coefficients of small circular orifices. *Civil Engineering*, Vol.7, No.7, (July), pp. 422-424

Mika, L. (2011). Energy losses of ice slurry in pipe sudden contractions. *Experimental Thermal and Fluid Science*, Vol.35, pp. 939-947.

Miller, D.S. (1978). *Internal flow systems*, Cranfield, BHRA Fluid Engineering

Ntamba Ntamba, B.M. (2011). *Non-Newtonian pressure loss and discharge coefficients for short square-edged orifice plates,* Unpublished M.Tech. thesis, Cape Peninsula University of Technology, Cape Town.

Ostwald, W. (1925). Über die geschwindigkeitsfuncktion der viskosität disperser systeme I., *Kolloid-Z.*, 36, pp. 99-117

Pal, R. and Hwang. C-Y. J. 1999. Loss coefficients for flow of surfactant-stabilised emulsions through pipe components. *Trans. IChemE*, Vol.77 (Part A): 685-691

Pienaar, V.G. & Slatter, P.T. (2004). Interpretation of experimental data for fittings losses; *12th International Conference on Transport and Sedimentation of Solid Particles*, pp. 537 - 546, ISBN 80-239-3465-1, Prague, Czech Republic September 20-24, 2004

Pienaar, V.G. (1998). *Non-Newtonian fitting losses,* Unpublished M.Tech. thesis, Cape Technikon, Cape Town

Roberts, G.P., Barnes, H.A & Mackie, C. (2001). Using the Microsoft Excel 'solver' tool to perform non-linear curve fitting, using a range of non-Newtonian flow curves as examples, *Applied Rheology*, September/October

Schlichting H (1960), *Boundary layer theory*, 4th edition, McGraw-Hill, New York

Sisavath, S., Jing, X., Pain, C.C. & Zimmerman, R.W. (2002). Creeping flow through axisymmetric sudden contraction or expansion. *Journal of Fluids Engineering,* (Trans. ASME), Vol.124, No.1, (March) pp. 273-278

Sisko, A.W. (1958), The flow of lubricating greases. *Ind Engng Chem.*, Vol.50, pp. 1789-1792

Slatter P T (1995), The laminar/turbulent transition of non-Newtonian slurries in pipes; *14th World Dredging Congress*, Amsterdam, 14-17 November, 1995, ISBN 90-9008834-2 pp. 31-48

Slatter P T (1996); Turbulent flow of non-Newtonian slurries in pipes; *J. Hydrol. Hydromech.*, Vol. 44, No.1, pp. 24-38

Slatter P T (1999); The laminar/turbulent transition prediction for non-Newtonian slurries, *Proceedings of the International Conference "Problems in Fluid Mechanics and Hydrology"*, Academy of Sciences of the Czech Republic, Prague, June 23-26, 1999.ISBN 80 238 3824 5 pp. 247 - 256

Slatter P T (1999b); The role of rheology in the pipelining of mineral slurries, *Min. Pro. Ext. Met. Rev.*, Vol 20, pp 281-300

Slatter PT (2011) The engineering hydrodynamics of viscoplastic suspensions. *J. Particulate Science and Technology*, Vol. 29, No.2, pp. 139-150

Slatter PT and Fester VG (2010); Fittings losses in Paste Flow Design, R.J.Jewel and A.B.Fourie (eds); *Paste 2010,13th International Seminar on Paste and Thickened Tailings*, Toronto, Canada;3-6 May 2010. pp. 303-310. ISBN 978-0-9806154-0-1

Steffe, J.F. (1996), *Rheological methods in Food Process Engineering*, 2nd ed., Freeman Press, Michigan

Streeter, V.L. & Wylie, B. (1985). *Fluid Mechanics*, 6th edition, McGraw-Hill Inc., USA

Turian, R.M., Ma, T.W., Hsu, F.L.G., Sung, M.D.J. & Plackmann, G.W. (1998). Flow of concentrated slurries: 2. Friction losses in bends, fittings, valves and venturi meters. *International Journal of Multiphase Flow*, Vol.24, No.2, pp. 243-269.

Ward-Smith, A.J. 1976. Component interactions and their influence on the pressure losses in internal flow systems. *Heat and Fluid Flow © ImechE*, Vol.6, No. 2, pp. 79-88

Whitehouse, R.C. (1993). *The Valve and Actuator User's Manual*, British Valve Manufacturers' Association, Mechanical Engineering Publications, London.

8

Measurement and Impact Factors of Polymer Rheology in Porous Media

Yongpeng Sun, Laila Saleh and Baojun Bai*
Petroleum Engineering Program,
Missouri University of Science and Technology, Rolla, Missouri,
USA

1. Introduction

Crude oil is found in underground porous sandstone or carbonate rock formations. In the first (primary) stage of oil recovery, oil is displaced from the reservoir into the wellbore and up to the surface under its own reservoir energy, such as gas drive, water drive, or gravity drainage. In the second stage, an external fluid such as water or gas is injected into the reservoir through injection wells located in the formation that have fluid communication with production wells. The purpose of secondary oil recovery is to maintain reservoir pressure and displace hydrocarbons towards the wellbore. The most common secondary recovery technique is waterflooding. Once the secondary oil recovery process has been exhausted, about two thirds of the original oil in place (OOIP) is left behind. Enhanced oil recovery (EOR) methods aim to recover the remaining OOIP. Enhanced oil production is critical today when many analysts are predicting that world oil peak production is either imminent or has already passed while demand for oil is growing faster than supply. For example, the United States has a total of 649 billion barrels original oil in place, and only about 220 billion barrels are recoverable by primary and secondary recovery methods. EOR methods offer the prospect of recovering as much as 200 billion barrels of more oil from existing U.S. reservoirs, a quantity of oil equivalent to the cumulative oil production to date.

Polymer flooding is one of the most successfully methods to enhance oil recovery. The polymers that are mainly used in oilfields are water soluble polyacrylamide (HPAM), Xanthan gum (Xc), and associative polymer (AP). Polymer solutions, in contrast to water, exhibit non-Newtonian rheological behaviors, such as shear thinning and shear thickening effects, which lead to different viscosity properties in a reservoir, as compared with those in water flooding. The results from labs, numerical simulations and fields all demonstrate that the rheology of a polymer solution not only impacts the polymer injectivity but also dominates the oil production rate and final recovery of a polymer flooding project. However, the rheological properties of polymer solutions measured by conventional rheological instruments which strive to produce pure shear flow could not be used directly to predict the pressure-to-flow relationship in porous media.

---

* Corresponding Author

When a polymer solution is injected into a reservoir from an injection well, the flow velocity, which is related to shear rate, will change from wellbore to in-depth of a reservoir; therefore, the polymer solution viscosity will also change from near wellbore to in-depth of a reservoir correspondingly. Polymer rheology in a porous media is also affected by polymer type, molecular weight, concentration, water salinity in the reservoir, and reservoir permeability. The flow of polymer solutions through a porous media is a complex topic, governed by polymer rheology and retention behavior, and is still not understood very well.

This chapter presents the fundamental terms/concepts used to characterize polymer rheology behaviors, the mathematical models that are used to calculate shear rate, the experimental procedures to measure polymer rheology in porous media, and the factors that impact the in situ rheology of polymers.

2. Definition of terms

2.1 Apparent viscosity

Darcy's law is one of the most well-known theories that describe fluid flow through a porous media as shown as follow:

$$Q = k\frac{A\Delta P}{\mu L} \tag{1}$$

Where Q is the liquid volume at a certain time interval in cm^3/s; A is the cross-sectional area in cm^2; L is the length of the sample in the macroscopic flow direction in cm; μ is the viscosity of a fluid flowing through a porous media in $mPa \cdot s$; ΔP is the pressure drop across the porous media in atm; k is the absolute permeability in *Darcy* for the porous media, and a measure of the conductivity of the porous media.

A polymer solution used in EOR is a non-Newtonian fluid; therefore, the viscosity term μ in Equation 1 is usually not constant. A macroscopic in situ 'apparent viscosity', η_{app}, is often used in polymer flooding, and is defined as follow:

$$\eta_{app} = k\frac{A\Delta P}{QL} \tag{2}$$

During a polymer flooding process, a polymer solution itself is a non-Newtonian fluid. A polymer is also retained in pore surfaces and reduces the permeability of the porous media. Thus, in Equation 2, η_{app}, is not constant and is affected by flow rate and rock permeability.

2.2 Effective viscosity

The apparent viscosity mentioned above is from Darcy's law, used to describe macroscopic rheology of a polymer fluid flow in a porous media. Effective viscosity η_{eff} is another term which is also often used in polymer flooding. It is from Poiseuille's law, refers to the observed effective viscosity in a single capillary channel, which is in microscopic. The effective viscosity is expressed as follows:

$$\mu_{eff} = \frac{\tau}{\gamma} \tag{3}$$

Where τ is the shear stress, N/m^2; γ is the shear rate, s^{-1};

Shear stress is the force per unit area required to sustain a constant rate of fluid movement. It can be defined as:

$$\tau = \frac{F}{A} \tag{4}$$

Where F is the force applied, N; A is the cross-sectional area of material in the area parallel to the applied force vector, m^{-2}.

Shear rate is defined as the velocity change through the thickness:

$$\gamma = \frac{v}{h} \tag{5}$$

Where v is the velocity, m/s; h is the thickness, m.

The distinction between effective and apparent viscosity should be clearly maintained, especially when considering mathematical porous media models based on capillary bundles, which as discussed below. The overall viscosity in macroscopic of the polymer fluid in the capillary bundles is η_{app}, whereas the viscosity in each of the capillaries may be different and is η_{eff}. In the latter case, η_{app} is like an 'average' value of the η_{eff} in the individual capillaries (Sorbie 1991).

2.3 Resistance factor

In order to characterize the behavior of different polymer solutions in response to pressure build-up during polymer injection, the resistance factor F_r is often used. A resistance factor is defined as the ratio of mobility of water to the mobility of a polymer solution (Littmann 1988):

$$F_r = \frac{M_w}{M_p} = \frac{\frac{k_w}{\mu_w}}{\frac{k_p}{\mu_p}} \tag{6}$$

Where M_w and M_p are the mobility of water and polymer, respectively; k_w and k_p are the water and polymer permeability in μm^2, respectively; μ_w and μ_p are the viscosity of water and polymer in $mPa \cdot s$, respectively.

The resistance factor describes the effective viscosity of a polymer solution in porous media.

3. Calculation of shear rate in porous media

Since most polymer solutions are non-Newtonian fluids, their shear rate vs. viscosity is not a linear relationship. For a given viscosity, a shear rate should be defined in advance. An average shear rate in a porous media is related to its permeability, porosity and fluid actual velocity. This leads to an expression of an apparent shear rate, γ_{app}, to describe the interstitial shear rate in the porous media. Table 1 provides some representative models which can be used to calculate the shear rate in a porous media. The application conditions and assumptions for each model have been listed in the Table.

No.	Equations	Prerequisite conditions and Assumptions	Source
1	$$\gamma_{app} = \frac{4u}{\phi r} \cdot \frac{L_e}{L} \quad (7)$$ where u is the superficial flow rate, *cm/s*; ϕ is porosity; r is the average pore radius, *cm*; L_e is length of tortuous flow path, *cm*; L is the porous media length, *cm*;	1. Porous media is considered as a bundle of capillary tubes with a length that is greater than the porous media by a tortuous factor; 2. Bundle of straight capillaries; 3. Capillaries are of uniform sized; 4. Capillaries are not interconnected; 5. Flow rate in capillary is constant (at constant pressure drop).	(Kozeny 1927) (Carman 1937)
2	$$\gamma_{app} = \left(\frac{3n+1}{4n}\right) \frac{12G}{\rho\sqrt{150k\phi}} \quad (8)$$ where n is the Power law parameter; ρ is fluid density, *g/cm³*; k is permeability, *cm²*; ϕ is porosity; G is mass velocity, *g/(cm² ·sec)* $$G = \rho(\frac{k}{H}\frac{\Delta P}{L})^{1/n} \quad (9)$$ where ΔP is the pressure drop, *dyne/cm²*; k is permeability measured by a Newtonian fluid, *cm²*; L is the unit length, *cm*; H is the non-Newtonian bed factor, *dyne ·secⁿ ·cm⁻¹⁻ⁿ* $$H = \frac{K}{12}(9+3/n)^n(150k\varphi)^{(1-n)/2} \quad (10)$$ where K is the Power law parameter, *dyne ·secⁿ ·cm⁻²*	The permeability of the porous media is the same for all identical packed bed configurations, independent of flow conditions in the bed.	(Christopher and Middleman 1965)
3	$$\gamma_{app} = \left(\frac{3n+1}{4n}\right)^{n/(n-1)} \frac{12u}{\sqrt{150k_p\phi_w}} \quad (11)$$ where n is the bulk power law index; u is the	1. Based on capillary bundle model for non-Newtonian fluid flow; 2. The polymers such as partially hydrolyzed	(Hirasaki and Pope 1974)

	Darcy velocity, *cm/s*; k_p is polymer permeability, *cm*^{2}; Φ is porosity; Φ_w is the pore space occupied by water, $\Phi_w = \Phi S_w$	polyacrylamides, the permeability reduction due to the adsorption of the polymer is considered; 3. When applied to Xanthan biopolymer through rock cores (at residual oil), very good agreement was obtained between calculated and experimental values; 4. A tortuosity of 25/12 has been assumed.	
4	$$\gamma_{app} = C\left(\frac{3n+1}{4n}\right)^{n/(n-1)} \frac{u}{\sqrt{k_w S_w \phi}} \quad (12)$$ where k_w is water permeability, *cm*^{2}; S_w is water saturation; n is the bulk power law index; C is shear rate coefficient, not constant, but a function of the network parameters; φ is porosity; u is Darcy velocity, $u=Q/A$, *cm/s*;	Equation is developed to relate the flow of Xanthan solutions in cores having different permeabilities, lithologies, and oil saturations.	(Cannella, Huh et al. 1988)
5	The average pore radius in the pack is estimated from the capillary bundle model of the porous medium. The equation: $$r = \left(\frac{8k}{\varphi}\right)^{1/2} \quad (13)$$ is shown to give a good estimate of the average hydrodynamic pore radius in homogeneous unconsolidated porous media. The equation for the shear rate at the pore wall in such media is taken as: $$\dot{\gamma} = \alpha\frac{4v}{r} \quad (14)$$ Where k is permeability, *cm*^{2}; ϕ is porosity; a is a shape parameter characteristic of the pore structure, a=1.7 for packs of large spheres having same diameter, a=2.5 for packed beds of angular grains; v is the superficial velocity, *cm/s*; r is the average pore radius, *cm*;	1. Calibrated glass beads having different diameters are packed to conduct experiments; 2. The porous media is assumed to have similar pore shapes but different pore sizes. 3. The shear rate is a maximum wall shear rate in the average pore throat diameter.	(Chauveteau and Zaitoun 1981) (Zitha, Chauveteau et al. 1995)

Table 1. Representative mathematical models to calculate apparent shear rate

4. Experimental apparatus and procedures

4.1 Apparatus

The polymer rheology in porous media is often measured by coreflooding experiments. Figure 1 shows a typical coreflooding apparatus. It is mainly composed of a pump used for pumping distilled water, piston accumulator used for polymer storage, core holder or sandpack, and pressure gauge or transducer to record injection pressure.

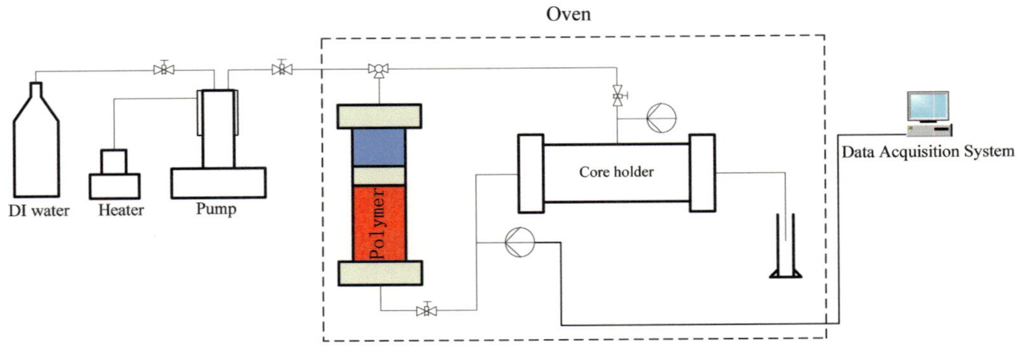

Fig. 1. Schematic of polymer flooding experiment.

4.2 Experimental procedures

1. Prepare a core or sandpack, and measure their weight;
2. Vacuum the core/sandpack for half an hour to one day, depending on permeability;
3. Saturate core with 4% KCl solution or a synthesized formation water;
4. Measure the weight of saturated core and calculate the effective porosity of the core;
5. Inject same brine to measure the core/sandpack permeability;
6. Saturate oil for the core, and calculate original oil saturation;
7. Inject same brine to displace oil until no oil comes out, and calculate residual oil saturation;
8. Inject polymer solution until a stabilized pressure can be seen at a given flow rate;
9. Continue injecting polymer solution at different flow rates and stabilized pressures as a function of injection rate are recorded.

Steps 6 and 7 can be ignored if the purpose of an experiment is only to test the rheology behavior of a polymer solution in porous media.

5. Factors impact the flow behaviors of polymers in porous media

5.1 Polymer type

Hydrolyzed Polyacrylamide (HPAM), Xanthan gum and associative polymers are the most frequently used EOR polymers, and they are all water soluble rather than hydrocarbon soluble.

5.1.1 Hydrolyzed Polyacrylamide (HPAM)

A HPAM solution is the most widely used polymer in oil field application. HPAM is a synthetic linear copolymer of acrylic acid and acrylamide (nonionic) monomers, with negative charges in the carboxylate groups. Some amide groups ($CONH_2$) would be replaced by carboxyl groups (COO^-) during the hydrolysis process and, thus, have strong interactions with cations. Through reacting polyacrylamide with a base, such as sodium, potassium, etc., the adsorption on a solid surface can be reduced. The degree of hydrolysis is the mole fraction of amide groups that are converted by hydrolysis. It ranges from 15% to 35% in commercial products. The chemical structure of HPAM is demonstrated in Figure 2.

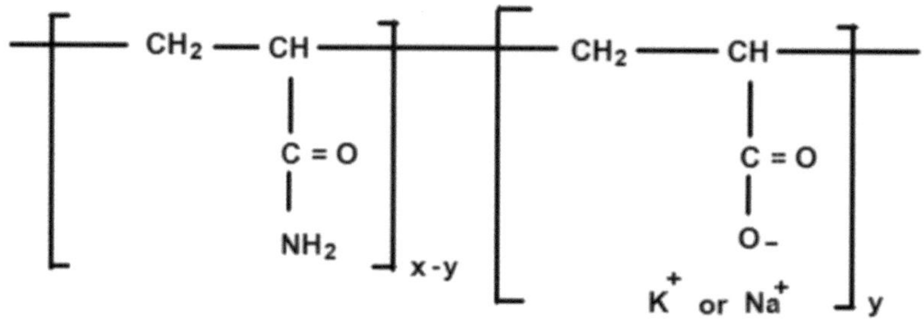

Fig. 2. Chemical structure of HPAM (Aluhwal and Kalifa 2008)

Hydrolysis of polyacrylamide introduces negative charges on the backbones of polymer chains that have a large effect on the rheological properties of a polymer solution (Sheng 2010). When HPAM is dissolved in distilled water or low salinity water, unshielded electrostatic repulsion exists between the anionic groups along the polymer chains, and the polymer molecules expand extensively. Therefore, the polymer chain and molecules are easy to stretch. This also explains that when HPAM solution is prepared in distilled water and flux at moderate to high shear rate, little or no shear thickening is seen (Seright, Fan et al. 2011). Once the HPAM molecular encounters the electrolyte solution, such as NaCl, the electrostatic repulsion is shielded by a double layer of electrolytes. Therefore, the stretch between the polymer chain and the molecules is decreased and a low viscosity is exhibited at a high salt concentration.

5.1.2 Xanthan gum

Xanthan gum, usually shortened to Xanthan, is another widely used polymer in oil fields. Xanthan is a microbial biopolymer, produced by the fermentation of glucose, sucrose, or lactose by the Xanthomonas campestris bacterium that are present in Xanthan in D-glucose, D-mannose, and D-glucuronic acid, as shown in Figure 3.

Xanthan exhibits high viscosity at a low shear force. Its shear thinning recovers rapidly once the shearing force is removed, and has a high resistance to mechanical shear degradation. Its good shear stability and thickening power at high salinity are major advantages of Xanthan over HPAM. The major disadvantages of this biopolymer are its high cost, the difficulty to prepare a uniform solutions and make it do not plug the porous media, and the viscosity loss during possible biochemical or chemical reactions (Wellington 1983).

5.1.3 Hydrophobically associative polymer

Water soluble hydrophobically associative polymers have developed very quickly in recent years, in order to possibly substitute them for HPAM and Xanthan polymers in oil field applications (Taylor and Nasr-El-Din 2007). Associative polymers have very similar structure with the conventional polymers used in the oil industry, which is usually water soluble, except that they have a small number of hydrophobic groups attached directly into the polymer backbone (carbon chain) as shown in Figure 4. These hydrophobic groups can significantly change polymer performance, even at levels of incorporation of less than 1 mol%. During polymer flooding, the associative polymers show more stable fronts, as compared with conventional polymer solutions.

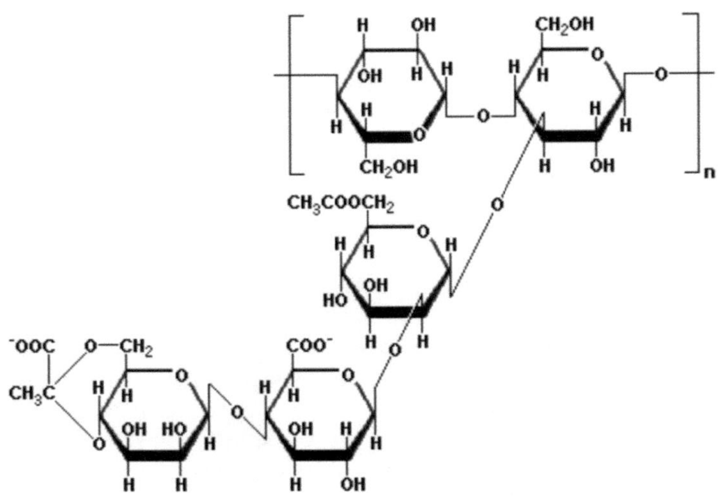

Fig. 3. Chemical structure of Xanthan Gum (Zaitoun, Makakou et al. 2011).

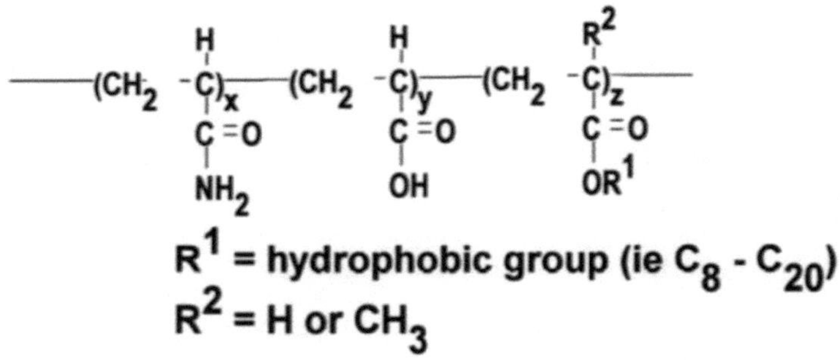

Fig. 4. Associating Acrylamide Copolymer (Taylor and Nasr-El-Din 2007)

Many hydrophobically associative polymers have been developed: SNF Floerger-S255, C1205; I-20; HMpolyDMAEMA; AP-P3; AP-P4, and etc. Different polymers have diverse characteristics. Lab tests and field applications show that one specific associative polymer may have one or two special characteristics that are better than those of HPAM or Xanthan gum. Take SNF Floerger-C1205 for instance, C1205 is an anionic-polyacrylamide-based tetra-polymer which has associative properties (Gaillard and Favero 2010). Typically, it has a hydrophobic monomer content range from 0.025 to 0.25 mol%, a 12 to 17 million g/mol

Molecular weights (Mw) and a 15 to 25 mol% total anionic content. Sulfonic monomer is present with less than 8 mol%. HPAM and C1205 (called AP in the Figure) are prepared with concentrations of 500, 900, 1,500, and 2,500 ppm in 2.52 % of the total dissolved solid (TSD) brine. Experiments are conducted in porous polyethylene cores. As shown in Figure 5, the resistance factors of the associative polymer (at a given flux) are substantially greater than those of HPAM. The associative polymer displays a similar trend with that of HPAM, with shear thinning at a low shear rate, and then changes to shear thickening at a high shear rate. The low flux resistance factors for associative polymer are around twice those of HPAM. Specifically, 500 ppm of the associative polymer (solid circles) behaves in a manner like that of 900 ppm of HPAM (open triangles). 900 ppm of the associative polymer (solid triangles) conducts like that of 1,600 ppm of HPAM (open squares) (Seright, Fan et al. 2011).

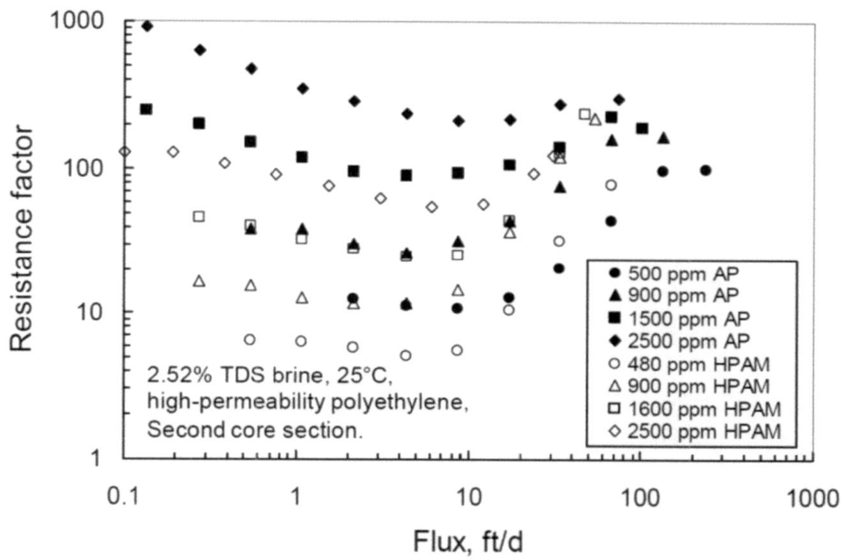

Fig. 5. Resistance factors in polyethylene: Associative polymer C1205 vs. HPAM (Seright, Fan et al. 2011).

However, when exposed to 235 psi/ft, the resistance factor losses for the associative polymer are significant—from 31 to 45%, while it is modest for HPAM—from 0 to 15%. This tells us that this associative polymer could behave in a manner similar to that of the HPAM with half concentration, but the resistance factor losses are relatively higher, as compared with those of HPAM.

Therefore, the polymer should be selected with consideration being given to its in situ geological and fluid properties, in order to make it fit and be cost effective.

5.2 Shear rate

When a polymer solution is injected into a reservoir from an injection well, the flow rate, which is related to shear rate, will change from wellbore to in-depth of a reservoir; therefore, the solution viscosity will also change from near wellbore to the in-depth of a reservoir correspondingly.

Masuda, Tang et al. (1992) conducted a series of experiments using 2.8 cm diameter, 47 cm long, unconsolidated cores packed with glass beads (70/100 mesh). Porosity and permeability are tested to be 37% and 25 μm², respectively. Two white mineral oils with viscosities of 25, 60 mPa·s, and a 200 ppm HPAM solution were employed as working fluids. In each experiment, polymer flooding was carried out after water flooding. Initial water saturation was controlled to be almost the same level at the start of each polymer flooding. The relationship of HPAM solution viscosity vs. shear rate measured in porous media and by viscometer is plotted in Figure 6.

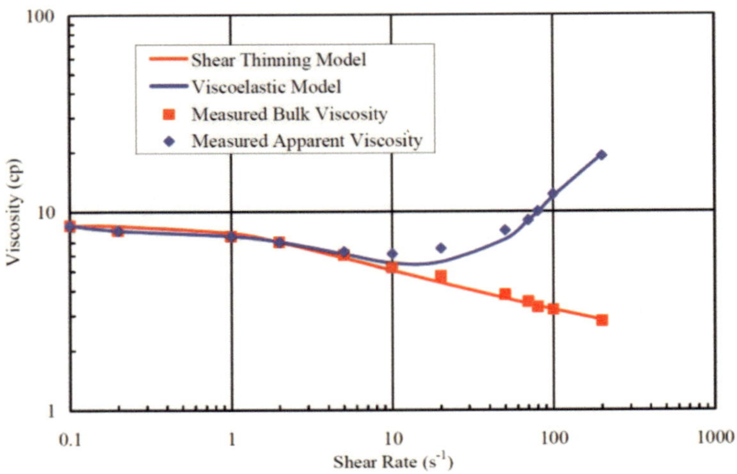

Fig. 6. Viscosity vs. shear rate of HPAM measured in porous media and by viscometer (Delshad, Kim et al. 2008).

The red box and blue diamond are bulk viscosity and apparent viscosity, acquired by viscometer and coreflooding experiments, respectively. Delshad, Kim et al. (2008) successfully matched their results with the shear thinning model and viscoelastic model as given in Figure 6.

At low velocities in a porous media, a Newtonian, near-Newtonian or mild shear thinning is observed, and the bulk viscosity has the same trend, as measured by a capillary viscometer. With the shear rate increasing, the bulk viscosity curve starts to decline, displaying an obvious shear thinning, while the apparent viscosity in the porous media starts to deviate from the declining bulk viscosity. When the shear rate continues extending, the bulk viscosity follows the same decreasing slope, but the apparent viscosity in porous media starts increasing and displays a shear thickening (Masuda, Tang et al. 1992). This behavior of HPAM solution is also confirmed by other authors (PYE 1964; Smith 1970; Jennings, Rogers et al. 1971; Hirasaki and Pope 1974; Seright 1983; Heemskerk, Rosmalen et al. 1984; Delshad, Kim et al. 2008; Seright, Seheult et al. 2008; Seright, Fan et al. 2011).

However, there is no consensus formula which could give the best prediction on the relationship of in situ viscosity and in situ shear rate. Generally, we need data both from viscometer and coreflooding experiments. Viscometer data is easy to obtain. If we do not have coreflooding data, we will need to match other available data by adjusting the shear rate coefficient: C in Equation 12. Gogarty (1967) stated that the shear rate coefficient is related to velocity, permeability, and porosity. Teeuw and Hesselink (1980) found that the shear rate coefficient is a function of the power law coefficient and exponent in the viscometer, or the geometry of the porous medium. Cannella, Huh et al. (1988) stated that the shear rate coefficient is directly related to the ratios of the effective capillary tube radii which used to model the non-Newtonian fluid in porous media. From their findings, it is still not clear which variables keep a major impact on the shear rate coefficient (Wreath, Pope et al. 1990).

5.3 Molecular weight

Molecular weight is the ratio of the mass of the polymer molecular to 1/12th of the mass of carbon-12 (Elias 2008), used to describe the constitutional size of a macromolecule substance. Its unit is Dalton, symbol: Da, g/mol is also used, 1 Da = 1 g/mol. Molecular weight can be determined using mass spectrometry.

Molecular weight is the main factor that affects the flow behavior of a polymer. The molecular weight of a polymer is directly related to its molecule size, which means the polymer with higher molecular weight has a larger molecular size. The polymer solutions with higher molecular weight usually have higher adsorption, higher resistance, and higher viscosity, as shown in Figure 7 (Dong, Fang et al. 2008).

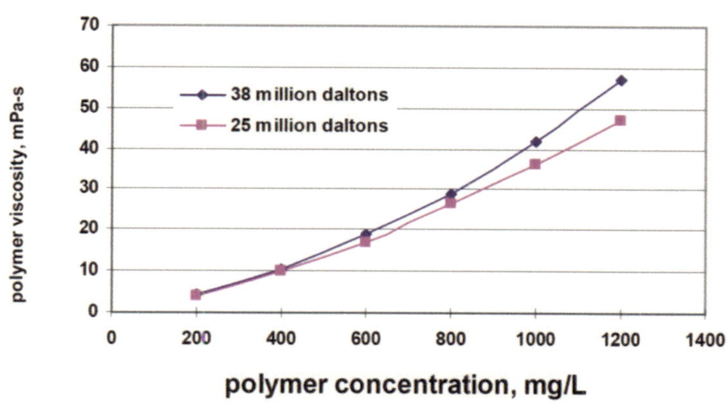

Fig. 7. A polymer's apparent viscosity as a function of molecular weight and concentration (Dong, Fang et al. 2008).

In many circumstances, a larger polymer Mw also leads to higher oil recovery. Dong et al (2008)'s coreflooding experiments shown in Table 2 verify this expectation. In these experiments, water flooding was first used, and their recoveries were almost the same (32 to 33%). Then different molecular weight polymers were applied to improve oil recovery. Total injected polymer mass was 570 mg/L·PV. Polymer concentration was 1,000 mg/L. Each core had three zones with a heterogeneity coefficient of 0.72. As shown in Table 2, the polymer with the molecular weight of 18.6 million Daltons had the highest ultimate recovery.

But for the reservoirs with middle to low permeability, higher molecular weight needs high injection pressure and also might cause more unswept pore volume, and thus resulting in lower recover factor shown as Figure 8 (Zhang, Pan et al. 2011).

Mw, 10^6 Daltons	Water flooding recovery, %	Polymer flooding recovery %	Ultimate recovery, %
5.5	32.7	10.6	43.3
11	32.9	17.9	51.8
18.6	32.2	22.6	54.8

Table 2. Effect of polymer molecular weight (Mw) on oil recovery (Dong, Fang et al. 2008)

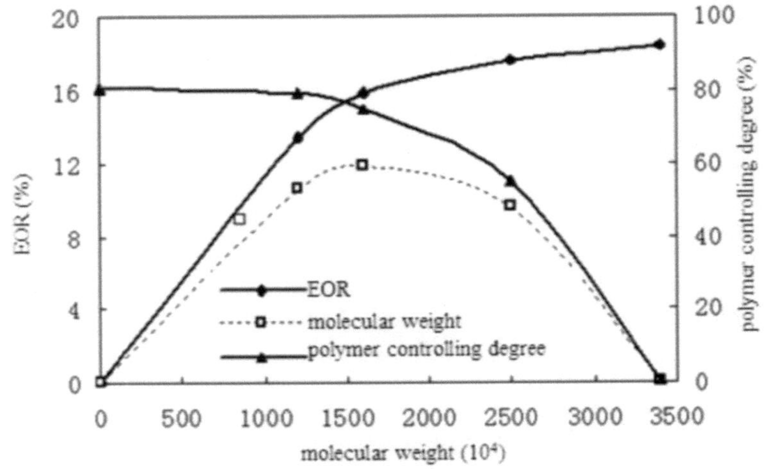

Fig. 8. Molecular weight as a function of enhanced recovery rate and controlling degree (Zhang, Pan et al. 2011).

Other previous studies have shown that the high Mw polymer solutions are known to be shear sensitive (Maerker 1976; Martin 1986). Due to the breakage of macromolecule chain, shear degradation is irreversible. Zaitoun (2011) tested six polymers and co-polymers with different Mw from 2 to 18.5 million Daltons through a 5 cm long stainless steel capillary tube with a 125 μm internal diameter. The results are shown that an increase in Mw induces higher shear sensitivity. Since a large coiled macromolecule is stretched more under the same deformation regime.

However, some factors should be taking into account when choosing the polymer molecular weight. Firstly, polymer with highest Mw could minimize the amount of polymer. Secondly, the relationship between Mw and permeability should be taken in account. Thirdly, the Mw must be small enough so that the polymer can enter and propagate effectively through the pore throat and channels (Dong, Fang et al. 2008), moreover the lower Mw may be a better option for field application because very high Mw polymers may lose most of their viscosifying power when submitted to shear degradation (Zaitoun, 2011).

5.4 Polymer concentration

The concentration of a polymer can either be expressed as parts per million (ppm), pound per thousand gallon (pptg), or mg/L, although ppm is the most frequently used term.

Aluhwal and Kalifa (2008) conducted series of experiments with various concentrations of HPAM solution through sandpack at 77 °C. Results are shown in Figure 9. They measured the pressure and calculated the apparent viscosity. In porous media, for a HPAM solution at low concentration (<1,000 ppm), the apparent viscosity slightly increases with the polymer concentration increasing. While at a high concentration (>1,500 ppm), however, the apparent viscosity is increasing at a faster rate.

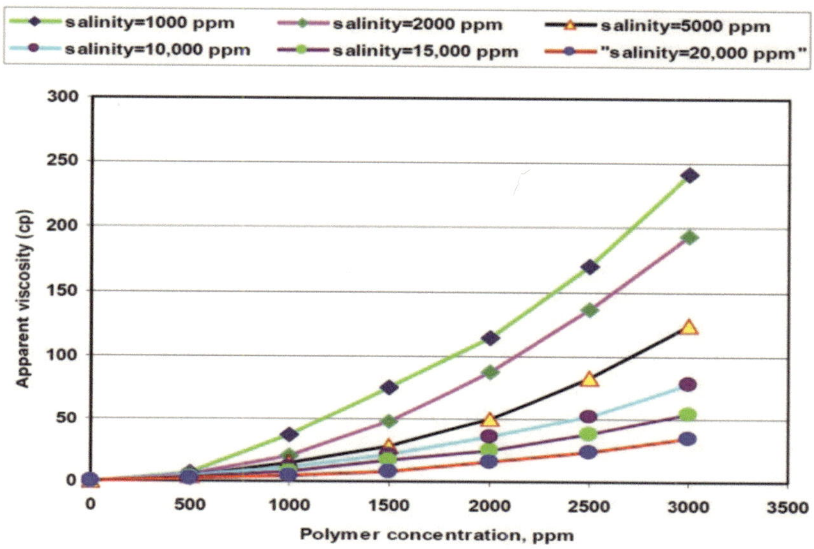

Fig. 9. A polymers' apparent viscosity as a function of concentration and salinity (Aluhwal and Kalifa 2008).

However, the rheological behaviors of associative polymers can be divided into a dilute region and a semi-dilute region by a critical concentration: c*. At concentrations below c*, it is in a dilute region where intramolecular hydrophobic associations within the polymer dominate the behavior of the polymer, and the viscosity decreases with the solution concentration increase. At a concentration above c*, it is in a semi-dilute region where intermolecular associations between polymers control the polymer rheology and the viscosity increases with the solution concentration increases (Taylor and Nasr-El-Din 2007).

Generally, the polymer concentration has a direct relationship with the apparent viscosity where an increase in the polymer concentration results in an apparent viscosity increase.

5.5 Salinity

Beside concentration, the salinity of water used to prepare a polymer solution also has a major impact on apparent viscosity. A change in the quantity of water directly affects the polymer solution's viscosity. As mentioned above, when a polymer, like HPAM, is exposed to different salinity water, its anionic and cationic in the water would have attraction/repulsion to the polymer chain, and make it compressed or stretched. When it exhibits to the distilled or low salinity water, the electrostatic repulsion between anionic groups along the polymer chains would be unshielded, make the polymer molecules expand extensively. Then it is relative hard to pass through a porous medium and generate a relative high viscosity. When it encounters the higher salinity water, electrostatic repulsion is shielded by a double layer of electrolytes. Therefore, the stretch between the polymer chain and the molecules is decreased and a low viscosity is produced at a high salt concentration.

The electrostatic interaction between the charged groups plays an important role in adsorption as well as the permeability reduction phenomena (Aluhwal and Kalifa 2008). Figure 9 above also illustrates that salinity has a relatively strong effect on the polymer's apparent viscosity. A higher salinity would generate a lower apparent viscosity and with an increase in polymer concentration, this effect is strengthened.

5.6 Permeability

Using two Berea sandstone cores (55 and 269 md permeability) and one porous polyethylene core (5120 md permeability with a pore structure similar to Berea sandstone), Seright (2011) conducted experiments with 600 ppm Xanthan in 2.52% TDS solution with a wide injection rate range (0.0174 to 1111 ft/D). Results are shown in Figure 10, which indicates that cores with high permeability have a higher resistance factor, and it continuously decrease with the extension of velocity, while the resistance factor difference between the high permeable cores and low permeable cores is narrowing. At a flux above 100 ft/D, the resistance factors of 55 md and 269 md sandstone cores show almost no difference.

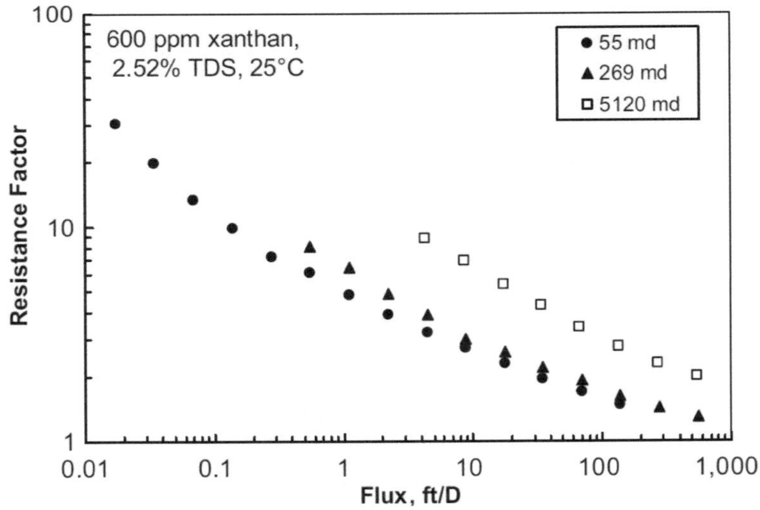

Fig. 10. Resistance factor vs. flux for 600 ppm of Xanthan with different permeability cores (Seright, Fan et al. 2011).

6. Summary

In this chapter, we have discussed some basic rheological properties of polymer solution flow in porous media, such as rheological terms, shear rate models, experiment measurement and impact factors. Apparent viscosity, effective viscosity and resistance factor are the basic terms that can be used to estimate the polymer flow behavior in a porous media. Based on capillary bundle model, five mathematical shear rate models are presented with their prerequisite conditions and assumptions. Then experimental method and flow chart are detailed. The factors that impact polymer solution rheology in porous media were reviewed, including polymer types, shear rate, molecular weight, concentration, salinity, and formation permeability.

7. Acknowledgement

Financial support from the China Scholarship Council, Libyan Ministry of Higher Education and Research Partnership to Secure Energy for America (RPSEA) is gratefully acknowledged.

8. References

Aluhwal, H. and O. Kalifa (2008). Simulation study of improving oil recovery by polymer flooding in a Malaysian reservoir. Master, Universiti Teknologi Malaysia.

Cannella, W. J., C. Huh, et al. (1988). Prediction of Xanthan Rheology in Porous Media. SPE Annual Technical Conference and Exhibition. Houston, Texas.

Carman, P. C. (1937). "Fluid flow through granular beds." Transactions of the Institution of Chemical Engineers 15: 150-166.

Chauveteau, G. and A. Zaitoun (1981). "Basic rheological behavior of xanthan polysaccharide solutions in porous media: Effects of pore size and polymer concentration." Enhanced oil recovery: proceedings of the third European Symposium on Enhanced Oil Recovery: 197-212.

Christopher, R. H. and S. Middleman (1965). "Power-Law Flow through a Packed Tube." Industrial & Engineering Chemistry Fundamentals 4(4): 422-426.

Delshad, M., D. H. Kim, et al. (2008). Mechanistic Interpretation and Utilization of Viscoelastic Behavior of Polymer Solutions for Improved Polymer-Flood Efficiency. SPE/DOE Symposium on Improved Oil Recovery. Tulsa, Oklahoma, USA, Society of Petroleum Engineers.

Dong, H., S. Fang, et al. (2008). Review of Practical Experience & Management by Polymer Flooding at Daqing. SPE/DOE Symposium on Improved Oil Recovery. Tulsa, Oklahoma, USA.

Elias, H. G. (2008). Macromolecules: Volume 3: Physical Structures and Properties, Wiley-VCH.

Gaillard, N. and C. Favero (2010). High molecular weight associative amphoteric polymers and uses thereof, Google Patents.

Gogarty, W. B. (1967). "Mobility Control With Polymer Solutions." SPE Journal 7(2).

Heemskerk, J., R. Rosmalen, et al. (1984). Quantification of Viscoelastic Effects of Polyacrylamide Solutions. SPE Enhanced Oil Recovery Symposium. Tulsa, Oklahoma.

Hirasaki, G. J. and G. A. Pope (1974). "Analysis of Factors Influencing Mobility and Adsorption in the Flow of Polymer Solution Through Porous Media." (08).

Jennings, R. R., J. H. Rogers, et al. (1971). "Factors Influencing Mobility Control By Polymer Solutions." SPE Journal of Petroleum Technology(03).

Kozeny, J. (1927). "Über kapillare Leitung des Wassers im Boden." Sitzungsber Akad. Wiss., Wien 136(2a): 271-306.

Littmann, W. (1988). Polymer flooding, Elsevier Publishing Company.

Maerker, J. M. (1976). "Mechanical Degradation of Partially Hydrolyzed Polyacrylamide Solutions in Unconsolidated Porous Media." (08).

Martin, F. D. (1986). "Mechanical Degradation of Polyacrylamide Solutions in Core Plugs From Several Carbonate Reservoirs." SPE Formation Evaluation(04).

Masuda, Y., K.-C. Tang, et al. (1992). "1D Simulation of Polymer Flooding Including the Viscoelastic Effect of Polymer Solution." SPE Reservoir Engineering(05).

PYE, D. J. (1964). "Improved Secondary Recovery by Control of Water Mobility." SPE Journal of Petroleum Technology(08).

Seright, R. S. (1983). "The Effects of Mechanical Degradation and Viscoelastic Behavior on Injectivity of Polyacrylamide Solutions." SPE Journal 23(3).

Seright, R. S., T. Fan, et al. (2011). "New Insights Into Polymer Rheology in Porous Media." SPE Journal(03).

Seright, R. S., T. Fan, et al. (2011). Rheology of a New Sulfonic Associative Polymer in Porous Media. SPE International Symposium on Oilfield Chemistry. The Woodlands, Texas, USA.

Seright, R. S., J. M. Seheult, et al. (2008). Injectivity Characteristics of EOR Polymers. SPE Annual Technical Conference and Exhibition. Denver, Colorado, USA.

Sheng, J. (2010). Modern Chemical Enhanced Oil Recovery: Theory and Practice, Gulf Professional Publishing.

Smith, F. W. (1970). "The Behavior of Partially Hydrolyzed Polyacrylamide Solutions in Porous Media." SPE Journal of Petroleum Technology(02).

Sorbie, K. S. (1991). Polymer-Improved Oil Recovery. Blackie and Son Ltd.

Taylor, K. C. and H. A. Nasr-El-Din (2007). Hydrophobically Associating Polymers for Oil Field Applications. Canadian International Petroleum Conference. Calgary, Alberta.

Teeuw, D. and F. T. Hesselink (1980). Power-law flow and hydrodynamic behaviour of biopolymer solutions in porous media. SPE Oilfield and Geothermal Chemistry Symposium. Stanford, California, 1980,. American Institute of Mining, Metallurgical, and Petroleum Engineers, Inc.

Wellington, S. L. (1983). "Biopolymer Solution Viscosity Stabilization - Polymer Degradation and Antioxidant Use." (12).

Wreath, D., G. Pope, et al. (1990). "Dependence of polymer apparent viscosity on the permeable media and flow conditions." In Situ;(USA) 14(3).

Zaitoun, A., P. Makakou, et al. (2011). Shear Stability of EOR Polymers. SPE International Symposium on Oilfield Chemistry. The Woodlands, Texas, USA.

Zhang, X., F. Pan, et al. (2011). A Novel Method of Optimizing the Moelcular Weight of Polymer Flooding. SPE Enhanced Oil Recovery Conference. Kuala Lumpur, Malaysia, Society of Petroleum Engineers.

Zitha, P., G. Chauveteau, et al. (1995). Permeability~Dependent Propagation of Polyacrylamides Under Near-Wellbore Flow Conditions. SPE International Symposium on Oilfield Chemistry. San Antonio, Texas.

Part 3

Drilling Fluids

9

Stability and Flow Behavior of Fiber-Containing Drilling Sweeps

Matthew George[1], Ramadan Ahmed[1] and Fred Growcock[2]
[1]University of Oklahoma,
[2]Occidental Oil & Gas Corp.,
USA

1. Introduction

In the oil and gas industry, drilling sweeps are used to improve borehole cleaning when conventional drilling fluid fails to or is suspected of failing to sufficiently clean the wellbore. They are often applied immediately prior to tripping operations to clean the wellbore and reduce excessive annular pressure losses. The sweeps remove cuttings that cannot be transported to the surface during normal fluid circulation while drilling and provide additional vertical lift to the cuttings. Sweeps can be performed in all well inclinations from vertical to horizontal, as required by wellbore conditions. Drilling sweeps are an effective tool for counteracting poor hole cleaning, which can lead to an increase in non-productive time and costly drilling problems such as stuck pipe, premature bit wear, slow rate of penetration, formation fracturing, and high torque and drag (Ahmed & Takach, 2008).

Multiple field-tested techniques have been introduced over the years to prevent and reduce the proclivity of drilled cuttings to settle within the wellbore, therefore enhancing cuttings transport and improving hole cleaning. Previous studies indicate that cuttings transport in directional wells is dependent on fluid rheology, wellbore inclination angle, rotary speed of the drillpipe, flow rate, wellbore geometry, and other drilling parameters (Valluri et al., 2006). Considering these factors, the easiest and most economical procedures to improve hole cleaning involve increasing annular velocity preferably into the turbulent flow regime or using sweeps of higher density or viscosity than the drilling fluid. Sweeps are commonly categorized as i) high-viscosity; ii) high-density; iii) low-viscosity; iv) combination; and v) tandem sweeps (Hemphill, 2002). Sweeps provide a number of benefits such as reducing cuttings beds – thus decreasing annular pressure loss – and reducing torque and drag at the surface. Increasing the flow rate also provides the sweeps with extra lifting potential, though this must be closely monitored, as the details depend on the characteristics of the well. Pressure losses along the wellbore, as well as the equivalent circulating density (ECD), must be considered when designing and applying sweep fluids. Hole cleaning fiber added to the sweep fluid is thought to improve its performance with a negligible increase in viscosity and pressure loss (George et al., 2011). Previous studies (Marti et al., 2005; Rajabian et al., 2005; Guo et al., 2005) showed similar trends, with the apparent viscosity of fiber-polymer suspensions increasing linearly but slightly with increasing fiber concentration up to an approximate critical fiber concentration. At this critical concentration, the apparent viscosity began to increase rapidly and exponentially with only small increases in fiber concentration. For field applications and for this study, fiber concentrations are maintained well below the critical threshold.

Reports from previous experimental studies (Ahmed & Takach, 2008) and field applications (Cameron et al., 2003; Bulgachev & Pouget, 2006) noted that adding a small amount of synthetic monofilament fiber (less than 0.1% by weight) improved solids transport and hole cleaning efficiency of the sweep over comparable non-fiber sweeps, with no noticeable change in fluid rheology. This favorable performance may be attributed to fiber-fiber and fiber-fluid interactions that create a stable network structure which can support cuttings. The fiber-fiber interactions can be in the form of direct mechanical contact and/or hydrodynamic interference among fiber particles. Mechanical contact among fibers improves the solids-carrying capacity of the fluid (Ahmed & Takach, 2008), while mechanical contact between the fibers and the cuttings beds helps to re-suspend cuttings deposited on the low side of the wellbore. As the fibers flow through the annulus, mechanical stresses develop between the fibers and settled cuttings. These mechanical stresses result in a frictional force which helps to re-suspend the cuttings, while the fiber networks carry the solids to the surface of the hole. Due to the fiber-fiber interaction, the bulk fiber network may move as a plug through the annulus. With the increased fluid velocity in the high side of the annulus, the fiber plug may move at a higher velocity relative to the local conditions at the low side of the annulus near the drillpipe. These fast moving fibers can transfer momentum to the deposited solids, overcoming the static frictional forces and initiating particle movement.

Although application of fiber suspensions in the oil and gas industry has been limited, they are common in other industries. Non-Brownian fiber suspensions have wide application in the pulp and paper industry, as well as in the manufacture of composite materials. These applications generally require uniform distribution of fiber particles throughout the suspending medium, which are often short and rigid and oriented parallel to the direction of motion or deformation. Similar to drilling fluid sweeps, these suspensions exhibit non-Newtonian characteristics such as the Weissenberg effect (Nawab & Mason, 1958; Mewis & Metzner, 1974), shear thinning (Goto et al., 1986), and viscoelasticity (Thalen & Wharen, 1964). The rheological properties of these non-Brownian fiber suspensions depend on the structure of the suspensions, which in turn is affected by features such as the fiber properties, interactions, suspending fluid properties, and the imposed flow field (Switzer & Klingenberg, 2003, 2004).

For drilling applications of fiber sweeps, some fiber flexibility is thought to be desirable, in order that the suspended fibers will not uniformly orient themselves while flowing up the wellbore annulus. In a sheared flow of highly-flexible, large-aspect-ratio fibers, the fibers flip, frequently collide with each other and the container walls, and orient themselves randomly. In fluids that contain a sufficient concentration of suspended fibers, the contacting, colliding fibers form heterogeneous structures, or networks, that flocculate and exhibit viscoelastic properties (Meyer & Wahren, 1964). Sozynksi & Kerekes (1988a,b) implicated mechanical contact as the principal mechanism in the formation of fiber "flocs." They suggested that fibers become locked in strained configurations due to their elasticity and frictional forces at fiber contact points, and referred to the fiber flocculation mechanism as "elastic fiber interlocking". A yield stress is also generated within the fiber suspension, implying that a sufficient force must be applied to initiate flow (Meyer & Wahren, 1964). Bennington et al. (1990) measured the yield stress of various wood and nylon fiber suspensions. The yield stress (τ_y) measurements were directly influenced by volume fraction (Φ) as $\tau_y \sim \Phi^{\beta}$. The measured values of β varied with the fiber elasticity and aspect ratio, which was especially evident in comparisons between stiff wood fibers and relatively flexible nylon fibers. These frictional, or adhesive, contacts between fibers were believed to give rise to shear thinning behavior of suspensions of short nylon fibers in silicone oils at very low shear rates (Chaouche & Koch, 2001). This speculation is consistent with the idea that shear thinning arises when the torque associated with adhesive fiber-fiber contacts is comparable with the hydrodynamic torque rotating the gross flow of the particles. It may be noted that fiber flexibility (bending) and the base fluid's non-Newtonian behavior exert greater influence at high shear rates and with high-viscosity solvents, while fiber-fiber adhesion is most influential at low shear rates and with low-viscosity solvents (Chaouche & Koch, 2001).

Several studies have focused on simulating flexible fiber suspensions. Ross & Klingenberg (1997) modeled flexible fibers as inextensible chains of rigid prolate spheroids connected through ball and socket joints. This model can represent large aspect ratio fibers. Schmid et al. (2000) suggested that flexible fibers interact as chains of spherocylinders connected by ball and socket joints. This study also demonstrated that inter-fiber friction, even in the absence of attractive forces, can produce fiber flocculation (**Fig. 1**). While this and a subsequent study (Schmid & Klingenberg, 2000) were primarily concerned with the flow of non-Brownian fiber suspensions such as that thought to occur with pulp slurries in the paper-pulp industry, it is useful to understand the mechanisms which lead to flocculation. An important conclusion of those simulation studies (Schmid et al., 2000), which contradicts our work, is that fibers can flocculate when sheared. While this flocculation may enhance wellbore clean-up, the suspensions we tested generally dispersed when sheared and only flocculated under quiescent conditions when the fibers were allowed to separate and come together at the top of the vessel.

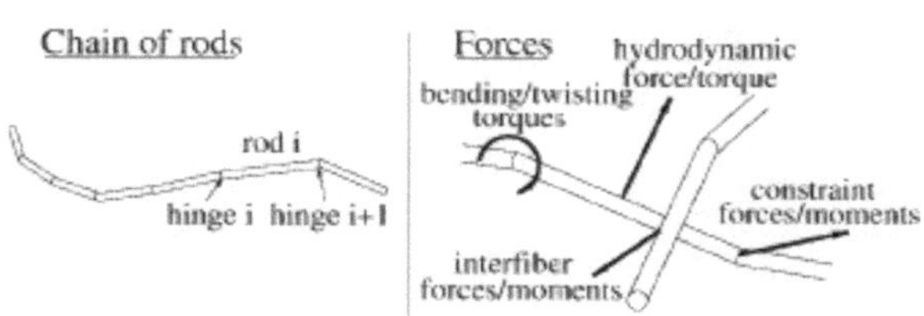

Fig. 1. Schematic of contacting fibers and forces (Schmid & Klingenberg, 2000)

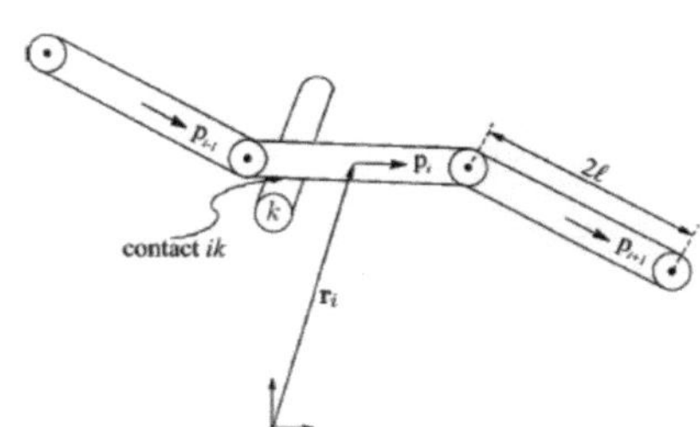

Fig. 2. Model of flocculated fiber particles that experience mechanical contact (Switzer & Klingenberg, 2003)

The fiber suspension modeling work was expanded by Switzer & Klingenberg (2003, 2004), who modeled flexible fiber suspensions as neutrally buoyant chains of linked rigid bodies immersed in a Newtonian fluid (**Fig. 2**). This model includes realistic features such as fiber flexibility, irregular shapes, and mechanical contact forces between fibers. The model also considers curvature, that is the degree of sustained flexibility of a fiber particle, which is measured from a straight, rigid baseline form. The study re-affirmed the conclusion of the prior simulation studies that flocculation can arise from friction alone, in the absence of other forces, and that fiber flocculation increases with decreasing fiber curvature.

This study is concerned with the rheology of fiber-containing fluids specific to the oil and gas industry, and the propensity of the fiber to separate within the suspending medium, both of which provide insight into the flow behavior and effectiveness of fiber sweeps in the wellbore. The literature suggests that at critical concentrations, the fiber particles come into contact with each other and form networks. The strength of these networks depend on the prevailing fiber orientation and fiber flexibility. The non-rigid fiber particles dispersed and oriented randomly throughout the suspension provide the maximum opportunity for fiber entanglement, therefore increasing the cuttings resuspension and carrying capacity in the wellbore.

2. Rheology of fiber containing fluid

Controlling the rheology of the drilling and sweep fluids is essential to maintain favorable wellbore hydraulics and hole cleaning efficiency. This is of utmost importance when drilling extended and ultra-extended reach wells in deep water, where the slight difference between the formation fracture pressure and pore pressure requires a minimum overbalanced wellbore pressure condition. In such environments, the pressure and temperature ranges rise to levels that are difficult to emulate in laboratory experiments and predict precisely the rheology of the fluids downhole.

To predict the transport properties and performance of fiber sweeps under downhole conditions, the rheological properties of the base fluid and suspension must be understood. The proposed formulations for such fiber sweeps will be most effective when the rheology has been accurately modeled and fine-tuned for specific wellbore conditions. To begin to grasp how the fluid behaves, the relationship between shear stress and shear rate must be known. This is denoted as the shear viscosity profile, which is an aspect of the rheology of a fluid that is thought to control the hydrodynamics of flow. The most common shear viscosity models used in the oil and gas industry to characterize non-Newtonian drilling fluids include:

- Bingham-Plastic (BP): $\tau = YP + PV\,\gamma$
- Power Law (PL): $\tau = K\gamma^n$
- Yield Power Law (YPL) or Herschel Bulkley: $\tau = \tau_y + K\gamma^n$

where τ = shear stress at the wall, γ = shear rate, YP = yield point, PV = plastic viscosity, K = consistency index, n = fluid behavior index, and τ_y = yield stress.

As the drilling fluid must have sufficient cuttings-carrying capacity to keep cuttings from accumulating in the wellbore, the fluid must be capable of suspending the cuttings under quiescent conditions, i.e. its yield stress must exceed the settling velocity of the cuttings. The classic viscosity model used for drilling fluids is the Bingham-Plastic or viscoplastic model. Here the shear stress rises linearly with shear rate, with a slope given by PV. The intercept on the τ axis, YP, is often identified with the carrying capacity of the fluid but does not adequately describe the viscosity profile at very low shear rates. Most drilling fluids exhibit a non-linear shear stress-shear rate relationship, which is best described by the Yield Power Law model. The YPL model is useful in describing a wide range of polymer-based, oil-based, and synthetic-based drilling fluids, from low shear rate to high shear rate. The yield stress (τ_y) defines the critical shear stress that must be reached before flow can initiate.

Recently, the shear viscosity profiles of synthetic-based drilling fluids were measured from 27°C to 138°C and from 1 bar to 346 bar (Demirdal et al., 2007). The study showed the viscosity to be extremely sensitive to downhole conditions, with the yield stress and consistency index drastically changing with temperature and pressure. The overall trend was that these parameters decreased with increasing temperature, and increased with increasing pressure. The evaluation also showed that the Yield Power Law model continued to describe adequately the shear stress-shear rate relationship over the entire range of temperature and pressure tested regimes. In another laboratory study (Yu et al., 2007), cuttings transport efficiency of drilling fluid was measured at elevated temperatures and pressures (up to 93°C and 138 bar). The experimental trend showed that higher temperatures diminished the cleaning efficiency of the fluid. Both of these studies suggest that as the fluid thins with increasing temperature, the amount of momentum transferred to the cuttings diminishes. The thinner fluid suspensions may exhibit a greater phase separation tendency, as it loses its ability to maintain a uniform fiber concentration while flowing in the annulus. This phase separation diminishes the quality of the hole cleaning provided by the fiber.

In designing a fiber-fluid formulation for wellbore cleaning sweeps, certain viscosity parameters can serve as indicators of how well the sweep will perform. The yield stress of the fluid represents the amount of force required to deform the fluid. At the same time, if the fluid possesses adequate yield stress to counteract the natural buoyancy of the fiber, the fiber may not separate. The yield stress may provide a good indication of how well the sweep will maintain uniformity when circulating up the annulus. However, the yield stress of the fluid does not tell the whole story, nor does shear viscosity itself provide a comprehensive portrait of the fluid. The elastic, extensional and thixotropic properties of the fluid very likely affect a fluid's ability to suspend and transport cuttings. Furthermore, in suspensions with sufficient fiber concentration, the fibers themselves can affect the fluid's rheology by forming networks in which each fiber maintains multiple contacts with adjacent fibers. These networks exhibit mechanical strength and viscoelastic behavior (Thalen & Wahren, 1964), which are controlled by the cohesive nature of the contact points (Switzer & Klingenberg, 2003). These cohesive forces are a result of the friction generated by normal forces at the contact points between elastically bent fibers (Kerekes et al., 1985). Soszynksi & Kerekes (1988a,b) proposed that these networks are the result of mechanical contacts, which are created due to "elastic fiber interlocking", a result of fiber elasticity and the fiber contact friction forces.

Multiple studies of the viscosity of fiber suspensions have reported the significance of fiber properties. Guo et al. (2005) showed that the rheological properties of fiber suspensions increased with increasing fiber volume fraction and fiber aspect ratio. These effects also increased with decreasing shear rate. Numerical simulations by Switzer & Klingenberg (2003) reported the strong influence that particle shape and inter-fiber friction have on the viscosity of fiber suspensions. The clustering of these suspensions also has a marked impact on the viscosity: Chen et al. (2002) observed spikes in the shear stress-shear rate flow curves of fiber suspensions when they flocculated at low shear rates.

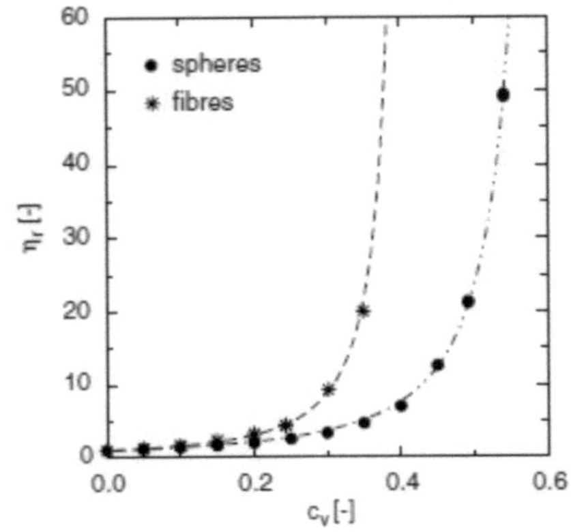

Fig. 3. Suspension viscosity vs. concentration for spheres and fiber (Marti et al., 2005)

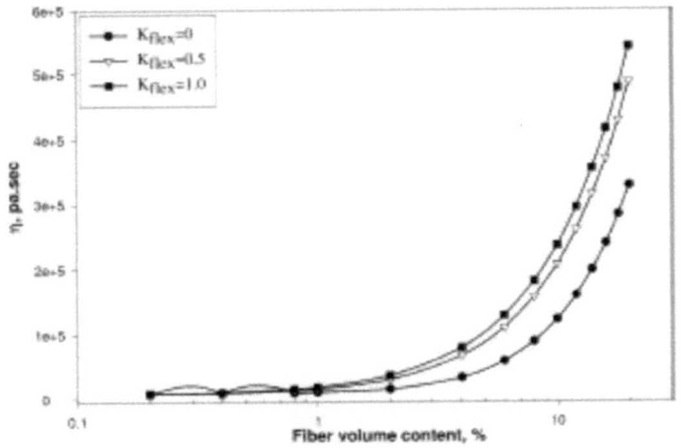

Fig. 4. Suspension viscosity vs. concentration for fibers of different flexibilities (Rajabian et al., 2005)

The ability of added fiber to improve hole cleaning without adversely impacting viscosity is the primary benefit in utilizing fiber sweeps. However, previous studies conducted on the viscosity of fiber suspensions (**Figs. 3 and 4**) have shown just how sensitive the fiber-fluid is to fluctuations in fiber concentration. Rajabian et al. (2005) investigated how the viscosity of fiber suspensions varied with increasing fiber concentration at varying degrees of fiber flexibility. In a similar study, Marti et al. (2005) looked into how the viscosity of suspensions of fibers and spheres, respectively, reacted with increasing particle concentration. While the data may not be highly correlated, both studies improve the basic understanding of the behavior of fibrous suspensions. The viscosity of the suspensions increases only slightly with increasing particle (fiber) concentration up to some threshold, above which it begins to increase rapidly and exponentially. These studies also indicated that the viscosity of the fiber-laden fluid is dependent on fiber flexibility and length. As observed in Fig. 4, the fibers with the highest degree of flexibility generated the highest viscosity at any given concentration.

This study focuses on identifying the behavior of fiber suspensions within the boundaries that are utilized in the oil and gas drilling industry. The scope of the experimental studies is limited to profiling the behavior and viscosity of a fiber suspension at concentrations not exceeding the upper practical limit at which this fiber has been used. This study will help to further verify the usefulness of fiber sweeps within the drilling industry.

2.1 Experimental investigations

Several base fluids were chosen to simulate the various drilling and sweep fluids utilized in the field (**Table 1**). A specially processed 100% virgin synthetic monofilament fiber was supplied for this research (**Table 2**), which was mixed with the base fluids at various concentrations of up to 0.08 % w/w.

	Base Fluid (kg / m³)	Weighting Agent	Fiber Concentration (%)
Water-Based Mud [WBM]	**XG** (1.00, 2.48, 4.99, 7.48)	None 998 kg/m³	0.00, 0.04, 0.08
Water-Based Mud [WBM]	**PAC** (1.00, 2.48, 4.99, 7.48)	None 998 kg/m³	0.00, 0.04, 0.08
Water-Based Mud [WBM]	**XG / PAC [50%/50%]** (1.00, 2.48, 4.99, 7.48)	None 998 kg/m³	0.00, 0.04, 0.08
Water-Based Mud [WBM]	**XG** (2.48, 4.99, 7.48)	Barite 1438 kg/m³	0.00, 0.04, 0.08
Water-Based Mud [WBM]	**PHPA** (0.49, 1.00, 1.49)	None 998 kg/m³	0.00, 0.04, 0.08
OBM	**Mineral Oil-base**	Barite 1462 kg/m³	0.00, 0.04, 0.08
SBM	**Internal-Olefin-base**	Barite 1450 kg/m³	0.00, 0.04, 0.08

Table 1. Test matrix of rotational viscometer measurements

Material	=	Polypropylene
Spec. Grav.	=	0.91
Length	=	10 mm (0.40 in)
Diameter	=	100 μm (0.004 in)
Melting Point	=	163°C – 177°C

Table 2. Fiber properties

The water-based fluids included those prepared with xanthan gum (XG) at two mud weights, polyanionic cellulose (PAC), partially hydrolyzed polyacrylamide (PHPA) and mixtures of XG and PAC. Formulations were prepared with a broad range of polymer concentrations. Two non-aqueous-based fluid systems were also tested: weighted mineral oil-based and internal olefin-based drilling fluids. These two fluid systems are designed and engineered to provide superior performance in locations not suitable for aqueous-based fluids. The rheology of these fluid systems are often adjustable, presenting the ability to custom-fit the fluid for certain environmental or in-situ conditions. Overall, the rheology of these fluid systems provide increased lubricity and yield stress over most conventional water-based fluids, but with reduced viscosity profiles. These characteristics can enable the fluids to prevent separation of the fiber within the suspension, and promote higher pump rates, both of which can reduce the size and occurrence of cuttings beds.

2.2 Experimental setup

The shear viscosity experiments were conducted using rotational viscometers (Chandler 35 and Fann 35A) with a thermocup. The Chandler 35 rotational viscometer has 12 speeds, and was modified to include a 1/5 spring. The weaker spring allows for more sensitive and accurate measurements in the low shear rate range and reports all shear stress dial readings 5x higher than their direct-reading counterparts. Both viscometers were calibrated and tested using multiple fluids to ensure readings were comparable.

2.3 Test procedure

The steps required to prepare the samples and record measurements are as follows:

Step 1. Preparation of Base Fluid: Bulk base fluid samples were prepared by mixing water, viscosifiers, and barite. Immediately after mixing, all water-based fluids were covered and left undisturbed for a minimum of 24 hours to ensure full hydration of the polymers. Each fluid was then re-agitated, and a uniform sample was obtained to determine its specific gravity using a mud balance.

Step 2. Preparation of Samples: After the base fluids were mixed and hydrated (where necessary), fiber was added at weight concentrations of 0.02%, 0.04%, 0.06%, and 0.08%. For the unweighted water-based fluids, 0.08% w/w fiber concentration corresponded to 0.80 kg of fiber/1.0 m³ of suspension, whereas for the weighted (~1440 kg/m³) water-based and non-aqueous fluids it corresponded to approximately 1.14 kg of fiber/1.0 m³ of suspension (**Table 3**).

Step 3. Viscometer Measurements at Ambient Temperature: After all samples were prepared, the viscosity profiles of the base fluids were measured using two rotational viscometers (Chan 35 and Fann 35A). If the viscosity of the fluid being measured exceeded the spring capacity of the Chan 35, the Fann 35A was utilized for the higher shear rate measurements.

Step 4. Viscometer Measurements at Elevated Temperature: Samples were placed in an oven for heating. The oven was set at approximately 82°C, and samples were agitated every 15 minutes to ensure uniformity. Once a sample was heated to 77°C as confirmed by a mercury thermometer, the sample was removed from the oven and mixed for 30 seconds using a stand mixer. This mixing time was deemed adequate to achieve uniform re-dispersion of the fibers. Immediately after mixing, a portion of the sample was poured into the thermocup. Using a mercury thermometer, the thermocup temperature was adjusted to achieve a constant fluid temperature of 77°C. The viscometer measurements were taken using the procedure described in Step 3.

%		$\rho \approx$ 998 kg/m³ (kg/m³)	$\rho \approx$ 1440 kg/m³ (kg/m³)
0.02	=	0.20	0.29
0.04	=	0.40	0.57
0.06	=	0.60	0.86
0.08	=	0.80	1.14

Table 3. Equivalent fiber concentrations (% w/w) for different fluid densities

2.4 Experimental results

The shear stress of each fluid was measured at rotational speeds of 1 rpm to 600 rpm (1.7 to 1022 sec^{-1}) at ambient temperature and 77°C. Using the measured shear stress values, the apparent viscosity was calculated at each viscometer speed for all the fluids tested.

When circulating through the annulus, the fiber sweep is likely traveling in the plug flow regime. Therefore, the low shear rate range is more significant when analyzing and predicting the behavior of these fiber sweeps under downhole conditions. However, to provide a general understanding of fiber sweeps, **Figs. 5** and **6** show the results of the viscometer measurements for the entire shear rate range.

Although the experiments were conducted with four (4) levels of fiber concentration (Step 2), to reduce data clutter only the intermediate (0.04%) and high (0.08%) fiber concentrations were included in the figures for the water-based drilling fluids. A detailed analysis of the various fluids' shear viscosity parameters was previously conducted (George et al., 2011).

2.4.1 Effect of fiber concentration

One goal of the research is to determine the effect that adding fiber and increasing the fiber concentration has on the viscosity of the fluid. In essence, the fiber did not substantially affect fluid viscosity in any of these cases. This is consistent with a previous study (Ahmed & Takach, 2008) which found that adding fiber to drilling fluids had an insignificant effect on the fluid's flow behavior. According to field results and supporting theories previously stated, adding fiber to the fluid may improve the fluid's hole cleaning performance without affecting its viscosity, which suggests that other rheological properties dictate hole-cleaning performance.

For most of the water-based drilling fluids, the addition of fiber to the base fluid resulted in only a slight increase in apparent viscosity, while in a few cases, the apparent viscosity appeared to decrease. Though small, these differences are considered greater than the expected experimental uncertainty.

As an example of the water-based drilling fluid results, let us evaluate the experimental results of the high-temperature weighted XG polymer fluids (not shown). The fiber fluid mixed with 2.48 kg/m^3 XG polymer showed the most common effect observed in this work, namely that the apparent viscosity increased with increasing fiber concentration, especially at low shear rates. Conversely, the fiber fluid mixed with 4.99 kg/m^3 XG polymer shows an opposing trend, with fiber reducing apparent viscosity throughout the shear rate range measured. At the shear rate 51 s^{-1}, the difference in shear stress between the base fluid and 1.14 kg/m^3 fiber fluid is 15%.

The effect of fiber on viscosity was found to be relatively insignificant for the non-aqueous based drilling fluids (Fig. 6). Even at low shear rates, the fiber at the highest concentration tested increased the viscosity or shear stress at 51 s^{-1} by only 4% to 6% in most cases, and 8.8% in the most extreme case (Fig. 6a). This finding is encouraging, as fiber can be added to sweeps to enhance hole cleaning without fear of increasing the ECD. Oil-based and synthetic-based muds are often used in harsh, not-easily accessible environments where there is concern for shale interaction and environmental impact. Fiber sweeps might be employed to reduce the cuttings beds in these extended reach wells where pressure loss along the annulus is a major concern.

In every case, the addition of fiber appeared to have no impact on the general shape of the apparent viscosity vs. shear rate plots. The approximate viscosity model used for the base fluid appears to be acceptable for accurately describing the behavior of the fiber fluid, at least up to a temperature of 77°C.

In a study conducted by Ahmed & Takach (2008), the hole-cleaning efficiency of fiber sweeps was compared to that of the base fluid. The experiments were carried out in a flow loop with varying inclination angles, measuring the cuttings bed height and frictional pressure loss during sweep circulation. For the same annular velocity, the fiber sweeps generally showed a reduced bed height in the flow loop annulus. Annular pressure loss was recorded as a function of time for various flow rates. In most cases, the frictional pressure loss was approximately the same for the base fluid and fiber sweep. In one instance, however, the fiber sweep pressure loss was less than that exhibited by the base fluid. Pipe viscometer experiments were also conducted comparing flow curves of the base fluid and fiber sweep. Viscometer pressure loss was measured as a function of flow rate. At low flow rates (laminar, plug flow regime), pressure loss for the base fluid and fiber sweep were equal, and the flow curves were similar.

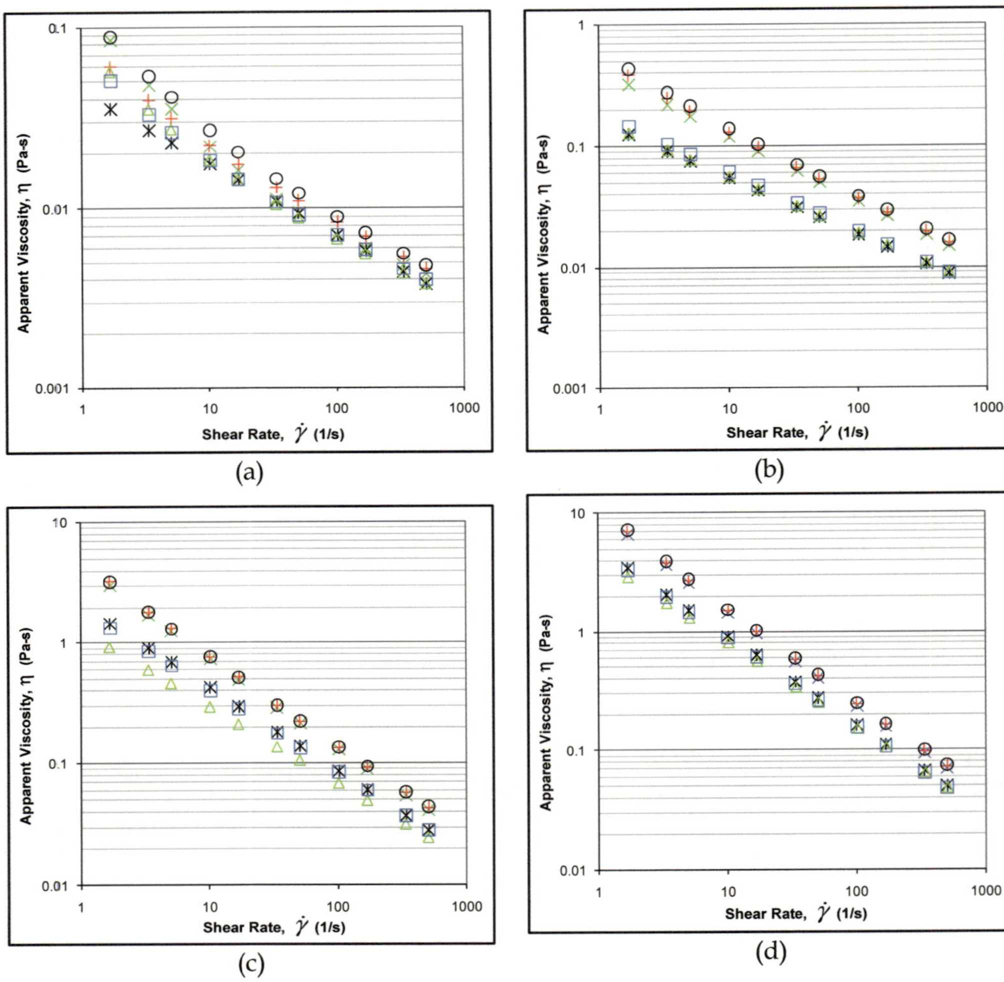

Fiber Concentration					
X Base Fluid	+ 0.40 kg/m³	○ 0.80 kg/m³	ж Base Fluid	Δ 0.40 kg/m³	□ 0.80 kg/m³
20°C			77°C		

Fig. 5. Rheology of XG based fluid at 20°C & 77°C varying fiber & polymer concentrations: a) 1.0 kg/m³ XG; b) 2.48 kg/m³ XG; c) 4.99 kg/m³ XG; & d) 7.48 kg/m³ XG

2.4.2 Effect of temperature

In order to more closely approximate the behavior of the fiber fluid under downhole conditions, the ambient temperature experiments were repeated at a somewhat higher temperature, as shown in Figs. 5 and 6. The general trend exhibited in all the fluids studied is that the fluid's ability to flow increases with temperature. The warmer temperature creates a "thinner" fluid that is more easily deformed and provides less resistance to flow.

As mentioned previously, adding fiber or increasing fiber concentration results in a slight general increase of the drilling fluid viscosity. In most cases, this same trend is observed in the elevated temperature measurements. However, in some fluids which exhibited an increase in viscosity at ambient temperature, no effect of fiber on viscosity was observed at the elevated temperature (Fig. 5b).

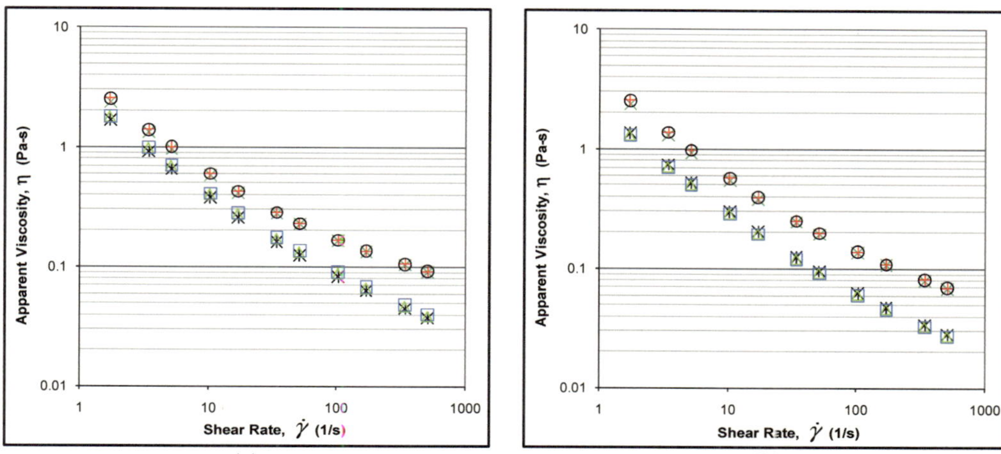

(a) (b)

Fiber Concentration					
x Base Fluid	+ 0.57 kg/m³	o 1.14 kg/m³	ж Base Fluid	Δ 0.57 kg/m³	□ 1.14 kg/m³
20°C			77°C		

Fig. 6. Rheology of weighted fluids at 20°C & 77°C and varying fiber concentrations:
a) Oil-based mud (OBM), 1452 kg/m³; b) Synthetic-based mud (SBM), 1450 kg/m³

For the oil-based and synthetic-based muds, regardless of temperature, fiber has an insignificant influence on viscometric measurements. None of the water-based fluids tested showed this lack of effect over the entire shear rate range at both temperatures.

Although no experiments were carried out elevated pressure, other works have studied pressure effects on base fluid viscosity. Zhou et al. (2004) conducted experiments to investigate aerated mud cuttings transport in a HPHT (high pressure, high temperature) flow loop. The effect of elevated pressure (up to 35 bar) was found to have minimal influence on cuttings concentration. Another study (Alderman et al. 1988) investigated HPHT effects on water-based drilling fluid viscosity. The viscous behavior of the fluids at HPHT showed a weak pressure dependence and an exponential temperature dependence. It was also shown that the fluid yield stress was essentially independent of pressure, but highly influenced by temperature. Inasmuch as pressure has little effect on viscosity of the base drilling fluids, it is expected that pressure will have little effect on fiber-laden drilling fluids.

2.5 Conclusions

This study was conducted to investigate the effects of temperature and fiber concentration on the rheology of fiber-containing sweeps. Rotational viscometers were used to conduct rheology experiments at ambient and elevated temperature (77°C). From these experiments, shear stress values were recorded and subsequent apparent viscosity flow curves were created. The shear viscosity profiles of fiber sweep fluids were compared using graphical and curve-fitting regression analyses. Based on the experimental results and data analysis, the following conclusions can be made:

- The addition of fiber up to 0.08 wt % has a minor effect on the fluid's shear viscosity profile, whether at ambient or elevated temperature. Some instances showed slight increases in viscosity, while others showed a decrease with increasing fiber concentration.
- Increasing the temperature decreases the non-Newtonian behavior of the fiber fluid, and decreases the viscosity throughout the shear rate range of 2 to 1022 s^{-1}.
- In most cases, as fiber concentration increased, the viscosity showed slightly increasingly non-Newtonian behavior: in the Yield Power Law model, n decreased while τ_y increased.
- Neither oil-based nor synthetic-based fluids exhibited any significant shear viscosity sensitivity to fiber concentration at ambient or elevated temperature. It may be possible for OBM or SBM sweeps to be utilized in the field with no increase in ECD.

3. Stability of fiber-containing fluid

Very limited experimental work has been accomplished in the study of fiber-containing drilling fluids. In particular, there is a paucity of data concerning the behavior of these long-aspect-ratio fiber particles as they move within the drilling fluid. As mentioned previously, these multi-phase complex fluid systems must possess certain rheological properties to enable homogeneous distribution of the fibers throughout the bulk of the fluid to maximize shearing force on the cuttings beds and provide transport up the annulus.

In contrast to the manufacturing industry, where fiber suspensions are common, the conditions in drilling operations to which the sweep fluids are subjected can be severe. These adverse conditions require that the drilling fluids be stable to temperatures at a minimum of 120°C (some times as high as 250°C) and possess a density in the range of 200 to 2200 kg/m^3. Most prior work on phase separation behavior of homogeneous, non-Brownian suspensions has been carried out with low-temperature, low-density fluids using particulates of density higher than the base fluid. In the case of drilling fluid fiber sweeps, the relative densities are reversed, as the specific gravity of the synthetic fiber is less than that of the base fluid. This relation encourages buoyancy, and the fibers tend to rise (separate) within the suspension. In spite of this difference, separation of fibers and fiber networks can be treated mathematically in a similar manner as settling of high-density particles.

The settling behavior of particles with density greater than the base fluid can be classified into four groups (Scholz, 2006):

- **Class I: Unhindered settling of discrete particles**. Non-interacting particles accelerate until a terminal settling velocity is reached, where the hydrodynamic drag and gravitational force are balanced. Stokes' Law is commonly used to describe this motion of spherical particles.

- **Class II: Settling of a dilute suspension of flocculant particles**. Randomly moving particles collide and form aggregates (flocs) which behave as larger particles and have increased settling velocities compared to a single particle.
- **Class III: Hindered and zone settling**. Particle concentration is increased to a point where particles interact and form a loose network that displaces the liquid phase and gives rise to upward flow of liquid. This motion of the liquid reduces the overall particle settling velocity, and is called hindered settling. In large-surface-area settling applications with high particle concentrations, the entire suspension may tend to settle as a "blanket" (zone settling).
- **Class IV: Compression settling (compaction and consolidation)**. At even higher concentrations of particulates, e.g. at the bottom of a column containing a suspension, , a layer of sludge forms that becomes compressed over time and ultimately contains little or no liquid.

At low concentrations, unhindered "settling" (or more simply and generally "separation") can be applied to describe the motion of synthetic fibers. The modeling study considers Class I motion, which simply describes the forces and rising velocity of a single fiber suspended in a fluid. This prediction can be extrapolated to determine the rising velocity of a dilute suspension of fiber particles. Flocculation is not expected, since the fibers in this study are thought to be chemically inert. On the other hand, another mechanism may occur with fibers that is not possible for granular or spherical particles, namely entanglement. At sufficiently high concentration and with high-aspect-ratio flexible fibers, entanglement could occur and lead to behavior not unlike flocculation. Thus, Class I and a modified form of Class II could occur. Hindered/zone separation is not likely to be a factor due to the low concentrations of fiber. Nor is compression a plausible scenario for buoyant fibers, since there is no driving force for compaction of the fibers at the surface of the liquid, as there is for particles at the bottom of a fluid column.

The settling motion of a spherical particle is simple. The separation motion of a fibrous particle is much more complex (Qi et al., 2011). In the absence of extraneous forces, a sphere settles in a purely vertical direction. A cylindrical low-density flexible fiber suspended in a fluid can exhibit profligate behavior in three dimensions, as well as drift horizontally during its vertical ascent (Herzhaft & Guazzelli, 1999). The rising velocity of these fibers also depends on the fiber concentration, and how the particles orient themselves. Experimental studies investigating dilute and semi-dilute cylindrical or spheroid particles indicate that concentration is vital in understanding the phase separation (Qi et al., 2011; Kuusela et al., 2001, 2003; Koch & Shaqfeh, 1989; Herzhaft & Guazzelli, 1996, 1999). The overall consensus of these studies regarding the behavior of fiber suspensions is that fiber flocs separate faster than an individual fiber. However, as fiber concentration increases (independent of flocculation), the fibers exhibit hindered separation, and the separation velocity decreases below that of a single fiber.

These prior studies were concerned with phenomena of a similar nature to this work. This study focuses on rising velocities of fiber particles within a suspension, based on theoretical models and experimental results.

3.1 Modeling rising velocity of fiber particles

A mathematical model was developed to describe the effect of viscosity on the stability of fiber-containing drilling fluid sweeps. In order for the fiber to perform as efficiently and effectively as possible as a hole-cleaning aid, it is important that the fiber be homogeneously distributed throughout the base fluid, i.e. minimal separation of the fiber occurs under downhole conditions. Thus, a theoretical study was conducted to determine the desirable base fluid properties to formulate sweep fluids that are stable under downhole conditions.

For the sake of simplicity, our analysis only considers a single fiber suspended in the fluid (unhindered separation). This assumption ignores the effect of fiber-fiber interaction and fiber concentration. The analysis considers two orientations of the fiber: perpendicular to the direction of motion (horizontal orientation) and parallel to the direction of motion (vertical orientation). These orientations represent the boundaries within which the fiber can theoretically orient. However, it has been shown that a single fiber will orient itself horizontally (Kuusela et al., 2001, 2003; Qi et al., 2011). This stable orientation is not dependent on fluid velocity, and the fiber will eventually return to the horizontal position if acted upon by an outside force (Qi et al, 2011). Liu and Joseph (1993) investigated how a rigid slender particle is affected by liquid properties, particle density, length, and shape. They found that only particle concentration and the end shapes influenced particle orientation. This is consistent with studies by Herzhaft et al. (1996, 1999), which concluded that orientation of a settling spheroid is almost independent of aspect ratio but correlated to suspension concentration. As the fiber concentration is increased, the hydrodynamic interactions between the fibers will upset the stable horizontal fiber. With increasing fiber concentration, the fiber will show greater tendency to orient parallel to the direction of motion (Herzhaft et al.; 1996, Qi et al., 2011). By analyzing both cases, we can predict the rheological properties of the base fluid that can keep the fiber in suspension for a sufficient length of time.

To determine the velocity at which the submerged fiber particles move upward to the surface of the liquid, the sum of the forces in the vertical direction (y-axis) are set equal to zero. As shown in **Fig. 7,** the forces acting on the fiber moving within the column of fluid are buoyancy (F_b), hydrodynamic drag (F_D) and gravity ($m \cdot g$). The projected surface area of a fiber particle, dependent upon particle orientation, is needed to compute the drag force. In this case, the fiber is horizontally oriented (perpendicular to the direction of motion of the fiber), and the generalized equation of the force balance in the vertical (y) direction is:

$$\sum F_y = F_b - F_D - mg = 0 \tag{1}$$

The forces acting on the fiber can be expanded as follows:

$$\sum F_y = V_p \rho_f g - \frac{1}{2} \rho_f C_{D,h} U_{p,h}^2 A_{p,h} - \rho_p V_p g = 0 \tag{2}$$

where $U_{p,h}$ is the rising velocity of the horizontally oriented particle. Rearranging the momentum balance equation to group similar terms:

$$\frac{1}{2} \rho_f C_{D,h} U_{p,h}^2 A_{p,h} = V_p \rho_f g - \rho_p V_p g \tag{3}$$

Both sides of Eqn. 3 are divided by (½ ρ_f) and the projected area for horizontal orientation ($A_{p,h} = L \times d$). Reducing like terms, expressing the fiber particle volume V_p explicitly (¼ $\pi d^2 L$) and rearranging gives the formula for the particle rising velocity:

$$U_{p,h} = \left[\frac{\pi d g}{2} \left(\frac{\rho_f - \rho_p}{\rho_f} \right) \frac{1}{C_{D,h}} \right]^{\frac{1}{2}} \tag{4}$$

A similar analysis for a vertically oriented particle with circular end area ¼ πd^2 gives

$$U_{p,v} = \left[2Lg \left(\frac{\rho_f - \rho_p}{\rho_f} \right) \frac{1}{C_{D,v}} \right]^{\frac{1}{2}} \tag{5}$$

The drag coefficients (C_D) of the fiber particle must be estimated to predict the rising velocities using the above equations. Hole cleaning fibers are more or less straight and stiff and they can be considered long cylinders for the drag force calculation. Drag coefficient correlations and charts for long cylinders are well documented in the literature. For cylinders oriented perpendicular to the flow (i.e. cross flow), Perry (1984) presented a correlation (**Fig. 8**) that can be approximated with the equation given below.

$$Log(C_{D,h}) = \frac{0.9842 - 0.554 Log(\mathrm{Re})}{1 + 0.09 Log(\mathrm{Re})} \tag{6}$$

This correlation is valid for Reynolds Number ranging from 10^{-3} to 1000. Due to the larger surface area of the horizontally oriented particle exposed to the direction of motion, the fibers moved at rates corresponding to very low Reynolds numbers. The Reynolds numbers in the lab experiments conducted here ranged from 10^{-17} to 10^{-3}.

The drag coefficient of a cylinder oriented in the direction of the flow is only a function of the aspect ratio (L/d). Based on available data in the literature (Hoerner, 1965), the following equation has been developed to estimate the drag coeffcieint C_{Dv}:

$$C_{D,v} = 0.825 + \frac{0.317}{1 + \left(\frac{l/d}{1.54} \right)^{4.23}} \tag{7}$$

As shown in Eqn. (6), the drag coefficient of a cylindrical fiber under cross flow condition is a function of the Reynolds Number, which is generally expressed as $\mathrm{Re} = \rho U_{p,h} d/\mu$ (i.e. the ratio of inertial force to viscous force). This definition holds true for Newtonian fluids, where shear stress $\propto$ shear rate. However, the fluids that are often utilized in fiber sweep applications are non-Newtonian. Hence, the Reynolds Number must be redefined using the apparent viscosity function as $\mathrm{Re} = \rho U_{p,h} d/\mu_{app}$. The viscosity for Newtonian fluids is independent of the shear rate. However, for non-Newtonian fluids, the apparent viscosity varies with shear rate. Applying the Yield Power Law (YPL) rheology model, the apparent viscosity is expressed as:

$$\mu_{app} = K(\dot{\gamma})^{n-1} + \tau_y(\dot{\gamma})^{-1} \tag{8}$$

The objective of this study was to determine the desirable Yield Power Law (YPL) fluid properties to keep the fiber in suspension in order to create efficient momentum transfer mechanisms between the sweep fluid and the cuttings bed. Only fluids with sufficient yield stress and/or low-shear-rate viscosity can be utilized to keep the fibers in suspension.

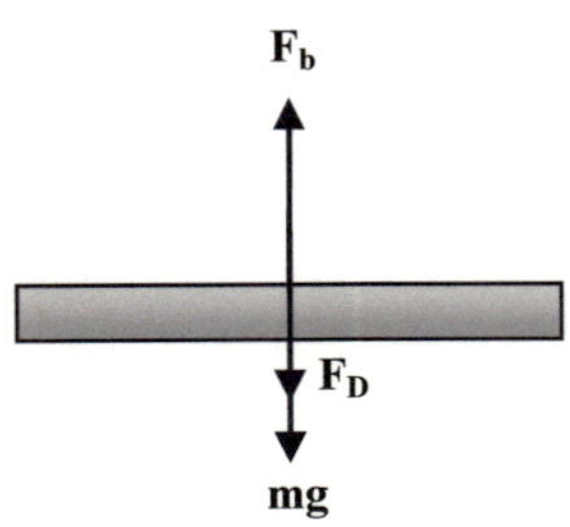

Fig. 7. Free body diagram of a cylindrical particle rising in static fluid

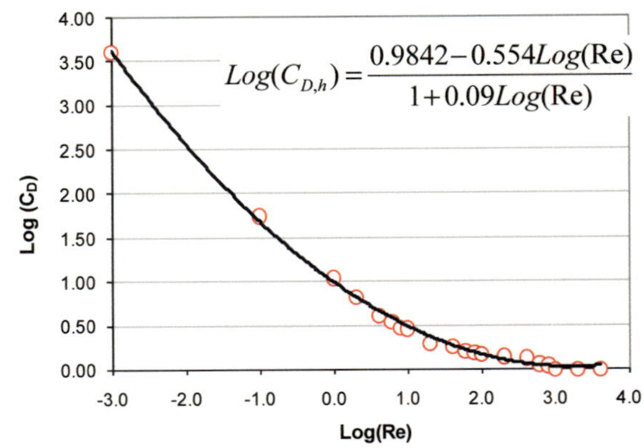

Fig. 8. Drag coefficient vs. Reynolds number for perpendicularly-oriented cylinders (Perry, 1984)

3.1.1 Non-rising particles under static conditions

In a static (quiescent) YPL fluid, a horizontally oriented cylindrical particle will not rise if the yield stress of the fluid is at least as high as its buoyancy (Dedegil, 1987). For a fully suspended particle, the momentum balance described by Eqn. (1) can be rewritten to include the static shear force acting on the particle in lieu of the drag force that is present under dynamic conditions. By taking a differential element of the cylindrical fiber, the vertical component of the maximum static shear stress (i.e., yield stress) acting on the fiber can be determined (**Fig. 9**). The direction of the shear stress acting on the cylinder depends on the location of the differential element as shown in Fig. 9. The stress acts on the area represented by the differential element shown in **Fig. 10**, which is expressed as:

$$dA = LRd\theta \tag{9}$$

Then, the vertical component of the shear force acting on the differential element is:

$$dF_{shear} = dA \cdot \tau_y \sin\theta = LRd\theta \cdot \tau_y \sin\theta \tag{10}$$

For a fiber-oriented vertically and horizontally, shear stresses act on the circumferential and end areas. However, the end areas are negligible when compared to circumferential area of the cylinder. This further simplifies the analysis. Neglecting the forces acting on the cylinder ends, the overall vertical component of the shear force is subsequently obtained by integrating Eqn. (10). After simplification, and considering that shear stress is exerted on opposing sides of the particle, the vertical component of the stress force acting on a horizontally oriented cylinder becomes:

$$F_{s,h} = 2dL\tau_y \tag{11}$$

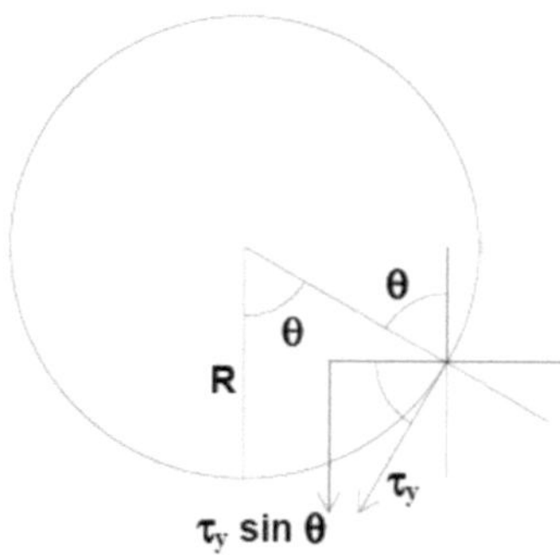

Fig. 9. Vertical component of shear force acting on a fully suspended cylinder

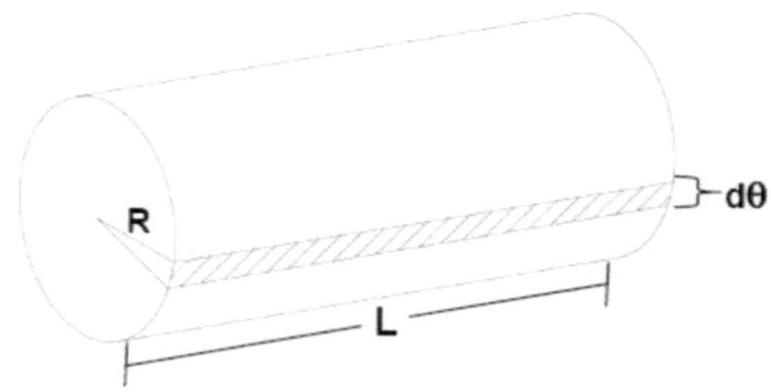

Fig. 10. Differential element of a cylinder subject to shear force

The above equation predicts the maximum value of the shear force acting on the cylinder. For the sake of simplicity, the analysis is only concerned with a single fiber suspended in fluid. The cylinder is considered to be non-rotating and constantly perpendicular to the direction of motion of the fiber. In addition, as stated earlier, the calculations ignore fiber-fiber interactions, which would influence the momentum balance. The momentum balance equation can be written:

$$\Sigma F_y = 0 = V_p\rho_f g - 2dL\tau_y - \rho_p V_p g \tag{12}$$

Replacing the particle volume V_p with (π $d^2 L/4$), and grouping like terms results in:

$$2dL\left[\frac{\pi dg}{8}\left(\rho_f - \rho_p\right) - \tau_y\right] = 0 \tag{13}$$

For a fiber particle oriented in the vertical direction (**Fig. 11**), the shear force can be written as:

$$F_{s,v} = \pi d\tau_y \int_0^L dh = \pi dL\tau_y \tag{14}$$

Rewriting the force balance equation to include the shear force acting on a vertical oriented particle and grouping like terms results in:

$$\frac{\pi dL}{4}\left[dg\left(\rho_f - \rho_p\right) - 4\tau_y\right] = 0 \tag{15}$$

For this study, the dimensions of the fiber are known and fixed. Therefore, in order to determine the fluid property that can hold the fibers is suspension, Eqns. (13) and (15) must be rewritten to solve for critical shear stress as a function of density difference and fiber diameter for a fiber oriented horizontally and vertically, respectively. For a fiber oriented horizontally, the critical yield stress is:

$$\tau_{y,h} = \frac{\pi dg}{8}\left(\rho_f - \rho_p\right) \tag{16}$$

For a vertically oriented fiber particle, the critical yield stress is:

$$\tau_{y,v} = \frac{dg}{4}\left(\rho_f - \rho_p\right) \tag{17}$$

For this analysis, the fiber size and density are known. Thus, the critical yield stress is essentially a function of the fluid density, and increases linearly with density. **Fig. 12** shows the yield stress required to keep a vertically and horizontally oriented fiber particle (with properties given in Table 2)) from rising within the base fluid. For any given fluid density, horizontally (perpendicular) oriented fibers require greater yield stress to counteract the fluid's natural buoyancy. At the highest fluid density considered (2800 kg/m³), only a small yield stress (less than 0.72 Pa) is needed to keep the fiber in suspension. Eqns. 16 and 17 represent the relationship between the fiber's dimensions and density difference to the yield stress required to keep the fiber from rising.

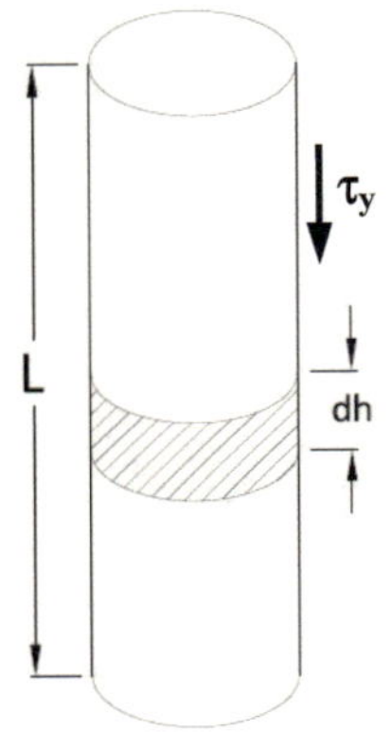

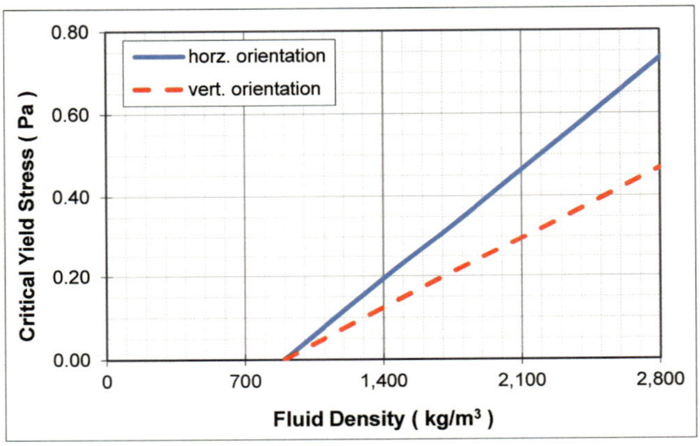

Fig. 11. Shear force acting on a vertically-oriented cylinder

Fig. 12. Critical yield stress as a function of fluid density for a non-rising particle

3.1.2 Non-rising particles under dynamic conditions

The models developed in previous sections are for fiber particles rising in a static fluid. They do not consider lateral motion and deformation of the fluid, as well as hydrodynamic diffusion effects. Eqns. (16) and (17) are appropriate for understanding rising behavior of fibers under static conditions. However, this study is to determine the hole cleaning efficiency of the fiber particles in real world situations such as fluid circulating up the annulus and drillstring rotation. The shearing motion of the fluid in the annulus will affect the apparent viscosity that subsequently influences the behavior of fiber particles in the base fluid. To accurately model the behavior of fiber under dynamic conditions, the overall shear rate must be computed from the primary and secondary flow shear rates, and the annulus is modeled as a narrow slot to obtain analytical solutions (**Fig. 13**).

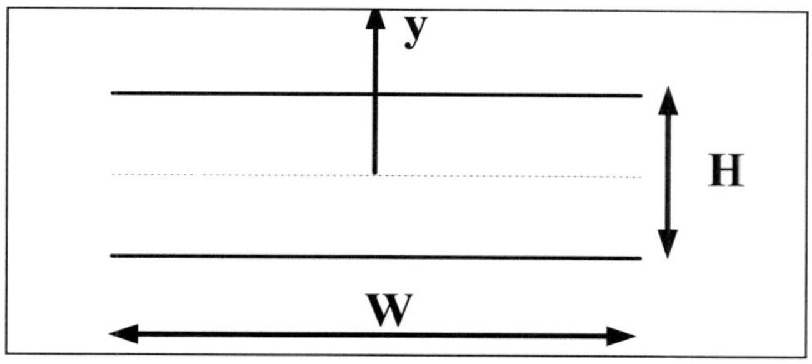

Fig. 13. Narrow slot representing annulus

As the sweep fluid is flowing in the annulus, it is subjected to primary and secondary flows. The primary flow is the gross flow of the fiber-fluid suspension in the annulus. For a fluid flowing in the annulus, the shear rate varies from zero to its maximum value, which occurs at the inner wall. Using the narrow slot approximation technique, the shear at any point in the annulus is given as (Miska, 2007):

$$\dot{\gamma}(y) = \frac{1}{K^n} \cdot \left[y\left(\frac{dp}{dL}\right) - \tau_y \right]^{\frac{1}{n}} \tag{18}$$

The above equation can be integrated to calculate the average shear rate as:

$$\dot{\gamma}_{ave} = \frac{1}{W(H/2)} \int_0^{H/2} \dot{\gamma}(y) W dy \tag{19}$$

For a YPL fluid, due to the presence of a plug zone, the average shear rate (i.e., primary share rate) calculation procedure is complex. However, in the plug zone the shear rate is zero and, therefore, can be neglected. For a power law fluid, Eqn. (19) yields:

$$\dot{\gamma}_{ave} = \dot{\gamma}_{primary} = \frac{1/K^n (dp/dL)^{1/n} (H/2)^{1/n}}{W(n+1/n)} \tag{20}$$

The slot width is $W = \pi (d_o + d_i)/2$, and the clearance is $H = (d_o - d_i)/2$. The primary shear rate is a function of flow geometry, properties of the fluid, and pressure gradient or annular velocity. The velocity gradient, or rising motion, of the particle induces the secondary flow, which is a function of the fiber particle rising velocity and the particle diameter:

$$\dot{\gamma}_{secondary} = \frac{U_p}{d_p} \tag{21}$$

Knowing that shear rate is the magnitude of the deformation tensor, the resultant shear rate scalar can be determined by the Eucledian norm:

$$\dot{\gamma}_{total} = \sqrt{\dot{\gamma}^2_{primary} + \dot{\gamma}^2_{secondary}} \tag{22}$$

3.2 Modeling results

In order to predict the behavior of the fiber under dynamic conditions, the typical values of annular velocity and hydraulic diameter shown in **Table 4** were used. For a dynamic condition, the rising velocity of the fiber particle can be determined applying the rising velocity equations in combination with the resultant shear rate.

Fiber Diameter	=	0.0001 m	$D_{hydraulic}$	=	0.0889 m
Fiber Length	=	0.01 m	K	=	1.00 $N\text{-}s^n/m^2$
Fiber Density	=	897.04 kg/m^3	n	=	0.52
$U_{annulus}$	=	0.9144 m/sec			

Table 4. Input Data

To predict the possible results of the subsequent bench-top experiments, sensitivity analysis was conducted using the model. By varying certain properties of the fluids and determining the resulting rising velocities, the behavior of the fibers in suspension were investigated. For the sensitivity analysis, the rising velocity of a horizontally oriented fiber was determined under dynamic conditions varying the yield stress, fluid behaviour index "n", consistency index "K" and fluid density **(Figs. 14 and 15)**.

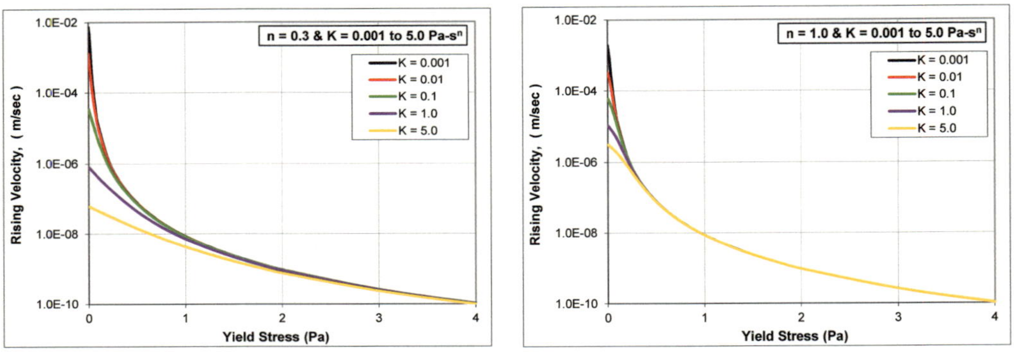

(a) (b)

Fig. 14. Rising velocity vs. yield stress for horiz. oriented fiber in 998 kg/m^3 fluid
a) n = 0.3; and b) n = 1.0

The rising velocity of the fiber decreases as the consistency index increases and the fluid behavior index decreases. But these effects are relatively minor compared to yield stress and fluid density. The rising velocity decreases with increasing yield stress, as expected, and it decreases essentially in exponential fashion. Furthermore, with increasing yield stress, the consistency index term in Eqn. (8) becomes increasingly irrelevant in determining the Reynolds Number and has only a marginal effect on the rising velocity of the fiber. Like yield stress, fluid density has a strong effect on rising velocity. However, rising velocity decreases with decreasing fluid density, again decreasing in exponential fashion.

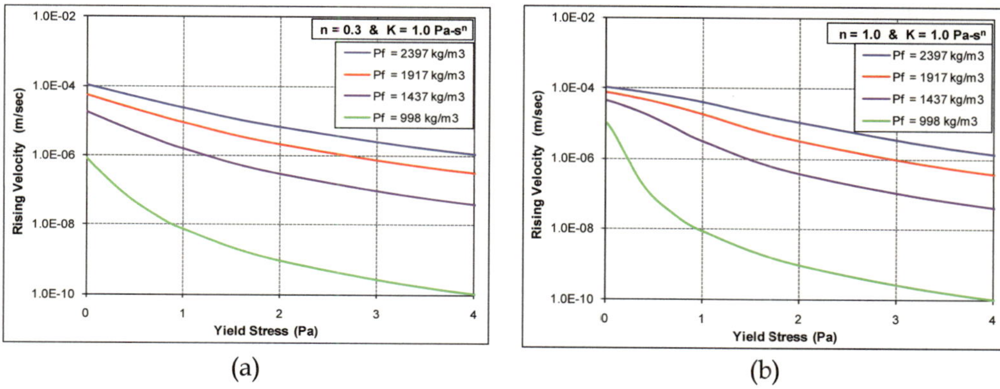

Fig. 15. Rising velocity vs. yield stress varying fluid density for horizontally oriented fiber particle, K = 1.0 Pa-s^n
a) n = 0.3; and b) n = 1.0

The yield stress and inherent "n" value of a specific fluid characterizes the degree to which the fluid behavior is non-Newtonian. Another trend worth investigating from Fig. 14 is the similarity of the rising velocity plots for different fluids with respect to their "n" values. It is observed that as "n" increasingly deviates from unity, the spread of the rising velocity plots at low yield stress values tends to increase. A careful examination of Fig. 14 reveals that the increase in the value of "n" substantially increases the rising velocity in fluids with high "K" values. Although this is unusual, analysis of Eqn. (10) shows that, at low shear rates (i.e., shear rates less than 1.0 s^{-1}), decreasing the values of "n" results in increased apparent viscosity.

To further explore the effect of yield stress on the upward motion of the fiber particles, the rising velocity of a horizontal fiber was determined as a function of yield stress, fluid density and "n" value. From Fig. 15, it can be seen that as the density of the fluid increases, the rising velocity increases. This is attributed to buoyancy, which increases as the difference in density between the fluid and fiber increases. This is consistent with the observation that the less dense fluids require smaller yield stresses to decrease the rising velocity or indefinitely suspend the fibers.

For a vertically oriented fiber particle, the rising velocity strictly relies on the fiber dimensions and density difference between the particles and the fluid. **Fig. 16** shows the effect of fluid density on the rising velocity of a vertically oriented fiber under dynamic conditions. This trend is similar to that observed for horizontally oriented fibers. As shown from Eqn. (7), when the fiber particle orients itself in the direction of motion, the drag coefficient becomes independent of the Reynolds Number and rheological properties of the fluid. Due to the high aspect ratio and low drag coefficient of the fiber, the rising velocity of a vertically oriented fiber particle is very high, and is about twice as high as when it is oriented perpendicular to the direction of motion (Herzhaft & Guazzelli, 1999). Although Koch & Shaqfeh (1989) found that a spheroid falls faster when its "thin" side is pointing in the direction of gravity, a single, unconstrained fiber will turn horizontal and oscillate around the stable horizontal orientation, separating slowly compared to vertical fibers (Kuusela et al., 2001, 2003; Qi et al., 2011). Due to the flexibility of the fiber and high mechanical and hydrodynamic interferences, a vertical configuration is difficult to maintain, though it is preferred under dynamic conditions; therefore, predicted values do not reflect the actual rising speeds.

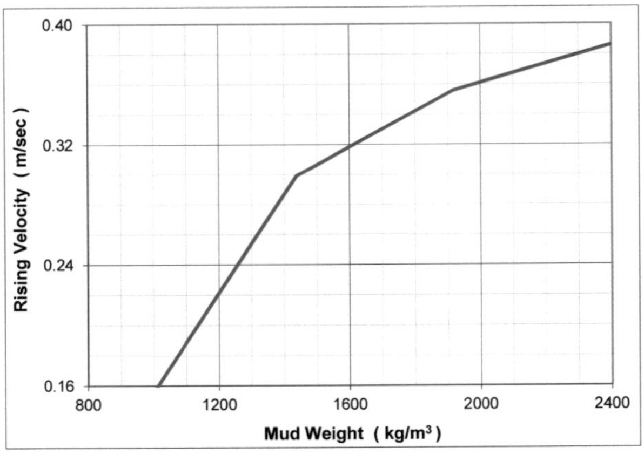

Fig. 16. Rising velocity vs. mud weight under dynamic conditions for vert. oriented fiber

3.3 Experimental study of fiber sweep stability

Several base fluids were chosen to simulate typical drilling and sweep fluids utilized in the field. The purpose of this investigation was to determine how well various base fluids would hold a fiber in suspension under ambient and elevated temperature conditions. The fibers used for these tests are described in Table 2. They have a specific gravity of approximately 0.91, which is less dense than the typical fluids in which they are suspended. Therefore, their natural tendency is to rise to the surface of the fluid and form fiber lumps. If the fibers rise while suspended in the fluid, the hole cleaning performance of the fluid diminishes and fiber lumps may plug some of the downhole tools. As discussed previously, fiber sweeps are mainly applied to minimize cuttings beds and reduce the concentration of cuttings in the wellbore. The fluids utilized for the sweep operations must possess properties conducive to maintaining a uniform fiber concentration throughout the bulk volume without increasing the ECD.

3.3.1 Experimental setup and procedure

Stability experiments were conducted using 250 mL graduated cylinders. Stand mixers were used to prepare the test fluid. Elevated temperature tests were carried out by heating test samples in a static oven. All fluids were prepared using the same process, unless otherwise specified by the fluids' respective product literature or laboratory preparation guidelines. Multiple polymeric fluids were tested, as well as oil-based and synthetic-based muds. Each polymeric fluid was prepared at the polymer concentrations shown in **Table 5**. The fluids and polymer concentrations utilized for this study were designed to simulate fluids used in drilling applications. **Table 6** lists the base % w/w concentration of the tested polymers, as well as their corresponding volumetric concentrations at different fluid densities.

	Base Fluid (kg / m³)	Weighting Agent	Fiber Concentration (%)
Water-Based Mud [WBM]	XG (1.00, 2.48, 4.99, 7.48)	None 998 kg/m³	0.00, 0.04, 0.08
Water-Based Mud [WBM]	PAC (1.00, 2.48, 4.99, 7.48)	None 998 kg/m³	0.00, 0.04, 0.08
Water-Based Mud [WBM]	XG / PAC [50%/50%] (1.00, 2.48, 4.99, 7.48)	None 998 kg/m³	0.00, 0.04, 0.08
Water-Based Mud [WBM]	XG (3.59, 7.19, 10.78)	Barite 1438 kg/m³	0.00, 0.04, 0.08
Water-Based Mud [WBM]	PHPA (0.49, 1.00, 1.49)	None 998 kg/m³	0.00, 0.04, 0.08
OBM	Mineral Oil-base	Barite 1462 kg/m³	0.00, 0.04, 0.08
SBM	Internal-Olefin-base	Barite 1450 kg/m³	0.00, 0.04, 0.08

Table 5. Test matrix for stability experiments

%		ρ ≈ 998 kg/m³ (kg/m³)	ρ ≈ 1440 kg/m³ (kg/m³)
0.05	=	0.49	0.72
0.10	=	1.00	1.44
0.15	=	1.49	2.16
0.25	=	2.48	3.59
0.50	=	4.99	7.19
0.75	=	7.47	10.78
1.20	=	11.98	17.26

Table 6. Equivalent polymer concentrations

The process used for preparing the water-based fluids followed these steps:

Step 1. **Preparation of Base Fluid**: The fluid samples were initially mixed with tap water in bulk using a stand mixer to begin hydration of the polymer. Hot water was used to accelerate the hydration time. The polymeric fluids were then placed in a blender and mixed for 30 minutes, and left to sit for 24 hours to ensure complete hydration.

Step 2. **Sample Preparation**: After sitting static for 24 hours, the fluids were re-agitated using a stand mixer to ensure uniformity of the samples. The bulk fluid was then divided into 300-mL samples (**Fig. 17**). Fiber was added to the samples by volumetric concentration in increments of 0.20, 0.40, 0.60, and 0.80 kg/m³ for unweighted, water-based fluid (approx. 0.02%, 0.04%, 0.06%, and 0.08% by weight for 998 kg/m³ mud, see Table 3).

Fig. 17. Fluid samples organized by polymer and fiber concentration (George et al., 2011)

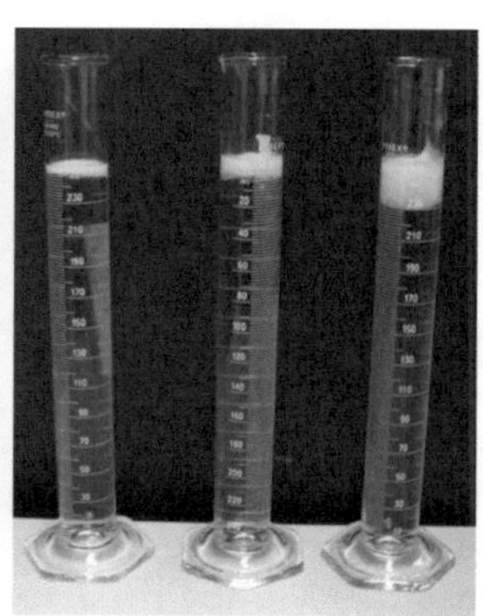

Fig. 18. Unstable test fluids after 1.0 hour test

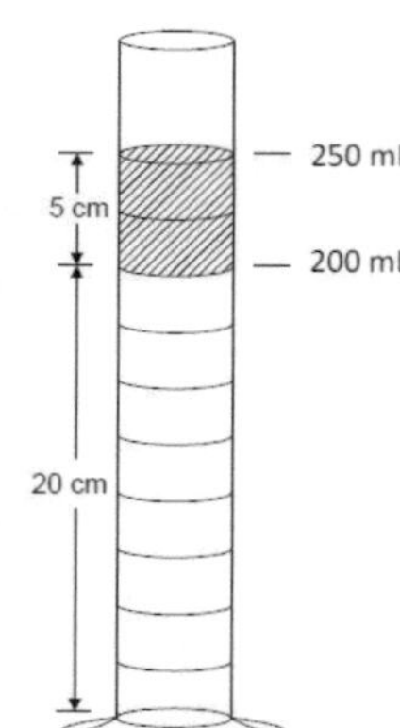

Fig. 19. Graduated cylinder used for stability experiment

Step 3. Heating the Samples: The samples were placed in the oven for approximately 10 minutes to preheat the fluid. They were removed from the oven and re-agitated with the stand mixer to ensure fiber uniformity. The fluid samples were then immediately transferred to 250-mL graduated cylinders, and placed in the oven for one hour.

Step 4. Extracting the Fiber: The graduated cylinders were promptly removed from the oven after one hour. Under quiescent conditions, buoyant fiber particles move toward the surface of the sample, increasing the fiber concentration near the surface of the liquid. In unstable fluids, most of the fiber particles reached the surface of the sample (**Fig. 18**) after one hour. Using a 60 cc syringe, an aliquot of the top 50 mL of the fluid (**Fig. 19**) was removed from each cylinder and placed in a beaker. Water and surfactant were mixed in with the fiber-fluid to aid in cleaning the fibers.

Step 5. Weighing the Fiber: The fibers were separated from the fluid using a screen, and remixed with water and surfactant to further clean the fibers. The fibers were then screened again, dried in an oven, and weighed.

3.4 Experimental results

The experiments were designed to look at the effects of fluid viscosity profile on rising velocity. For the water-based drilling fluids, the viscosity was varied by using different types and concentrations of polymers, as well as barite. The primary purpose of the barite is to control density, but it also affects viscosity. For the oil-based and synthetic-based drilling fluids, the viscosity profiles and density were varied solely by using barite. The stability of the fiber-fluid suspension was tested for various fluids at varying polymer concentration. A few stability experiments were also conducted at ambient temperature, but with essentially the same qualitative results. The tests were conducted with no shear applied to the fluid, so the only shear present was exerted by the fiber itself as it moved upward in the fluid. The fluids tested are shear-thinning, so that apparent viscosity decreases with increasing shear rate. Under downhole conditions with the fluid circulating up the annulus, the viscosity may be considerably lower than under static conditions.

3.4.1 Effect of base fluid rheology on stability of fiber sweep

Visual observations during the tests showed the formation of three fluid layers within the graduated cylinder: clear layer on the bottom, uniform concentration layer in the middle and fiber accumulation layer on top. As the fiber particles migrate, the clear layer with a distinct boundary forms at the bottom. In the absence of hindered separation, the boundary is expected to move upward at a constant velocity, same as the rising velocity of the particles. At the same time, the accumulation layer develops at the top. As particles continuously migrate from the bottom to the top, the concentration and thickness of the accumulation layer increases. Assuming initial uniform fiber dispersion throughout the entire test specimen, the fiber volume of the top layer is equivalent to 20% of the total fluid column. Therefore, the change in average concentration of the top layer can be used to quantify the level of instability in the fluid system. Applying a mass balance for the solid phase, the average fiber concentration at a given time (t) can be estimated as:

$$c(t) = \frac{\left(U_{p,h} * c_f * A_c * t\right) + \left(V_c * c_f * 20\%\right)}{V_c * 20\%} \tag{23}$$

where A_c and V_c are the cross-sectional area and volume of the cylinder, respectively. For a completely unstable fluid, all fibers will raise to the top layer, and the final concentration will be five times greater than the initial concentration.

Figs. 20 to 25 show the test results along with model predictions. The results are presented in terms of final fiber concentration of the top layer as a function of initial fiber concentration. In the plots, the unstable fluid line illustrates the maximum fiber concentration that would occur in the top layer if all the fiber particles migrate into this layer. The stable line shows the initial fiber concentration in the top layer that does not change with time because of complete stability. The results complement the viscosity profiles measured previously for these fluids. The fluids that possess yield stress generally showed greater stability. Xanthan gum (XG), a very common drilling fluid, exhibited expected behavior (Fig. 20a). All except the 0.10% XG fluid showed complete stability.

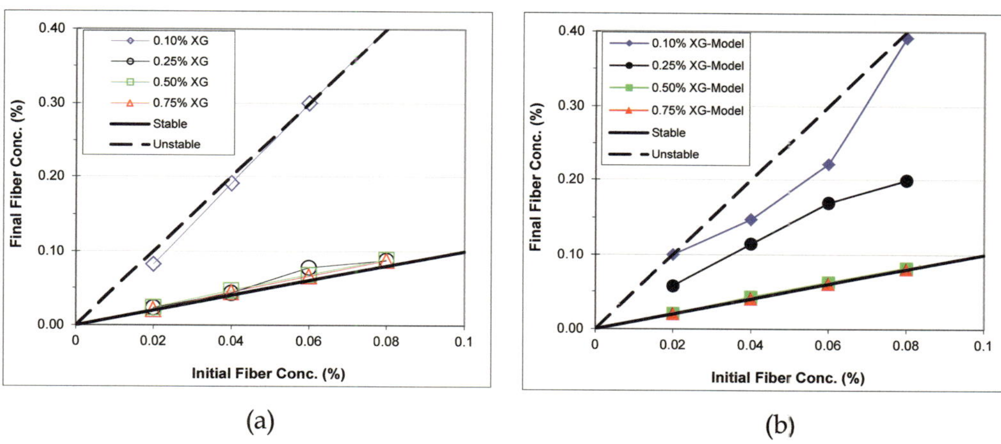

Fig. 20. One-hour stabilities of XG based fiber sweep at 77C: a) Lab experiment; b) Mathematical model

The PAC polymeric fluid showed no stability at 77°C (Fig. 21a). PAC is typically used as a fluid loss additive, not for its cuttings carrying capacity. However, the solution remains clear when mixed with water, and is a good tool for observing the fibers in suspension. In an attempt to determine if the PAC fluid had any potential as a fiber suspension fluid at 77°C, the duration of the experiment was shortened to 30 minutes. In this short amount of time, the fiber in the thicker fluid had not yet migrated to the surface, while the thinner fluids once again showed no stability. The PAC fluid mixed at 0.25% w/w fiber showed unexpected stability, but it appeared to be caused by an artifact. Upon visual inspection, the fiber appeared to have grouped in the middle of the cylinder, preventing the majority of the fiber from rising to the top part of the cylinder. The hydrodynamic and mechanical effects between the container wall and the rising fibers caused the formation of a fiber plug (lump) that did not move. This phenomenon persisted for the length of the experiment. Upon slight perturbation of the cylinder with the fiber-fluid, the fiber plug fell apart and the fiber rose to the surface in a matter of minutes.

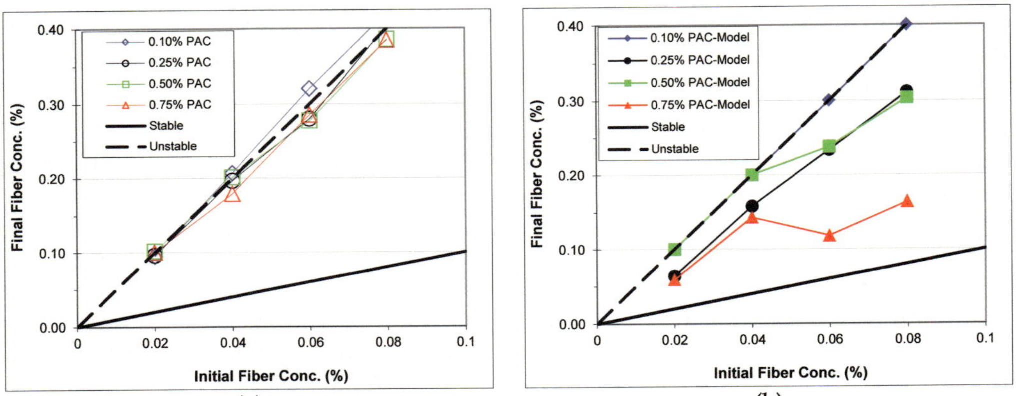

(a) (b)

Fig. 21. One-hour stabilities of PAC based fiber sweeps at 77°C: a) Lab experiments; b) Mathematical model

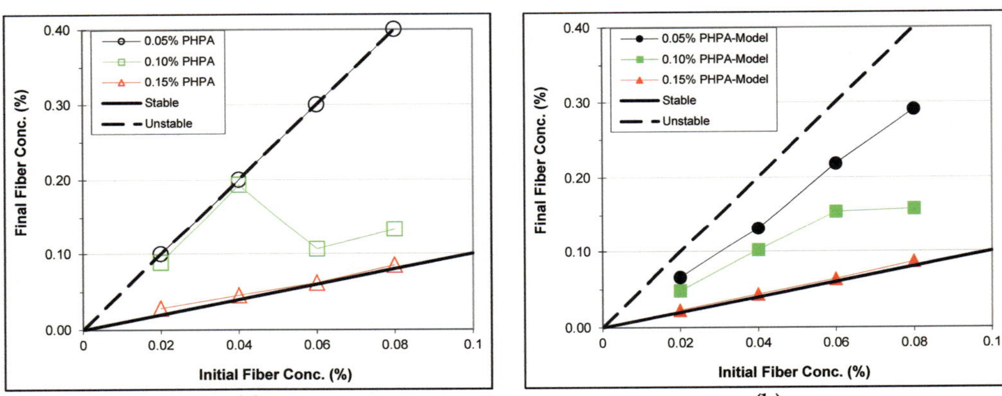

(a) (b)

Fig. 22. One-hour stabilities of PHPA based fiber sweeps at 77°C: a) Lab experiments; b) Mathematical model

Another clear fluid tested during the experiments was Partially Hydrolyzed Polyacrylamide (PHPA), which is often utilized as a shale stabilizer in drilling applications. When used at concentrations commonly used in the field, the thickest fluid (0.15% PHPA) showed stable behavior (Fig. 22a) while the thinner fluid (0.05% PHPA) became unstable. The intermediate viscosity fluid (0.10% PHPA) was essentially unstable, too. At high fiber concentrations, the experimental results showed stable behavior, but this is probably attributable to the formation of a fiber plug, as in the case described above for PAC. Another polymer tested was a 1:1 blend of XG and PAC, which was pursued to determine whether the combination of these two would provide as much stability as XG alone. The two thicker fluids exhibited stability, while the two thinner fluids did not (Fig. 23a). Thus, the blend did not perform as well as pure XG, and the addition of XG resulted in an opaque fluid, so that visual clues to the mixed fluid's stability were absent.

Drilling fluids are usually weighted in order to control formation pressure and support the borehole wall. Weighting is usually carried out through the addition of a weighting agent like barite. As the fluid density is increased, so does the density difference between the fluid and the fiber, which increases the buoyancy force. It is expected that this in turn will increase the rising velocity of the fiber. On the other hand, increasing the concentration of weighting material usually results in an increase in viscosity, including the yield stress. Figs. 24 and 25 depict the results of the weighted fluid stability experiments. Despite the increased buoyancy force acting on the fiber particle, the experiments showed that the weighted fiber fluids were just as stable as the unweighted ones (ref **Fig. 20**), evidently because the viscosity increase of the weighted fluid kept pace with the density increase. The weighted and unweighted oil-based mud (OBM) and synthetic-based mud (SBM) demonstrated similar stability behavior (Fig. 25). In these two fluid systems, increasing the density by the addition of barite had no detrimental effect on the stability of the suspension. The fluids are formulated to provide properties advantageous to drilling applications. They exhibit relatively low overall viscosities but have sufficient yield stress to prevent the fibers from rising.

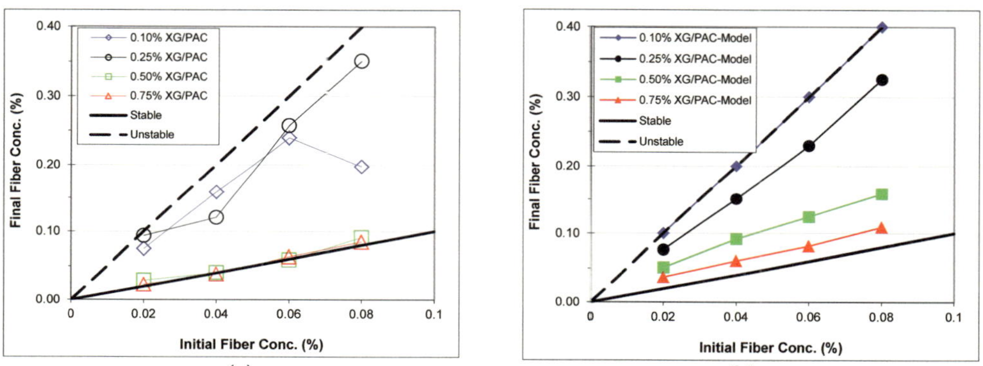

(a) (b)

Fig. 23. One-hour stabilities of XG/PAC based fiber sweeps at 77°C: a) Lab experiments; b) Mathematical model

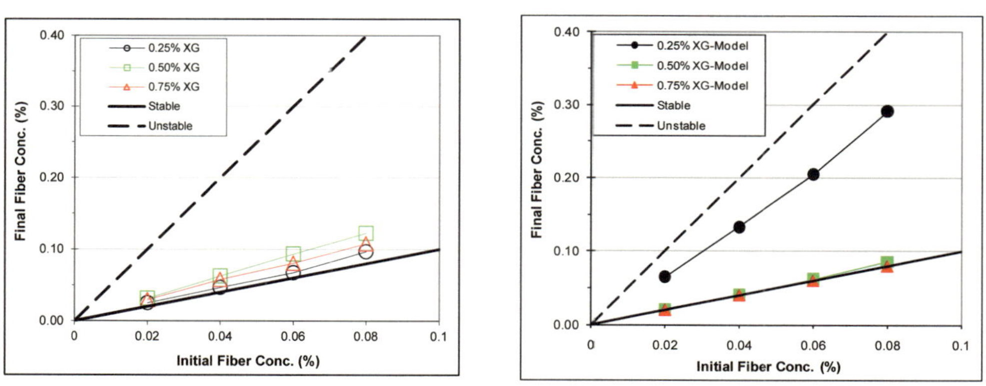

(a) (b)

Fig. 24. One-hour stabilities of XG based weighted (1438 kg/m^3) fiber sweeps at 77°C: a) Lab experiments; b) Mathematical model

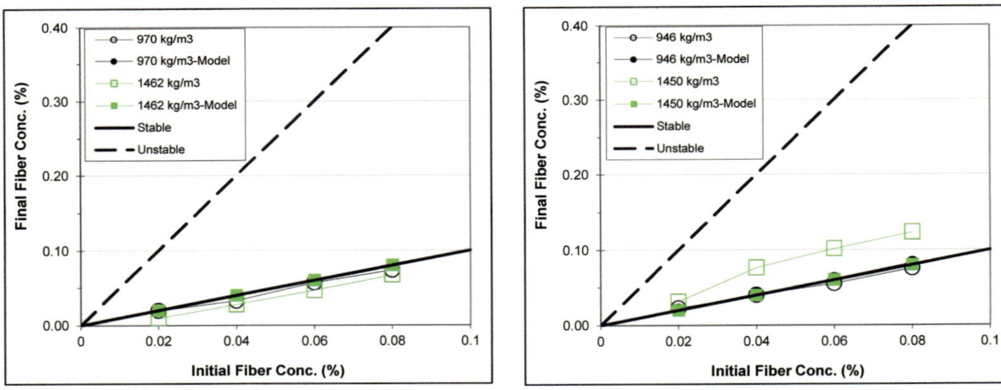

(a) (b)

Fig. 25. Measured and predicted one-hour stability of non-aqueous based fluids at 77°C: a) Oil-based mud (OBM); b) Synthetic-based mud (SBM)

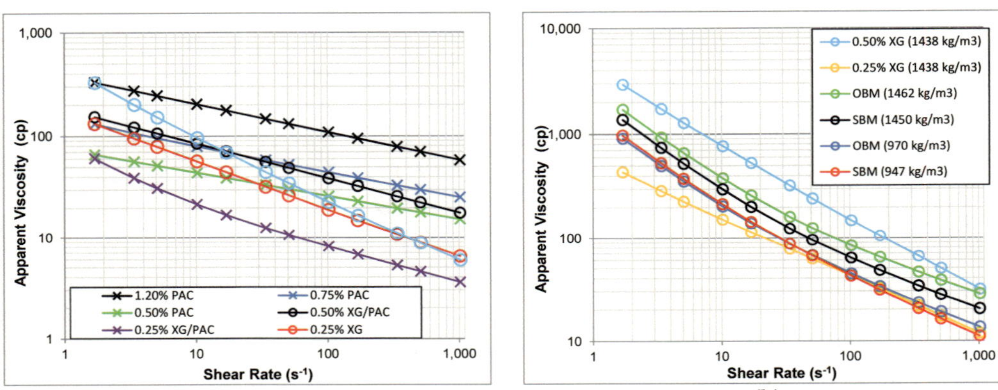

(a) (b)

Fig. 26. Apparent viscosity vs. shear rate of water and non-aqueous based fluids at 77°C: a) Low-viscosity fluids; b) High-viscosity fluids

Intuitively, the magnitude of the fluid's resistance to deformation or viscosity should provide a good indication of its internal ability to resist flow. Therefore, an increase in a fluid's viscosity would result in an increase in the ability to hold particles in suspension. As it pertains to the current study, an increase in polymer concentration generally results in higher viscosity, which in turn should yield a more stable fiber suspension. Visual observation of the fluids while mixing can also provide insight into the fiber fluid's probable stability. The rheological study on fibrous fluids that was undertaken previously was reconciled with the stability experiment results to compare the apparent viscosities of the stable and unstable fluids (**Fig. 26**). Fig. 26a shows a few low viscosity fluids, some of which were unstable. This data contradicts the original hypothesis that the fluid's viscosity or rheological properties would determine its ability to maintain stability. Easily apparent is the viscosity profile of the 1.20% PAC fluid, overall the most viscous fluid present in the figure. Despite its relatively high viscosity, this fluid was unstable (Fig. 26a). Also present in this figure are three stable fluids (0.25% XG, 0.50% XG/PAC, and 0.15% PHPA). It is important to note that these fluids exhibited lower apparent viscosity at low shear rate than the 1.20% PAC fluid. The 0.25% XG fluid even exhibited lower viscosity than the 0.75% PAC fluid, which was also unstable.In general, XG based fluids showed lower fiber rising velocities than other polymeric fluids. Xanthan Gum polymer may have a structure than can easily tangle with the fiber particles. Fig. 26b depicts various stable high-viscosity fluids that were tested, including invert emulsion OBM and SBM.. With a dispersed phase (i.e., water phase) ranging from 20% to 30%, the OBM and SBM exhibit strong structure that can hinder the movement of fiber particles. Therefore, even though rheological properties play a great role in maintaining the fiber in suspension, other properties of the fluid such as the type of polymer and the presence of fluid structure may have some influence on the ease with which fiber fluids undergo phase separation.

3.4.2 Effect of temperature on stability of fiber sweep

As stated previously, most of the experiments were carried out at 77°C, though a few were conducted at ambient temperature (20°C). As expected, most of the suspensions tested at ambient temperature showed essentially no separation of the fiber. At elevated temperature, however, most of the fluids showed some instability. Although this effect of temperature appears to be associated with viscosity, the influence of temperature declines as the fluids becomes more viscous and increasingly non-Newtonian in nature.

3.5 Comparison of model predictions with experimental results

The rising velocity model presented in Section 3.1 was developed to predict the stability of fiber suspensions and determine fluid properties necessary to prevent the separation of fibrous particles. These models take into account various forces acting on each individual particle while suspended in the fluid, such as the buoyancy force created by the difference between fiber and fluid density,. The model was formulated to predict rising velocities for particles oriented horizontal and vertical. This was done to simulate the two extreme cases of particle orientation within the fluid. The rheological parameters of the fluids obtained from viscometric measurements are inputs into both models to compare model predictions with the actual experimental measurements. As the tested fluids exhibit non-Newtonian behavior, the rising velocity of the fiber particles becomes a function of rheological parameters of the fluid, which were determined in Section 2.4. The Yield Power Law (YPL) model is a suitable constitutive equation for describing the majority of drilling fluids currently used in the industry. From the viscometric measurements, the parameters of the test fluids were determined using regression analysis and used in the models. By determining the distance the fiber particles rise in one hour and the amount of fiber that entered into the top layer in that period, we were able to estimate the final concentrations of the top layer. Model predictions shown from Figs. 20 to 25 were obtained using the horizontal orientation model. These results generally compare favorably with the test results. However, in some cases discrepancies are apparent, especially with PAC-based fluids. This indicates that some other measure of the internal structure of fiber-laden fluids is lacking.

The predictions of the vertical orientation model were also compared with fiber separation measurements, but the results did not compare favorably. Indeed, that model predicted all fluids to be unstable, which contradicts the experimental results wherein many of the fluids proved to be relatively stable. When comparing the two mathematical models, the dimensional terms differ. For both models, the area projected to the fluid flow depends on the orientation of the fiber. If the fiber is oriented perpendicularly (horizontally) to the flow, its rectangular profile (length x diameter = 1.0×10^{-6} m^2) provides the governing dimensions, whereas for vertical orientation the profile is the circular end area ($\frac{1}{4} \pi d^2 = 7.85\times10^{-9}$ m^2). However, geometry of the projected area only partially explains the difference in results. The drag coefficient (C_D) is also different for the two cases. C_D is inversely proportional to rising velocity, as it counteracts the buoyancy force. For a vertically oriented particle, drag force resisting the rising fiber is exclusively related to the fiber aspect ratio (Hoerner, 1965), independent of the fluid properties. Conversely, for a horizontally oriented particle, the drag force is implicitly related to the Reynolds Number (Perry, 1984). The rather large aspect ratio in comparison to the rectangular projected area and fluid property dependency results in a vertically oriented particle rising velocity that is anywhere from three to five orders of magnitude higher than that for horizontally oriented particles.

Fluid density is one of the controllable model parameters that has a marked influence on the stability of fiber sweeps. However, operationally this is dictated by wellbore stability issues. Rising velocity is a direct function of the difference between fluid density and particle density. However, the increased viscosity associated with weighted fluid hinders settling of the barite particles For instance, the model prediction for the stability of the weighted XG based fluids (Fig. 24b) shows both favorable and unfavorable results, while the experimental results indicated stable fluids with a slight departure from complete uniformity for the two thicker fluids. With its current formulation, the model considers a single particle rising in the fluid. It does not account for the hindering effect of barite particles.

The oil-based and synthetic-based muds (Fig. 25) showed remarkable stability, both experimentally and mathematically. Both the weighted and unweighted OBM and SBM exhibited high yield stress, and all fiber-fluid combinations tested showed stable behavior. This was reinforced with the mathematical model predictions of similar results.

3.6 Conclusions

This study was undertaken to investigate the stability of fiber sweeps at ambient and elevated temperature conditions. Experiments were conducted using different base fluids (several water-based and non-aqueous fluids) with varying fiber concentrations. Fibers were extracted from the samples after the test and weighed to determine the final fiber concentration in the top layer. This data was used to determine if the fiber had risen while in the fluid sample, or if uniformity had persisted throughout the length of the experiment. These measurements were compared with stability predictions obtained from the mathematical model. After analyzing and comparing all the data to date, the following inferences can be made:

- Horizontally oriented particle model predictions are in general concurrence with the experimental data, and reasonable real-time application performance can be predicted using the model. The vertically oriented particle model overestimates rising velocity of fibers in all fluids tested, which does not reflect experimental results and is not expected to provide accurate predictions.

- Despite the dominant effect of fluid viscosity on the phase separation of fiber sweeps, other properties of the fluid such as elasticity and extensional viscosity should be examined.
- Selecting the type of polymer used for drilling sweep applications is critical in designing fluids that have good stability under downhole conditions. XG polymer appears to be a better choice than PAC or PHPA.
- Oil-based and synthetic-based fluids possess high resistance to separation of fibers from the base fluids. This could be attributed to the high yield stress that they exhibit and the presence of internal structure perhaps associated with their invert emulsion character.

4. Nomenclature

A_c = Cross-sectional area (m^2)
$A_{p,h}$ = *Projection area of horizontally oriented particle (m^2)*
$A_{p,v}$ = *Projection area of vertically oriented particle (m^2)*
BP = Bingham Plastic
c = *Average fiber concentration*
$C_{D,h}$ = *Drag coefficient for a horizontally oriented particle*
$C_{D,v}$ = *Drag coefficient for a vertically oriented particle*
c_f = *Initial fiber concentration*
d = *diameter of fiber particle (m)*
d_i = *inner diameter of annulus (m)*
d_o = *outer diameter of annulus (m)*
D_h = *hydraulic diameter, $d_o - d_i$ (m)*
ECD = Equivalent circulating density
F_B = *Buoyancy force (N)*
F_D = Drag force (N)
F_{shear} = Shear force (N)
g = gravitational acceleration
K = Consistency index (Pa-s^n)
L = Fiber particle length (m)
N = Flow behavior index
n = Fluid behavior index
OBM = Oil-based mud
PAC = Polyanionic Cellulose
PHPA = Partially-Hydrolyzed Polyacrylamide
PL =Power Law
R = Radius (m)
R^2 = Coefficient of Determination
Re = Reynolds Number
V_p = Volume of fiber particle (m^3)
XG = Xanthan Gum
YPL = *Yield Power Law*
s = second
SBM = Synthetic-based mud
t = time
$U_{p,h}$ = *Rising velocity of horizontally oriented particle (m/s)*
$U_{p,v}$ = *Rising velocity of horizontally oriented particle (m/s)*

Greek Letters

$\dot{\gamma}$ = shear rate (s_{-1})
$\dot{\gamma}_{ave}$ = Average shear rate (s^{-1})
$\dot{\gamma}_{primary}$ = Primary shear rate (s^{-1})
$\dot{\gamma}_{secondary}$ = Secondary shear rate (s^{-1})
$\dot{\gamma}_{total}$ = Total shear rate (s^{-1})
θ = Angle
μ = Fluid viscosity (Pa-s)
μ_{app} = Apparent fluid viscosity (Pa-s)
ρ = density (kg/m^3)
ρ_f = fluid density (kg/m^3)
ρ_p = particle density (kg/m^3)
τ = Shear stress (Pa)
τ_y = Yield stress (Pa)
$\tau_{y,h}$ = Critical yield stress of horizontally oriented particle (Pa)
$\tau_{y,v}$ = Critical yield stress of vertically oriented particle (Pa)

5. References

Ahmed, R.M. & Takach, N.E. 2008. Fiber Sweeps for Hole Cleaning. Paper SPE 113746 presented at the SPE/ICoTA Coiled Tubing and Well Intervention Conference and Exhibition, The Woodlands, Texas, 1-2 April.

Alderman, N.J., Gavignet, A., Guillot, D. & Maitland, G.C. 1998. High-Temperature, High-Pressure Rheology of Water-Based Muds. Paper SPE 18035 presented at the SPE Annual Technical Conference and Exhibition, Houston, Texas, 2-5 October.

Bennington, C.P.J., Kerekes, R.J., & Grace, J.R. 1990. The Yield Stress of Fibre Suspensions. *Can. J. Chem. Eng.* 68 (5): 748-757

Bivins, C.H., Boney, C., Fredd, C., Lassek, J. Sullivan, P., Engels, J., Fielder, E.O. et. al. 2005. New Fibers for Hydraulic Fracturing. *Schlumberger Oilfield Review* 17 (2): 34-43.

Bulgachev, R.V. & Pouget, P. 2006. New Experience in Monofilament Fiber Tandem Sweeps Hole Cleaning Performance on Kharyaga Oilfield, Timan-Pechora Region of Russia. Paper SPE 101961 presented at the SPE Russian Oil and Gas Technical Conference and Exhibition, Moscow, Russia, 3-6 October.

Cameron C. 2001. Drilling Fluids Design and Management for Extended Reach Drilling. Paper SPE 72290 presented at the IADC/SPE Middle East Drilling Technology conference, Bahrain, 22-24 October.

Cameron, C., Helmy, H., & Haikal, M. 2003. Fibrous LCM Sweeps Enhance Hole Cleaning and ROP on Extended Reach Well in Abu Dhabi. Paper SPE 81419 presented at the SPE 13th Middle East Oil Show and Conference, Bahrain, 5-8 April.

Chaouche, M. & Koch, D.L. 2001. Rheology of non-Brownian Rigid Fiber Suspensions with Adhesive Contacts. *J. Rheol.* 45 (2): 369-382

Chen, B, Tatsumi, D., & Matsumoto, T. 2002. Floc Structure and Flow Properties of Pulp Fiber Suspensions. *J. Soc. Rheol., Jpn.* 30 (1): 19-25

Dedegil, M.Y. 1987. Drag Coefficient and Settling Velocity of Particles in Non-Newtonian Suspensions. *Journal of Fluids Engineering* 109 (3): 319-323.

Demirdal, B., Miska, S., Takach, N.E. & Cunha, J.C. 2007. Drilling Fluids Rheological and Volumetric Characterization Under Downhole Conditions. Paper SPE 108111 presented at the SPE Latin American and Caribbean Petroleum Engineering Conference, Buenos Aires, Argentina, 15-18 April.

Drilling Fluid Rheology. 2001. Kelco Oil Field Group, Houston, Texas (Rev. Sep 2005).

George, M.L., Ahmed, R.M. & Growcock, F.B. 2011. Rheological Properties of Fiber-Containing Drilling Sweeps at Ambient and High Temperature Conditions. Paper AADE-11-NTCE-35 presented at the AADE National Technical Conference & Exhibition, Houston, Texas, USA, 12-14 April.

Goto, S., Nagazono, H., & Kato, H. 1986. The Flow Behavior of Fiber Suspensions in Netwonian Fluids and Polymer Solutions. I. Mechanical Properties. *Rheologica Acta* 25 (2): 119-129

Guo, R., Azaiez, J., & Bellehumeur, C. 2005. Rheology of Fiber Filled Polymer Melts: Role of Fiber-Fiber Interactions and Polymer-Fiber Coupling. *Polymer Eng. and Sci.* 45 (3): 385-399.

Hemphill, T. & Rojas, J.C. 2002. Drilling Fluid Sweeps: Their Evaluation, Timing, and Applications. Paper SPE 77448 presented at the SPE Annual Technical Conference and Exhibition, San Antonio, Texas, 29 September-2 October.

Herzhaft, B., Guazzelli, E., Mackaplow, M., & Shaqfeh, E. 1996. Experimental Investigation of a Sedimentation of a Dilute Fiber Suspension. *Phys. Rev. Lett.* 77 (2): 290-293

Herzhaft, B. & Guazzelli, E. 1999. Experimental Study of Sedimentation of Dilute and Semi-Dilute Suspensions of Fibres. *J. Fluid Mech.* 384: 133-158

Hoerner, S.F. 1965. *Fluid-Dynamic Drag; Practical Information on Aerodynamic Drag and Hydrodynamic Resistance.* Midland Park, New Jersey: Hoerner Fluid Dynamics

Koch, D. & Shaqfeh, E. 1989. The Instability of a Dispersion of Sedimenting Spheroids. *J. Fluid Mech.* 209: 521-542

Kuusela, E., Hofler, K., & Schwarzer, S. 2001. Computation of Particle Settling Speed and Orientation Distribution in Suspensions of Prolate Spheroids. *J. Eng. Math.* 41 (2-3): 221-235

Kuusela, E. & Lahtinen, J. 2003. Collective Effects in Settling of Spheroids Under Seatdy-State Sedimentation. *Phys. Rev. Lett.* 90 (9): 1-4

Liu, Y.J. & Joseph, D.D. 1993. Sedimentation of Particles in Polymer Solutions. *J. Fluid Mech.* 255: 565-595

Maehs, J., Renne, S., Logan, B. & Diaz, N. 2010. Proven Methods and Techniques to Reduce Torque and Drag in the Pre-Planning and Drilling Execution of Oil and Gas Wells. Paper SPE 128329 presented at the IADC/SPE Drilling Conference and Exhibition, New Orleans, Louisiana, 2-4 February.

Majidi, R. & Takach, N. 2011. Fiber Sweeps Improve Hole Cleaning. Paper AADE-11-NTCE-36 presented at the AADE National Technical Conference & Exhibition, Houston, Texas, USA, 12-14 April.

Marti, I., Hofler, O., Fischer, P. & Windhab, E.J. 2005. Rheology of Concentrated Suspensions Containing Mixtures of Spheres and Fibres. *Rheologica Acta* 44 (5): 502–512.

Metzner, A.B. & Reed, J.C. 1955. Flow of Non-Newtonian Fluids - Correlation of the Laminar, Transition, and Turbulent-flow Regions. *A.I.Ch.E. Journal* 1 (4): 434-440.

Mewis, J. & Metzner, A.B. 1974. The Rheological Properties of Suspensions of Fibers in Newtonian Fluids Subjected to Extensional Deformations. *J. Fluid Mech.* 62 (3): 593-600

Meyer, R., Wahren, D. 1964. On the elastic properties of three-dimensional fibre networks. *Sven. Papperstidn.* 67: 432-436

Miska, S. 2007. Advanced Drilling, Course Material, University of Tulsa.

Nawab, M.A. & Mason, S.G. 1958. The Viscosity of Dilute Suspensions of Thread-Like Particles. *J. Phys. Chem.* 62 (10): 1248–1253

Perry, R.H. & Green, D.W. 1984. *Perry's Chemical Engineering Handbook. 6th Edition.* Japan: McGraw-Hill.

Power, D.J., Hight, C., Weisinger, D. & Rimer, C. 2000. Drilling Practices and Sweep Selection for Efficient Hole Cleaning in Deviated Wellbores. Paper SPE 62794 presented at the IADC/SPE Asia Pacific Drilling Technology conference, Kuala Lumpur, Malaysia, 11-13 September.

Qi, G.Q., Nathan, G.J., & Kelso, R.M. 2011. Aerodynamics of Long Aspect Ratio Fibrous Particles Under Settling. Paper AJTEC2011-44061 presented at the ASME/JSME 8th Thermal Engineering Joint Conference, Honolulu, Hawaii, USA, 13-17 March.

Rajabian, M., Dubois, C., & Grmela, M. 2005. Suspensions of Semiflexible Fibers in Polymeric Fluids: Rheology and Thermodynamics. *Rheologica Acta* 44 (5): 521–535.

Ravi, K.M. & Sutton, D.L. 1990. New Rheological Correlation for Cement Slurries as a Function of Temperature. Paper SPE 20449 presented at the SPE Annual Technical Conference and Exhibition, New Orleans, Louisiana, 23-26 September.

Robertson, N., Hancock, S., & Mota, L. 2005. Effective Torque Management of Wytch Farm Extended-Reach Sidetrack Wells. Paper SPE 95430 presented at the SPE Annual Technical Conference and Exhibition, Dallas, Texas, 9-12 October.

Ross, D.F. & Klingenberg, D.J. 1997. Dynamic Simulation of Flexible Fibers composed of Linked Rigid Bodies. *J. Chem. Phys.* 106 (7): 2949-2960

Schmid, C.F. & Klingenberg, D.J. 2000. Mechanical Flocculation of Flowing Fiber Suspensions. *Phys. Rev. Lett.* 84 (2): 290-293

Schmid, C.F., Switzer, L.H., & Klingenberg, D.J. 2000. Simulations of Fiber Flocculation: Effects of Fiber Properties and Interfiber Friction. *J. Rheol.* 44 (4): 781-809

Scholz, M. 2006. *Wetland Systems to Control Urban Runoff.* Amsterdam, The Netherlands: Elsevier.

Sozynksi, R.M. & Kerekes, R.J. 1988a. Elastic Interlocking of Nylon Fibers Suspended in Liquid. Part I. Nature of Cohesion Among Fibers. *Nord Pulp Paper Res. J.* 3: 172-179

Sozynksi, R.M. & Kerekes, R.J. 1988b. Elastic Interlocking of Nylon Fibers Suspended in Liquid. Part II. Process of Interlocking. *Nord Pulp Paper Res. J.* 3: 180-184

Swerin, A. 1997. Rheological Properties of Cellulosic Fibre Suspensions Flocculated by Cationic Polyacrylamides. *Colloids Surf. A: Physicochem. Eng. Aspects* 133 (3): 279-294

Switzer, L.H. & Klingenberg, D.J. 2003. Rheology of Sheared Flexible Fiber Suspensions via Fiber-Level Simulations. *J. Rheol.* 47 (3): 759-778

Switzer, L.H. & Klingenberg, D.J. 2004. Flocculation in Simulations of Sheared Fiber Suspensions. *Intl. J. Mult. Flow* 30 (1): 759-778

Thalen, N. & Wahren, D. 1964. Shear Modulus and Ultimate Shear Strength of Some Paper Pulp Fibre Networks. *Sven. Papperstidn.* 67 (7): 259-264

Valluri, S.G., Miska, S.Z., Ahmed, R.M. & Takach, N.E. 2006. Experimental Study of Effective Hole Cleaning Using "Sweeps" in Horizontal Wellbores. Paper SPE 101220 presented at the SPE Annual Technical Conference and Exhibition, San Antonio, Texas, 24-27 September.

Xu, A.H. & Aidun, C.K. 2005. Characteristics of Fiber Suspension Flow in a Rectangular Channel. *Intl. J. Multiphase Flow* 31 (3) 318–336.

Yamamoto, S. & Matsuoka, T. 1993. A Method for Dynamic Simulation of Rigid and Flexible Fibers in a Flow Field. *J. Chem. Phys.* 98 (1): 644-650

Yu, M., Takach, N.E., Nakamura, D.R. & Shariff, M.M. 2007. An Experimental Study of Hole Cleaning Under Simulated Downhole Conditions. Paper SPE 109840 presented at the SPE Annual Technical Conference and Exhibition, Anaheim, California, 11-14 November.

Zhou, L., Ahmed, R.M., Miska, S.Z., Takach, N.E., Yu, M., & Pickell, M.B. 2004. Experimental Study & Modeling of Cuttings Transport with Aerated Mud in Horizontal Wellbore at Simulated Downhole Conditions. Paper SPE 90038 presented at the SPE Annual Technical Conference and Exhibition, Dallas, Texas, 26-29 September.

Zhu, C. 2005. Cuttings Transport with Foam in Horizontal Concentric Annulus Under Elevated Pressure and Temperature Conditions. Ph.D. Dissertation, University of Tulsa, Tulsa, Oklahoma.

Part 4

Food Rheology

10

Solution Properties of κ-Carrageenan and Its Interaction with Other Polysaccharides in Aqueous Media

Alberto Tecante[1,*] and María del Carmen Núñez Santiago[2]
[1]*Department of Food and Biotechnology, Faculty of Chemistry, National Autonomous University of Mexico, Mexico, D.F.,*
[2]*Centre for Development of Biotic Products, National Polytechnic Institute, Yautepec Morelos,*
[1,2]*México*

1. Introduction

Carrageenans are an important class of hydrophilic sulfated polysaccharides used as thickening, gelling and stabilizing agents in a great number of foods such as sauces, meats and dairy products. In frozen foods its high stability to freeze-thawing cycles is very important. They are also responsible of the smoothness, creaminess, and body of the products to which they are added. Its combination with starch allows different textures to be obtained and reductions in fat content of up to 50%.The commercially important forms are kappa (κ), iota (ι) and lambda (λ). κ-carrageenan, consists of an alternating linear chain of (1→3)-β-D-galactose-4SO_3^--(1→4)-3,6, anhydro-α-D-galactose. It is soluble in hot water (> 75 °C) and even low concentrations (0.1 to 0.5%) yield high viscosity solutions. Viscosity is stable over a wide pH range, because the semi-ester sulfates are always ionized even under strongly acidic conditions. κ-carrageenan can adopt different conformations in solution, e.g. random coil and double helix. Therefore, its rheological behavior is strongly affected by the total ionic concentration, temperature and ion content of the system.

In this chapter we review the solution properties of κ-carrageenan and its interaction with starch and non-starch polysaccharides in aqueous media. We stress the importance of the sol-gel transition and gelation mechanism of κ-carrageenan; particularly the effect of temperature, polysaccharide concentration and external counterions on the transition and the interaction of κ-carrageenan with other polysaccharides. Given the economic importance of κ-carrageenan, we also discuss its viscoelastic behavior and microstructure as well as actual and potential applications in foods.

2. Origin of carrageenans and their classification

Carrageenans are found in marine red algae of the family *Rhodophyceae* (Snoeren, 1976, Chen et al., 2002). The main source is the *Chondrus crispus*, collected along the coast of North

* Corresponding Author

America from Boston to Halifax and the *Eucheuma* species, which is obtained mainly from cultivation in shallow waters around the Philippines (Whistler & BeMiller, 1997). Carrageenans constitute from 30 to 80% of the cell wall of these algae, and their functionality depends on the species, season, and growing conditions. They are composed of linear chains of D-galactopyranosyl units linked via alternated (1→3)-β-D-and (1→4)-α-D-glucoside (Chandrasekaran, 1998), in which sugar units have one or two sulfate groups. Some units contain a 3,6-anhydro ring. This provides a sulfate content of 15 to 40% (Fennema, 2002). Depending on the amount and position of the SO_3^- groups carrageenans are classified as μ, ν, λ, ξ, κ, ι, and θ types (Stanley, 1987) (Figure 1). The chemical and functional properties of each type of carrageenan are different due to the presence of sulfate groups. For example, ι and κ-carrageenan do form gels in the presence of counterions, whereas λ-carrageenan does not. The commercially available carrageenans are utilized to prepare a wide variety of gels; clear or cloudy, rigid or elastic, hard or soft, thermo-stable and with or without syneresis (Whistler & BeMiller, 1997). Carrageenan gels do not require refrigeration since they do not melt at room temperature.

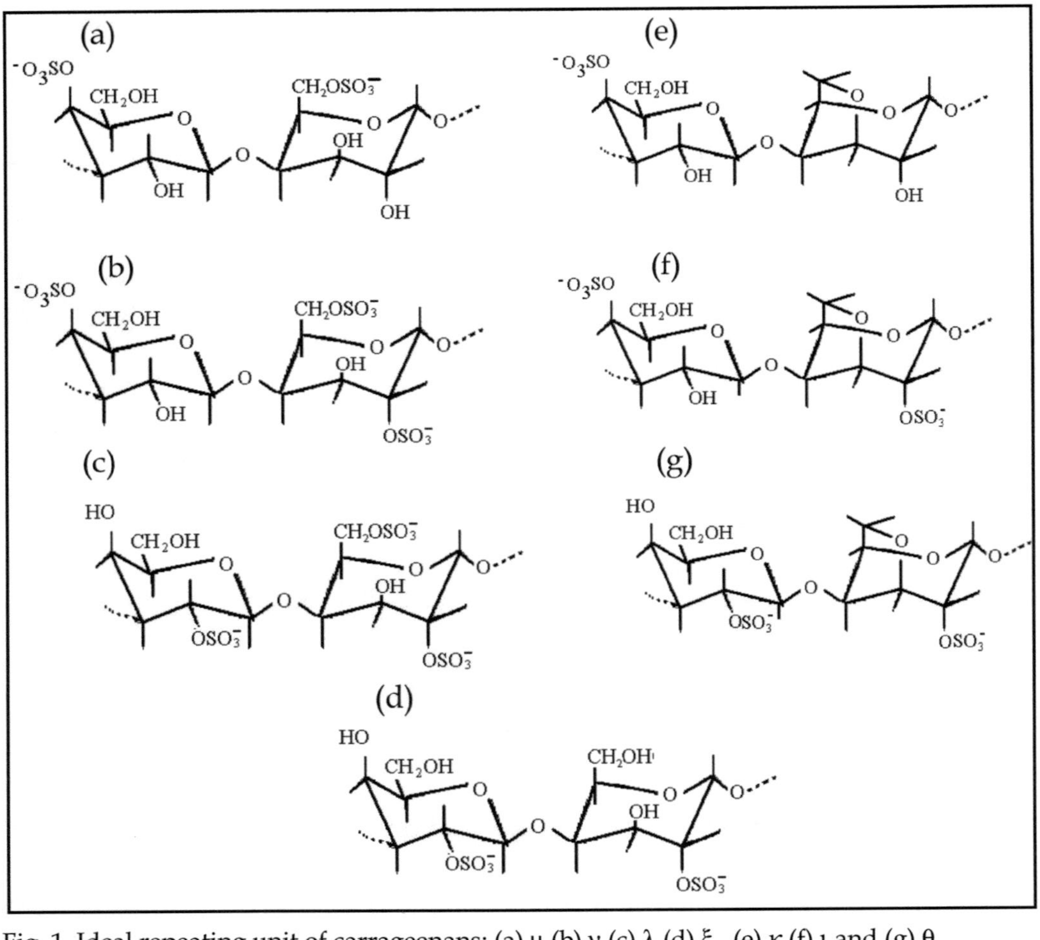

Fig. 1. Ideal repeating unit of carrageenans: (a) μ (b) ν (c) λ (d) ξ , (e) κ (f) ι and (g) θ carrageenan (adapted from Stanley, 1987).

Due to their hydrophilic nature, carrageenans are used as thickening agents (mainly λ-carrageenan), gelling agents (κ and ι–carrageenan), stabilizers or combinations of these functions in a number of food products, standardized with the necessary amounts of sucrose, glucose, salts or gelling aids, such as KCl.

3. Crystal structure

As κ-carrageenan molecules have a net negative charge, the polysaccharide is sensitive to ionic interactions with anions and cations. The interaction of κ-carrageenan with various counterions, particularly with potassium, has been studied by X-ray diffraction (Chandrasekaran, 1998). κ-carrageenan strands are less oriented and less crystalline that the salts of ι-carrageenan. However, like the iota form, κ-carrageenan adopts a double helix conformation. As shown in Figure 2, the double helices are stable parallel right-handed strands forming twisted in-phase structures so that when an anhydrous galactose residue of one strand is in front of a sulfate group of the other strand, the arrangement is stabilized by a hydrogen bond. A twist between the residues of the anhydrous galactose molecule has three hydrogen bonds that stabilize the helix (dotted lines in Figure 2). The sulfate groups are located on the periphery of the double helix, and are interacting with other ions.

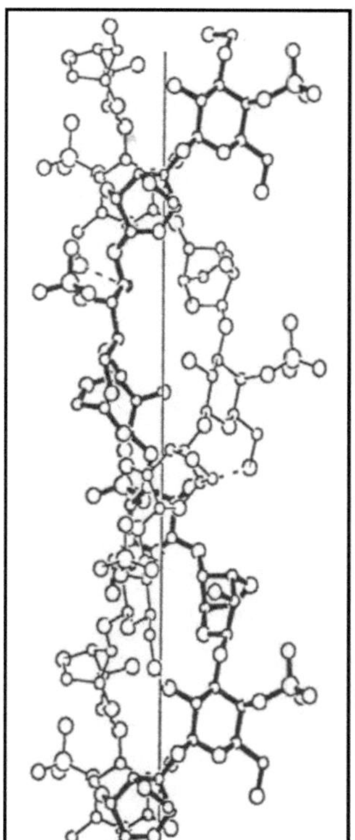

Fig. 2. Crystal structure of κ-carrageenan. One strand is clear and the other dark to distinguish them from each other. The dotted lines represent hydrogen bonds (adapted from Chandrasekaran, 1998).

4. Conformation in solution in the presence of counterions

The type of external counterions plays an important role in the solution properties of κ-carrageenan as a result of the decrease in the effective charge density of the polysaccharide (Takemesa & Nishinari, 2004). In sodium chloride solutions, κ–carrageenan adopts a disordered conformation (Snoeren, 1976) while in sodium iodide solutions the polysaccharide adopts an ordered helical conformation (Slootmaekers et al., 1988) at 25 °C in both cases. In the disordered state κ–carrageenan exists as a random coil, expanded as a result of the effect of the excluded volume and electrostatic repulsions between chain segments (Snoeren, 1976; Vreeman et al., 1980) with a high water absorption capacity (Harding et al., 1996). In this conformation, the chains are flexible and sensitive to the presence of ions. For some time, there was controversy on whether the ordered helical conformation consisted of one or two helices (Slootmaekers et al., 1988), however, light diffraction observations showed that the ordered conformation has twice the molecular mass of a random coil, a fact that is interpreted as the adoption of a double helix conformation (Viebke et al., 1995). Thus, the transition of κ–carrageenan is described from a disordered state to an ordered state as 2 random coils ↔ 1 double helix.

Meunier et al. (2001) showed that formation of double helices does not yield exactly twice the molecular mass of a random coil and that the relationship between the molecular masses and radii of gyration between the two conformations depends on the range of molecular mass of the random coils. Studies of κ-carrageenan in sodium iodide solution show that the ordered state is a rigid structure (Chronakis et al., 2000) with a high charge density (Takemesa & Nishinari, 2004). The disorder-order transition of κ-carrageenan occurs at a given temperature called the transition temperature, $T_{d\text{-}o}$ (Rochas, 1982). This temperature depends on the nature of the counterion and the total ionic concentration. The same behavior is shown by other polyelectrolytes, e.g. gellan (Milas et al., 1990) and ι-carrageenan (van de Velde et al., 2002).

κ-carrageenan has different affinities for some ions. Considering the ability to stabilize the ordered conformation, determined from measured values of $T_{o\text{-}d}$, the decreasing order of affinity is (Rochas, 1982):

- monovalent cations: $Rb^+ > Cs^+ > K^+ > NH_4^+ > (CH_3)_4N^+ > Na^+ > Li^+$
- divalent cations: $Ba^{2+} > Ca^{2+} > Sr^{2+} > Mg^{2+} > Zn^{2+} > Co^{2+}$

According to this sequence, the monovalent cations Rb^+, Cs^+ and K^+ induce the transition at lower temperatures. Unlike other polyelectrolytes, the disorder-order transition of κ-carrageenan is also sensitive to the presence of iodide as a result of electrostatic repulsion between the anion and the sulfate group of κ–carrageenan without aggregation of double helices to form a gel network. Because of this particular feature, the iodide ion has been used to study the mechanism of gelation of this polysaccharide (Slootmaekers et al., 1988; Viebke et al., 1995, 1998; Meunier et al., 2001; Takemesa & Nishinari, 2004).

5. The sol-gel transition diagram

Rochas (1982) obtained the sol-gel transition diagram of the potassium salt of κ-carrageenan using polarimetry and ionic conductivity during heating and cooling treatments. The diagram shows the relationship between the total ionic concentration and the inverse of the absolute temperature of transition (Figure 3). κ-carrageenan gels are thermoreversible, i.e. they can be formed upon cooling hot solutions and melted upon heating. Above a critical concentration C_{crit} the melting temperature (T_{melt}) is higher than the gelling temperature (T_{gel}). This phenomenon, known as thermal hysteresis, is a consequence of aggregation of the helical structures during gel formation. Below the critical concentration, which for the potassium salt of κ-carrageenan is about 0.007 mmol/dm³ (Figure 3), thermal hysteresis does not exist. The total ionic concentration, C_T, is given by:

$$C_T = C_S + \gamma C_P \quad (1)$$

where C_S is the concentration of added salt, C_P is the concentration of charged groups in the polysaccharide and γ is the average activity coefficient; 0.55 for the potassium salt of κ-carrageenan.

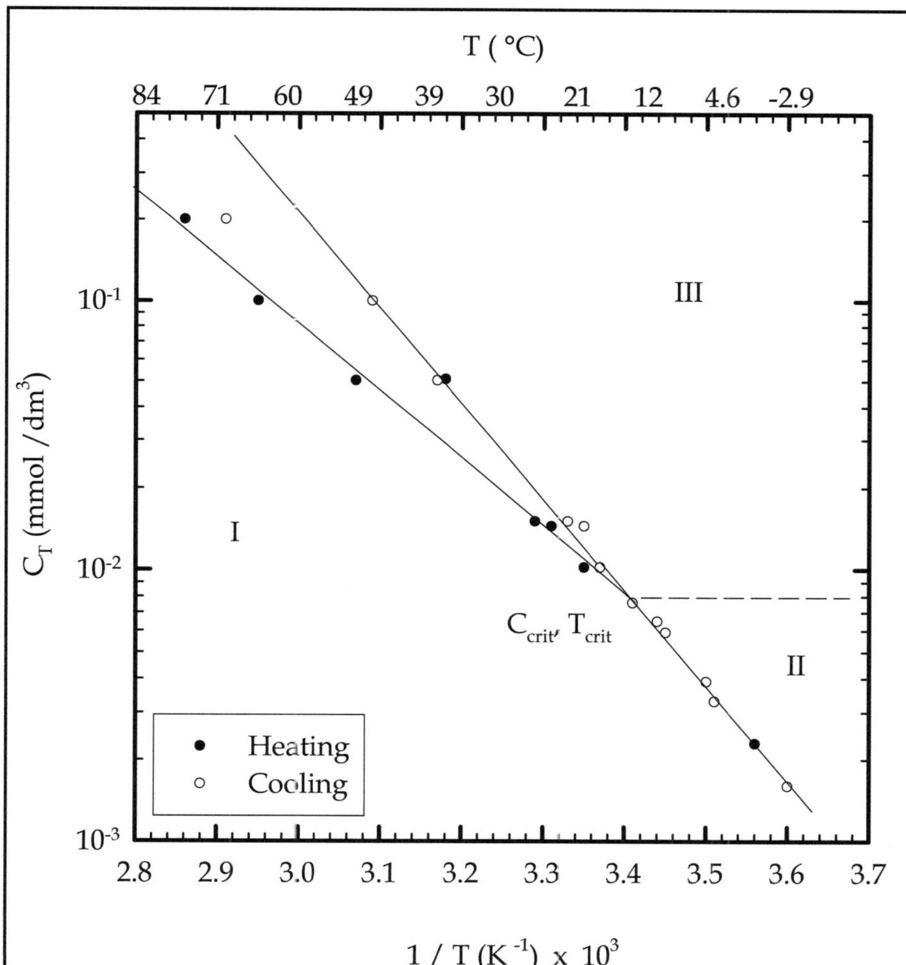

Fig. 3. Sol-gel transition diagram of the potassium salt of κ-carrageenan (adapted from Rochas & Rinaudo, 1982). The dotted line represents the division between zone I and II.

According to the sol-gel transition diagram, gels can be obtained from low concentrations of κ-carrageenan and high concentrations of potassium ion and viceversa; these gels have, however, different rheological behaviors.

Three regions can be identified on the transition diagram (Rochas, 1982). In each one of these regions κ-carrageenan adopts different conformations. The boundary between regions II and III is only approximate (Fig. 3). In region I, the polysaccharide exists as a random coil for temperatures above T_{melt}. In region II, dimers are formed without aggregation for temperatures below T_{gel} and C_T lower than C_{crit}. In region III, helical conformation exists and aggregation leads to a three-dimensional network for C_T greater than C_{crit} and T lower than T_{gel}. Therefore, it is possible to maintain conditions leading to the formation of double helices without aggregation. When the polysaccharide chains are in this condition, formation of double helices does not depend on polysaccharide concentration (Mangione et al., 2003), as long as C_T is lower than C_{crit}. In this type of polysaccharides, the transition is characterized by a linear relationship between the inverse of T_{gel} and log C_T (Rinaudo, 2001), but it is particular to each polysaccharide and the salt added to promote gelation. The diagram is very useful to set the conditions for a specific application.

6. Mechanism of gelation

The mechanism of gelation of κ-carrageenan is reported as a process divided into several stages (Figure 4). Starting from the random coil conformation, the polysaccharide is in solution at a concentration below C_{crit} and at a temperature above T_{melt} (**a** in Figure 4); this corresponds to region I in the sol-gel transition diagram (Figure 3). A decrease in temperature below T_{gel} keeping a constant concentration or an increase in concentration at constant temperature leads to the formation of a helical dimer (**c** in Figure 4); this corresponds to region II in Figure 3. For a given temperature, increasing C_T above C_{crit} leads to aggregation of helical dimers (**d** in Figure 4) and formation of a three-dimensional network (**e** in Figure 4) with possible dependence on time (Rochas, 1982; Chen et al., 2002; Takemesa & Chiba, 2001; Yuguchi et al., 2002). This condition corresponds to region III in Figure 3. It is possible to change from this condition to one in which a three-dimensional network does not exist, instead a semi diluted solution of random coil is present (**b** in Figure 4); it is necessary to increase the temperature or to decrease the concentration provided that it is kept above C_{crit}. Thus, a further decrease in C_T will lead to the starting point (**a** in Figure 4).

According to Bayley (1955) (cited by Rochas, 1982) the $-SO_3^-$, of κ-carrageenan interacts with K^+ through ionic bonds. Specifically, the potassium ion interacts simultaneously with two $-SO_3^-$ groups of anhydrous galactose.

In recent years, the development of new equipments and techniques has allowed the mechanism of gelation of natural polymers to be studied. In the case of κ–carrageenan, the transition from random coil to double helix has been studied by techniques including rheology (Takemasa & Chiba, 2001, Chen et al., 2002; Nishinari & Takahashi, 2003, Mangione et al., 2003), polarimetry (Mangione et al., 2003), light scattering (Mangione et al., 2003), photon transmission (Kara et al., 2003), spectrophotometry (MacArtain et al., 2003), low amplitude X-ray scattering(Yuguchi et al., 2002, 2003), differential scanning calorimetry (Nayouf, 2003) and laser dispersion during deformation (Takemasa & Chiba, 2001). All these techniques have confirmed the association of two linear κ–carrageenan strands to form a double helix during gelation, reaffirming the mechanism proposed by Rochas (1982). The transition from random coil to double helices is extremely fast (Norton et al., 1983), whereas subsequent aggregation of double helices into small domains to form a three dimensional network (Mangione et al., 2003) is slower. Gel cure experiments monitoring the change with time of the storage modulus, G', have revealed that the time for complete gel formation, i.e. aggregation of κ–carrageenan is between 12 (Meunier et al., 1999) to 15 hours (Tecante & Doublier, 1999).

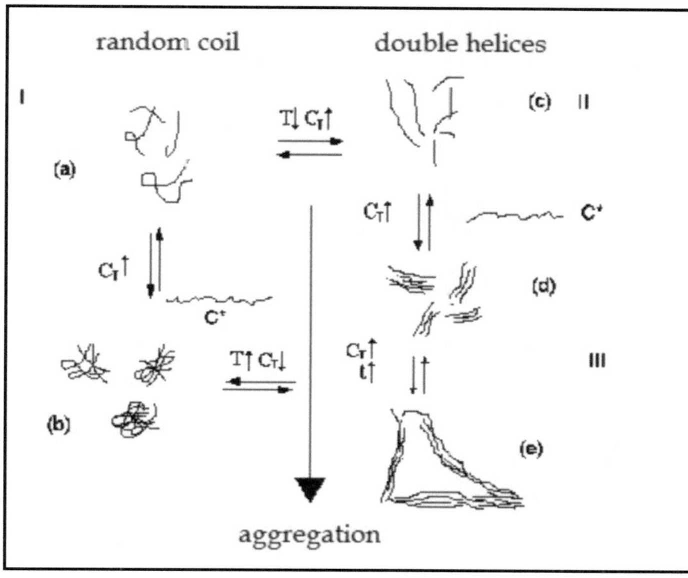

Fig. 4. Gelation model of κ-carrageenan (Rochas, 1982).

7. Rheological behavior of κ-carrageenan solutions

Like other polyelectrolytes in solution, e.g. gellan (Miyoshi & Nishinari, 1999), κ–carrageenan in the disordered conformation exhibits high sensitivity to counterions (Núñez-Santiago & Tecante, 2007; Núñez-Santiago et al., 2011). The viscoelastic behavior at 40 °C for 0.5% κ-carrageenan solutions with 0 - 30 mmol/dm^3 KCl is shown in Figure 5. As shown in Fig. 5a, the loss modulus, G", was greater than the storage modulus, G', and the dependence of both moduli with frequency ($G' \alpha\, \omega^{1.7}$, $G'' \alpha\, \omega^{0.93}$) approached that of a viscoelastic liquid ($G' \alpha\, \omega^2$, $G'' \alpha\, \omega^1$) (Ferry, 1980). Over the range of frequency, G" decreased with the increase in KCl concentration.The viscoelastic behavior was governed by the viscous component.The complex viscosity of the solutions, Fig. 5b, exhibited a plateau followed by a decrease at high frequencies.

The variation of the plateau value of the complex viscosity, i.e. the zero-shear complex viscosity, $|\eta^*|_0$, as a function of KCl concentration at 40 °C is shown in Fig. 6. Addition of KCl reduced the complex viscosity from about 14.7 to 9.7 mPa·s from 5 to 15 mmol/dm^3 KCl and then slightly increased to about 13.7 mPa·s. Therefore, κ-carrageenan is in the sol state over the KCl concentration range of 0-15 mmol/dm^3 at 40 °C. When the total ionic concentration approaches the sol-gel transition at 40 °C, incipient gelation probably occurs and consequently the zero-shear complex viscosity rises as shown. Addition of KCl reduced $|\eta^*|_0$; however, when the total ionic concentration approaches the sol-gel transition at 40°C, incipient gelation probably occurs and consequently the zero-shear complex viscosity rises as observed in Fig. 6. In polyelectrolytes, counterions shield the electrostatic repulsions between polymer chains and coil dimensions decrease as the counterions concentration increases. This shielding effect is well documented in the literature on the basis of viscosity measurements in dilute solution. The intrinsic viscosity decreases as the counterion concentration increases and a linear relationship between the intrinsic viscosity and the reciprocal of the ionic strength is generally reported; the slope depends on the rigidity of the polymer chain.

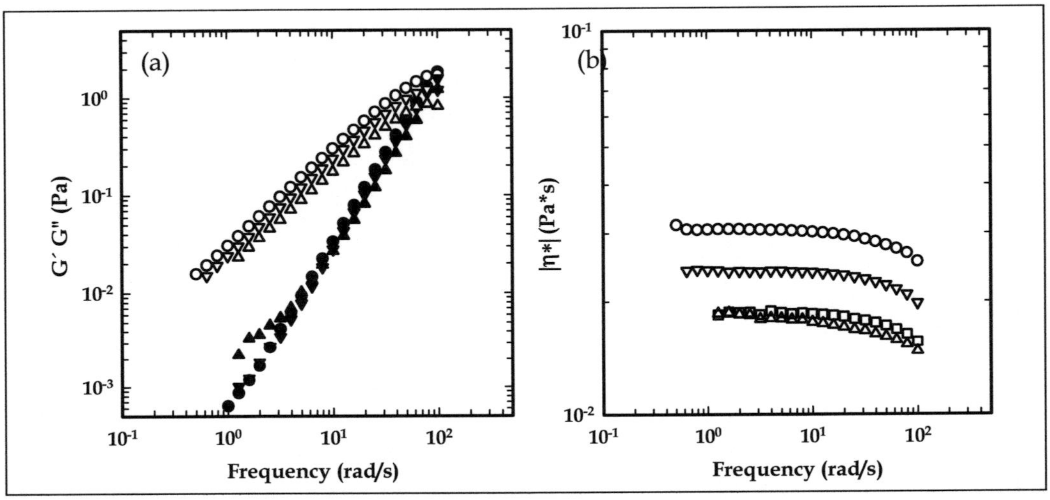

Fig. 5. Variation of G′ (black symbols), G″ (white symbols) and [η*] with angular frequency at 40 °C for 0.5% κ-carrageenan with different KCl concentrations (mmol/dm^3): (a) 0 (circles), 5 (inverted triangles), 15 (square) and 30 (triangles). Experimental conditions correspond to region I in Fig. 3.

This effect has been investigated for κ-carrageenan and values of rigidity have been determined (Snoeren, 1976; Vreeman et al., 1980). The present results (Fig 5.) for higher carrageenan concentrations can be interpreted by the same shielding effect. However, as the total ionic concentration approximates that for the sol-gel transition to occur, formation and possibly incipient aggregation of helical chains is promoted, increasing the number of junction zones driving the chains to network formation.

κ-carrageenan not only interacts with cations. In the presence of I^-, repulsion between negative charges compel κ-carrageenan to form intermolecular associations by adopting an ordered double helix conformation (Takemesa & Nishinari, 2004). "Weak gels" are formed when κ-carrageenan is used in high concentrations (> 0.9%) in the presence of I^- (Chronakis et al., 1996). This condition is characterized by the predominance of G′ over G″, with both moduli depending on frequency (Chen et al., 2002). However, additional creep tests

(Chronakis et al., 2000) and monitoring of dynamic moduli with temperature (Ikeda & Nishinari, 2001) shows that it is not a "weak gel" but a concentrated polysaccharide solution with low relaxation rates.

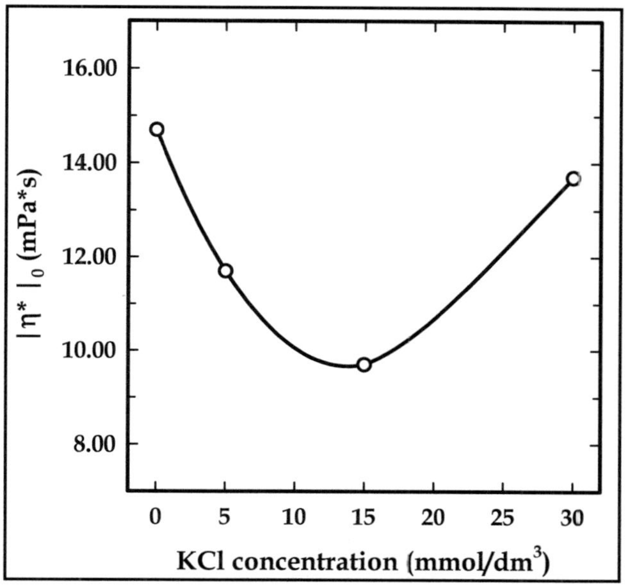

Fig. 6. Variations of the zero-shear complex viscosity with potassium chloride concentration at 40 °C. Experimental conditions correspond to zone I in Fig. 3.

Rheological and calorimetric studies of κ-carrageenan for conditions in which double helices do not aggregate in the presence of K^+ (region II in Figure 3) have been made (Núñez-Santiago et al., 2001). Figure 7a shows the DSC thermograms and the evolution of dynamic moduli with temperature for 0.3% κ-carrageenan (C_T = 4.3 mmol/dm^3) without KCl; for this condition, $C_T < C_{crit}$ ($C_{crit} \approx$ 7 mmol/dm^3). The change in temperature crosses the transition line. As a result, one "exo"and one "endo" peaks are observed on cooling and heating, respectively, with a difference between them of less than 2 °C. This confirms that helical chains formed upon cooling are not aggregated and hence do not form a three-dimensional network. The variations of G′ and G″with temperature illustrate the effect of the disorder-order on cooling and the order-disorder transition on heating although gelation does not occur as shown by the signals of G′ and G″. The dynamic moduli do not over-cross on cooling and thermal hysteresis is not observed on cooling or heating. These results show that by monitoring the change in dynamic moduli with temperature one can determine the order-disorder transition temperature either rheologically or with micro-DSC. Unlike gelation, the order-disorder transition is rapid, is time-independent and the transition enthalpy does not change with polysaccharide concentration as a consequence of the lack of aggregation of helical chains.

Fig. 7b shows the rheological behavior of 0.3% κ-carrageenan at 9 °C without KCl. The loss modulus is greater than the storage modulus over most of the frequency window and both dependon frequency ($G' \propto \omega^{0.96}$; $G'' \propto \omega^{0.74}$). This overall viscoelastic behavior reveals the fluid-like character of the system with the viscous character predominating over the elastic one. At 9 °C κ-carrageenan molecules are in the ordered state but neither aggregation nor gelation takes place. Moreover, κ-carrageenan molecules being in the ordered state are obviously stiffer than in the disordered one.

A decrease in the intrinsic viscosity, [η], with ionic strength is the classical behavior of a polyelectrolyte. In the ordered state, κ-carrageenan at 9 °C, for C_T = 1.4 mmol/dm^3 ($C_T < C_{crit}$ and $T < T_{crit}$, region I in Fig. 3), [η] = 45.5 dL/g. This value is particularly high and corresponds to a highly expanded coil because the salt concentration is so low that polysaccharide chains are not sufficiently shielded. For C_T values of 2.9, 3.6 and 4.3 mmol/dm^3 ($C_T < C_{crit}$ and $T < T_{crit}$, region II in Fig. 3), [η] was about 30 dL/g, indicating the presence of more compact chains in which aggregation is not likely to occur (Núñez-Santiago et al., 2011). The value abruptly decreases to 23 dL/g when C_T increases to 5.8 mmol/dm^3. A further slight decrease down to 20 dL/g is noticed when C_T reaches values in region III ($C_T < C_{crit}$). Moreover, for the two latter C_T values the Huggins coefficients ($k_i \approx 0.8$) are relatively high indicating a tendency to aggregation (Núñez-Santiago et al., 2011). It is noteworthy that the [η] values are more than twice higher than those obtained in NaCl (0.1 mol/dm^3) for this sample: 10.5 dL/g.

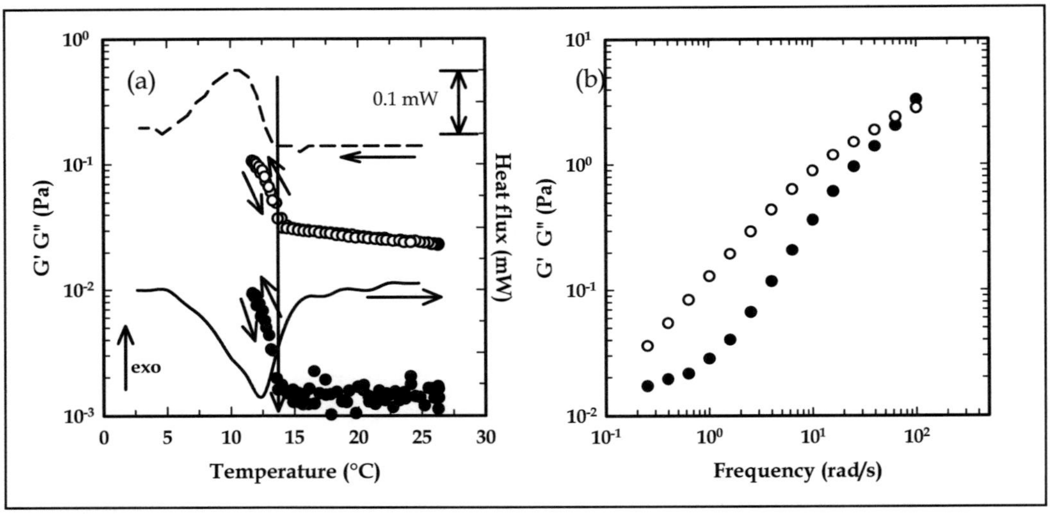

Fig. 7. (a) DSC thermograms for cooling (dotted lines) and heating (continuous lines) and variations of G′ (black symbols) and G″ (white symbols) with temperature for 0.3% κ-carrageenan, without KCl. The vertical arrow indicates T_{onset} for the transition. Frequency = 1 rad/s and strain = 50% (b) Variation of G′ (black symbols) and G″ (white symbols) with frequency for 0.3% κ-carrageenan without KCl at 9 °C for 50% strain. Experimental conditions correspond to region II in Fig. 3. Adapted from Núñez-Santiago et al. (2011).

8. Rheological behavior of κ-carrageenan gels

The solid character of κ-carrageenan gels arises from the formation of a three-dimensional network, which extends continuously through the entire volume and entraps the dispersed medium within its structure. The three-dimensional network is formed by non-covalent bonds, which have low energy and a finite lifetime. The interactions include van der Waals forces, hydrogen bonds, charge transference, ionic, hydrophilic and hydrophobic. Consequently, κ-carrageenan gels are considered physical gels as they are not formed by chemical cross-linking. The non-covalent bonds can be one or more of those mentioned above, combined with specific and complex mechanisms involving interaction zones (Kavanagh & Ross-Murphy 1998; Morris, 1990). Counterions play an important role in the gelation of polyelectrolytes. Among the monovalent cations that induce gelation of κ-carrageenan are K^+, Rb^+, Cs^+ and high concentrations of Na^+ and Li^+ (MacArtain et al., 2003). However, the gels formed with K^+ are the strongest and most stable (Chen et al., 2002).

Figure 8 shows the effect of the K^+ ion on the viscoelastic behavior of κ-carrageenan (1%) gels under conditions corresponding to region III in Figure 3. From 5 to 80 mmol/dm³ KCl, the behavior is characteristic of a gel (Ferry, 1980) with both moduli independent of frequency (i.e. $G' \alpha \omega^{0.03}$) in the range of 0.1 to 100 rad/s and G′ >> G″. In general, the rigidity of the gels, expressed by G′, increases with addition of potassium chloride. For concentrations below 80 mmol/dm³ KCl, Fig. 8a, addition of the salt yields a rapid increase in gel rigidity, whereas for concentrations above 80 mmol/dm³, such increase is limited as shown in Fig. 8b. Once a limiting concentration is reached, further addition of KCl does not increase gel rigidity. Additionally, the limiting gel rigidity depends on κ-carrageenan concentration (Núñez-Santiago & Tecante, 2007). The existence of a limiting salt concentration beyond which the rigidity of the gels reaches a constant value suggests that the polysaccharide is oversaturated with potassium ions reaching a maximum aggregation of helical chains in which no more space is available for further interactions.

Divalent cations such as Ca^{2+} and Cu^{2+} also have the ability to induce gelation of κ-carrageenan with the resulting gels having the same level of rigidity than those of monovalent cations (Michel et al., 1997). In such systems the calcium ion to sulfate group ratio is one to one (MacArtain et al., 2003). Regarding the effect of concentration of cations on the stiffness of gels, rheological (Chronakis et al., 1996, Chen et al., 2002) and uniaxial compression tests (MacArtain et al., 2003) show a progressive increase of the moduli with increasing salt concentration. The increased stiffness of the gels results from neutralization of the electrostatic charges of the polysaccharide chains which promotes their association. However, the reverse effect also exists when there is a high concentration of ions, as in the case of Ca^{2+} (Lai et al., 2000). This decrease is attributed to the fact that Ca^{2+} induces a large increase in the number of branches formed by the aggregates during gelation of κ-carrageenan, resulting in long aggregates and precipitation, which may explain the decrease of moduli when the concentration of the external ion in the medium increases (MacArtain et al., 2003).

In general, rheological studies of κ-carrageenan have been conducted with high concentrations of polysaccharide (0.5 to 2%) and concentrations of K^+ sufficiently high to ensure gel formation (> 10 mmol/dm³ KCl). However, rheological studies of κ-carrageenan in moderate concentrations (0.7 to 1.4%) in the absence of ions show that "weak gels" can be formed; their stiffness depends on temperature (Chen et al., 2002). Although these gels do not withstand their own weight (Chen et al., 2002) they are able to recover their structure when they are re-cooled. Under these conditions, although G′ > G″, both moduli depend on frequency. Thus, the term "weak gel" is used in various systems invariably when G′ > G″ without making or including additional tests to define this type of gels, for example, the analysis of the dependence of the phase angle with frequency. On the other hand, a larger amount of κ-carrageenan produces strong gels due to the presence of a larger proportion of aggregated helices (Núñez-Santiago et al 2011). However, it is possible that in spite of the presence of a sufficient amount of κ-carrageenan to induce gelation, potassium ions are necessary to aggregate the polymer.

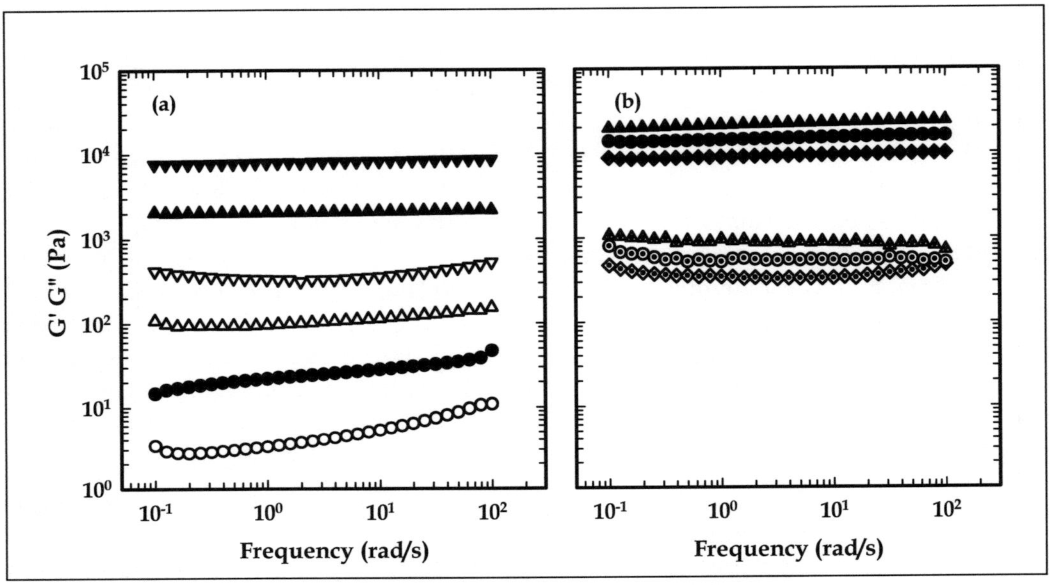

Fig. 8. Variation of G′ (black symbols) and G″ (white symbols) with angular frequency for different KCl concentrations (mmol/dm³) for 1.0% κ-carrageenan gels at 25 °C. (a) 5 (circles), 40 (triangles), 80 (inverted triangles). (b) 90 (diamonds), 100 (dotted circles) and 200 (dotted triangles). Experimental conditions correspond to region III in Fig. 3.

9. Microstructure of κ-carrageenan

Microscopy is one of the best tools to understand the rheological behavior of polysaccharides and materials in general. Various techniques, such as scanning electronic microscopy (SEM) and atomic force microscopy (AFM) allow the microstructure of materials to be observed. Other techniques, such as confocal laser scanning microscopy, CLSM, allow the ultrastructure and internal structure to be observed without breaking the material. On the other hand, most foods are mixtures of several components. Therefore, microscopic examination of the structural arrangement of the different components helps to understand the possible interactions among them. Sample preparation is the main problem to observe foodstuffs (Heertje & Pâkes, 1995). In SEM, there are physical and chemical methods to prepare the samples. Chemical methods are used to dehydrate and solidify liquid or semi-liquid samples through the use of chemical agents that can induce changes in the structures of the sample. Physical methods are used to solidify the samples by freezing and there is always the possibility of formation of small crystals that can break down their structure.

AFM has been used to study individual molecules and their interactions, particularly polysaccharides of complex structures. For example, the effect of salts on κ-carrageenan gelation (Ikeda et al., 2001). The images corresponding to an ionic concentration high enough for κ-carrageenan to form gels (100 mg/mL of κ-carrageenan in 100 mmol/dm^3 KCl), show the existence of local networks composed of faint filaments with some degree of branching resulting presumably from side to side aggregation of the helices, or by a thick bundle of aggregated helices, known as "supercritical spiral concentration". Ikeda et al. (2001) suggest that κ-carrageenan in the presence of NaI can form a local network without side to side aggregation of helices, however, this network is not as solid nor as rigid as the one formed in the presence of added salts.

In the case of CLSM, fluorescent staining is necessary to observe the materials under study. κ-carrageenan has been covalently labeled with rhodamine B isothiocyanate (RITC, excitation/emission wavelengths 543/575 - 640 nm) (Núñez-Santiago et al., 2011) and examined at 25 °C. The CLSM images of gels formed with 40 mmol/dm^3 KCl are shown in Fig. 9. With this amount of salt, 0.5% κ-carrageenan forms gels with a dense three-dimensional network and a homogeneous distribution of the polysaccharide as can be seen in Fig. 9a. As the polysaccharide concentration decreases to 0.25, and 0.01, Fig. 9b and 9c respectively, the network is still continuous but with perceptible dark zones of about 10-20 μm, Fig. 9d, the lack of continuity of the network becomes apparent in spite of a certain degree of aggregation. The three-dimensional images in CLSM confirm the presence of a three-dimensional network for polymer concentrations as low as 0.05%.

However, when polysaccharide concentration is very low, even though κ-carrageenan exists as aggregated helical chains, such aggregates are not enough to occupy all the volume and one obtains a salt solution in which short and fragmented κ-carrageenan networks are dispersed in the continuous phase.

As mentioned previously, κ-carrageenan has the ability to form a three-dimensional network in the presence of monovalent and divalent ions. However, the analysis of the structural conformation in the presence of two different cations, K^+ and Ca^{2+}, shows that the network structure is different. Figure 10a shows the image of κ-carrageenan in the presence of 10 mmol/dm^3 KCl (Dunstan et al., 2001) and Figure 10b is the image of κ-carrageenan in the presence of Ca^{2+} in a molar calcium to sulfate ratio of one to one (MacArtain et al., 2003). In both SEM images, the concentration of polysaccharide is 1.0%. The image with K^+ shows a dense structure, similar to the cross section of various tubular structures in the gel with a characteristic diameter of the spaces between 10 and 20 μm. The network with Ca^{2+} is fine with thin filaments of κ-carrageenan linked together to form a continuous network. The images are useful to understand that the gels formed with K^+ are stronger because they have denser and more continuous structures than those formed with Ca^{2+}.

Roh & Shin (2006) used SEM to observe the structure of mixtures, to analyze the replacement of alginate by κ-carrageenan and the effect of two agents to crosslink the polysaccharides. The images show that association between the two polysaccharides occurs by interpenetration between the two structures. Contrary to what happens with locust bean gum, replacing alginate with κ-carrageenan produces a structure with smaller pore sizes in relation to the network of alginate alone, which results in improved mechanical properties of the network. A similar behavior is observed in alginate-gellan mixtures, in which interpenetration occurs and the rheological behavior is governed by the polysaccharide present in higher concentration (Amici et al., 2001).

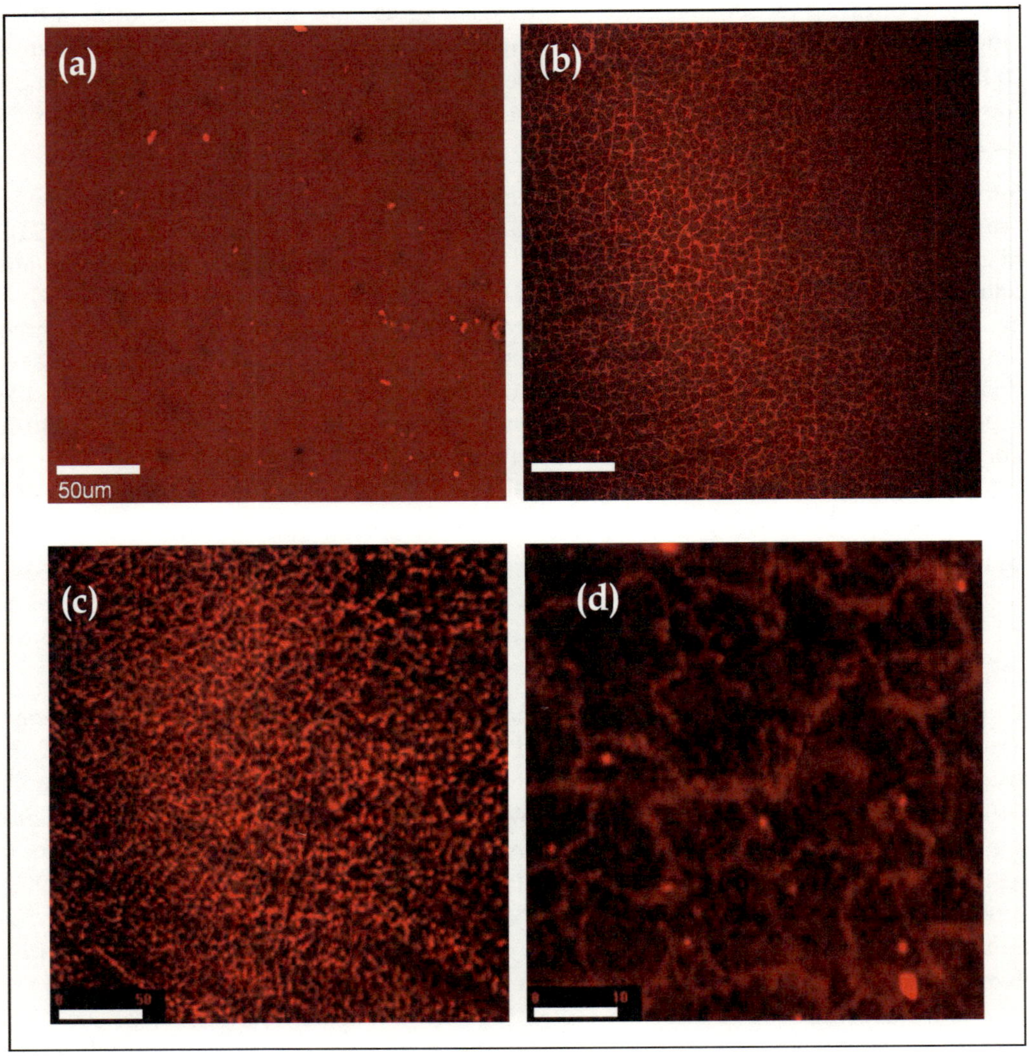

Fig. 9. CLSM images of covalently-labeled κ-carrageenan in the presence of 40 mmol/dm^3 KCl at λ= 580 nm for concentrations (%) of: 0.5 (a), 0.25 (b), 0.1 (c) and 0.05 (d). The scale bar is 50 μm in (a), (b) and (c), and 10 μm in (d).

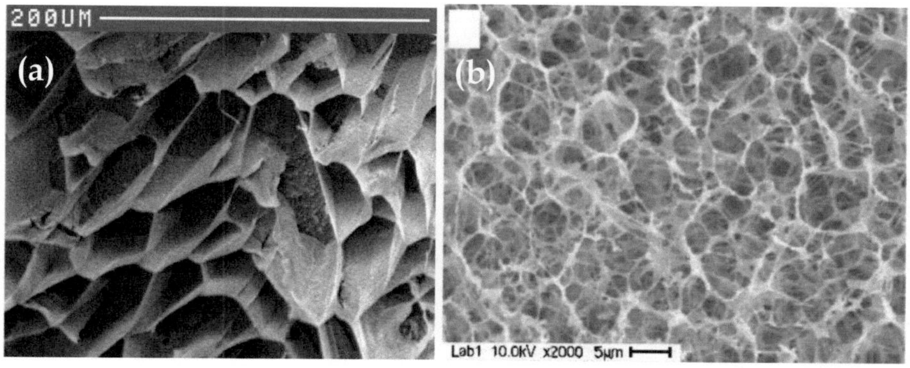

Fig. 10. SEM images of 1.0% κ–carrageenan networks. (a) with 10 mmol/dm^3 KCl (Reprinted from Dunstan, et al., (2001), with permission from Elsevier), (b) with Ca^{2+} in a 1:1 ratio (Reprinted from MacArtain et al., (2003), with permission from Elsevier).

10. Interaction between κ-carrageenan and other macromolecules

In industrial applications polysaccharides are not generally used alone because blends extend the range of possible different rheological and textural properties and improve the properties of the systems. Some binary mixtures of gelling hydrocolloids produce more resilient gels than the individual components or gels with the same resistance but with lower total concentrations, with an obvious economic advantage. Therefore, this has increased the interest in the study of mixtures of κ-carrageenan with starch, other non-starch polysaccharides or proteins in addition to systems containing only κ-carrageenan.

10.1 Non-starch polysaccharides

The simplest mixtures are those of two components. One of them can be a gelling agent such as κ-carrageenan, whereas the other a non-gelling macromolecule. These blends produce stronger gels with reduced syneresis and fragility. In this context, several rheological studies using oscillatory shear tests have shown increases in the storage modulus, G′, due to the presence of galactomannans such as locust bean gum (Stading & Hermansson, 1993; Lundin & Hermanson, 1997; Dunstan et al., 2001; Chen et al., 2001), or an increase in the maximum fracture stress of the gels (Dunstan et al., 2001). Other mixtures studied include κ-carrageenan-guar gum (Damasio et al., 1990) and κ-carrageenan-cactus mucilage (Medina-Torres et al., 2003). However, this phenomenon, sometimes called rheological synergy, occurs only under very specific conditions. Examination of the viscoelastic behavior (Stading & Hermansson, 1993) of systems containing potassium, sodium or calcium salt of κ-carrageenan and replacing the carrageenan with guar to a total polysaccharide concentration of 1.0%, shows that only the potassium salt of κ-carrageenan produces a synergistic effect in 100 mmol/dm^3 KCl. Compression and deformation tests (Chen et al., 2001) suggests more an "interaction" than a synergistic effect in mixtures of calcium salt of κ-carrageenan - locust bean gum. The synergistic effect between κ-carrageenan and locust bean gum can be reversed for high concentrations of the galactomannan, leading to lower stiffness from certain concentrations or ratios between the polysaccharides in the mixture. Dunstan et al. (2001) kept a constant total polymer concentration (1.0%) and gradually replaced κ-carrageenan with locust bean gum. They found a maximum value of rupture stress in compression tests for locust bean gum to κ-carrageenan ratios of 30/70 and 40/60. Medina-Torres et al. (2003) report on a synergistic effect with cactus mucilage observed in compression tests. They used a total concentration of 2.0% and found a fracture stress of 50 kPa in gels with a ratio of mucilage to κ-carrageenan of 80/20 with 120 mmol/dm^3 KCl.

Rheological or mechanical synergy has been observed but there is not a general criterion to explain it from a physicochemical point of view. One possible explanation is based on the presence of the volume exclusion effect between the polysaccharides that causes an increase in their effective concentration, as well as the presence of electrostatic interactions between the polymers in solution. Another possible explanation is based on the existence of interactions between the components in the network causing a synergistic effect due to the association of locust bean gum with the double helices stabilizing the "rods" of stiff κ-carrageenan in the presence of ions Na^+ and Ca^{2+} (Lundin & Hermanson, 1997). An important factor in the existence of synergy between these macromolecules is the presence of sulfate groups in κ-carrageenan, the type of salt of the κ-carrageenan (potassium, sodium or calcium), the ratio of galactomannans when they are present, presence of external ions, temperature, the ionic contribution of the other polysaccharide, the water absorption capacity and the molecular mass.

Another type of mixture is that of κ-carrageenan with another gelling agent. The mixture of κ-carrageenan with alginate improves the structure of films formed with alginate-κ-carrageenan ratios from 6 to 4, respectively, resulting in smaller gaps, which in turn improve the mechanical properties and stability of the network (Roh & Shin, 2006). In such systems, the synergistic effect has been attributed to interpenetration between the two polysaccharides, such as in agar-gellan (Amici et al., 2001), in which the network of one macromolecule passes through the other at a minimum scale, comparable to its interstitial size, reinforcing the three-dimensional network; this results in G′ values higher than the individual components.

10.2 Starch

In their blends with starches, a synergistic effect on the rheological properties of the pastes and gels exist (Loisel et al., 2000; Verbeken et al., 2004). Starch pastes and gels are considered a biphasic system with the continuous phase consisting of the solvent and dissolved starch during gelatinization and a disperse phase consisting of swollen granules (Doublier, 1981; Bagley & Christianson, 1982; Doublier, 1987). This viewpoint emphasizes the presence of swollen particles in starch suspensions. The swollen granules are not only deformable, but also compressible and elastic.

Mixtures of cross-linked waxy corn starch and κ-carrageenan have been also studied (Tecante & Doublier, 1999). This type of starch is practically free of amylose and starch granules are rigid enough to produce a suspension of swollen particles in a continuous medium. The biphasic model proposed for starch dispersions (Doublier, 1981) and gels can be applied. In these mixtures, addition of κ-carrageenan to starch pastes increases the apparent viscosity and the median diameter, D [v, 0.5], compared with individual starch pastes at 60 °C. However, the presence of K^+ (20 mmol/dm^3) produces an important decrease in the apparent viscosity of the mixture. One possible explanation for this phenomenon is that as the median diameter increases, the apparent viscosity increases and hence the nominal volume fraction of the disperse phase (Rao & Tattiyakul, 1999; Loisel et al., 2000; Paterson et al., 2001) producing mixtures with high viscosities; the addition of K^+ produced a shielding effect over the κ-carrageenan. The elastic character of these mixtures decreases with the decrease of starch concentration (2.0%) and the presence of K^+ (20 mmol/dm^3) (Tecante & Doublier, 1999). The evolution with time of G'at 25 °C in mixtures with 20 mmol/dm^3 KCl and 0.5% κ-carrageenan with 2.0 to 4.0% cross-linked waxy maize starch, show that more rigid gels are produced upon increasing the concentration of starch. These gels can be considered as composites of swollen particles embedded in a macromolecular network (Tecante & Doublier, 1999). Thus, the rheological behavior of starch/carrageenan is governed by the volume fraction of the disperse phase and the viscosity of the continuous phase.

Other studies on starch with non-starch polysaccharides mixtures have focused on changes in mixture properties during starch gelatinization (Nagano et al., 2008; Techawipharat et al., 2008). These investigations have revealed that the presence of the non-starch component decreases the gelatinization temperature of starch, probably due to a decrease in the amount of free water that the non-starch polysaccharideleaves to starch gelatinization, however, it also has been found that retrogradation decreases.

It is also possible to observe incompatibility between κ-carrageenan and other polysaccharides. In this context, Lai et al. (1999) studied the effect of incorporating κ-carrageenan to starch during cooking and gelation, through the rheological properties, the degree of gelatinization, swelling and solubility of starch granules. These authors conclude that the type of interaction between κ-carrageenan and starch is governed by the excluded effect of the swollen granules and the incompatibility existing in the gel matrix formed by κ-carrageenan and amylose. In a mixture of the calcium salt of κ-carrageenan with high amylose (70%) starch, Lai et al. (1999) found that for low concentrations of κ-carrageenan (0.3%) G' increases due to the presence of amylose, but with high concentrations of κ-carrageenan (1.0%) the rheological properties do not change, which shows that at high concentrations of κ-carrageenan, this polysaccharide prevailed over starch and governed the rheological behavior.

These results are consistent with those obtained by Tecante & Doublier (2002), who by measuring the change in turbidity and G' with time during gelation of amylose and κ-carrageenan mixtures, observed the presence of phase inversion between them. This phenomenon depends on the proportion of both polysaccharides; for concentrations of κ-carrageenan below 0.3%, amylose forms the continuous phase and carrageenan the disperse phase. For concentrations higher than 0.5% κ-carrageenan, the continuous phase is made up of κ-carrageenan and the disperse phase of amylose. κ-carrageenan behaves as a macromolecular solution when KCl is not present and forms three-dimensional networks when this salt is added.

11. Industrial applications in foods

κ–carrageenan is one of the most abundant polysaccharides in nature that can be used in prepared foods and cosmetics as a gelling, stabilizing and thickening agent due to its biocompatibility, biodegradability, high capacity of water retention and mechanical strength of its gels. According to the sol-gel transition diagram, κ-carrageenan can adopt three different conformations: random coil, double helix and double helix aggregated forming a three-dimensional network. The dependence of the rheological properties of κ-carrageenan on external factors, such as temperature, counterion and polysaccharide concentration permits to control the way in which the functionality of the polysaccharide is desired for a given product.

For example, when the product is to be kept at room temperature or when the low concentration of κ-carrageenan and counterions permit the polysaccharide to be in the random coil conformation ($C_T < C_{crit}$ and $T > T_{crit}$). Under these conditions κ-carrageenan can be used as a thickening agent and as a stabilizer in products such as chocolate drinks (Prakash et al, 2010) in which the polysaccharide provides creaminess to milk and maintains cocoa in suspension for a long time. The polysaccharide can be used also as a foam stabilizer in cream of whipped milk (Kováčová et al., 2010) due to the increase in viscosity. Thus, κ-carrageenan used in low concentrations stabilizes dairy products as a result of its interaction with casein micelles, for this reason it can be used in other products as soy and condensed milks.

On the other hand, when the products are liquid and refrigerated (4 °C), the low concentrations of κ-carrageenan (< 0.5%) permit to have the macromolecule in an ordered state without forming aggregates. Some examples are yoghurt, flans, jellies, fruit filling for puddings, cold coffee beverages, among others. However, it is important to consider the concentration of Ca^{+2} in these systems, because in the presence of this ion, κ-carrageenan can form gels (Dunstan et al., 2001; MacArtain et al., 2003).

Due to its property of forming gels, κ-carrageenan has been used to increase the hardness of starch gels (Huang et al., 2007; Tecgawipharat et al., 2008), to prepare low-calorie products, such as turkey sausages (Ayadi et al., 2009), ham and cold meat in general, where the appearance is enhanced but the flavor is not modified. Another alternative widely studied is its use in diet desserts (Descamps et al., 1986) in which fats are substituted by starches and carrageenan obtaining a decrease of up to 50% in the amount of fats. In desserts like custards the matrix is formed by starches, sugars and milk (Depypere et al., 2009). In some cases the custard is flavored with aromatic compounds. Therefore, addition of κ–carrageenan permits to have a synergistic effect in the texture of the dessert because of the interactions with starch and milk proteins (Cayot, 2006); besides, its polymeric matrix permits to preserve the aromatic compounds and flavoring agents (Seuvre et al. 2008). In this case, mixtures of one polysaccharide as a primary stabilizer and κ-carrageenan as a secondary stabilizer are used as a cryo-protective agent, giving excellent results in the preservation of ice cream (Soukoulis et al., 2008).

Another use of κ-carrageenan is in the preparation of biodegradable and edible films to increase the shelf-life of strawberries (Ribeiro et al., 2007), storage of cheese (Kampf & Nussinovitch, 2000) and antimicrobial films (Choi et al., 2005). κ-carrageenan microgels allow foods and aromas in fruit preparation to be preserved (Savary et al., 2007; Hambleton, et al., 2009; Marcuzzo et al., 2010; Hambleton et al., 2011; Fabra et al., 2011), drugs to be released over extended periods to the desired dosage when microgels are spherical particles (Cha et al., 2002, 2003; Bonferoni et al., 2004; Karbowiak et al., 2007; Hu et al., 2009; Nessem et al., 2011) and extended release of nutraceutical products (Ellis et al., 2009). In this context, κ-carrageenan is considered an excellent alternative for active compounds release in the intestinal tract and not in the first stages of digestion.

One of the new applications of κ-carrageenan is in copolymerization. One example is the copolymerization of κ-carrageenan with acrylic acid with further acidification to convert the sulfate groups into sulfuric groups. This copolymer has been used as a catalyst during the hydrolysis of sucrose to glucose and fructose (El-Mohdy & Rehim, 2008).

12. Conclusions

The existence of a given conformation, random coil, non-aggregated helical chains and three-dimensional networks (gels), depends on the total ionic concentration and temperature and determines the functionality of κ-carrageenan. Conditions that promote the presence of random coils result in viscoelastic behaviors typical of macromolecular solutions. Below a critical ionic concentration and its corresponding temperature, the functionality is determined by the existence of helices without aggregation. The viscoelastic behavior of this state is similar to the random coil condition at high and low temperatures. Therefore, the functionality of the polysaccharide is not expected to change drastically by a change in temperature. Above the critical conditions and low temperature, 25 °C, κ-carrageenan forms three-dimensional structures of different degrees of aggregation that result in very rigid gels. However, this rigidity remains practically unchanged above a limiting salt concentration. In making gels of this polysaccharide, it is important to bear in mind that the amount of κ-carrageenan and added KCl must be enough to form a continuous three-dimensional network over the entire aqueous medium.

13. References

Amici, E., Clark, A.H., Normand, V. & Johson, N.B. (2001). Interpenetrating network formation in agarose-sodium gellan gel composites. *Carbohydrate Polymers,* Vol. 46, No. 4, pp. 383-391, ISSN 0144-8617.

Ayadi, M.A., Kechaou, A., Makni, I. & Attia, H. (2009). Influence of carrageenan addition on turkey meat sausages properties. *Journal of Food Engineering,* Vol. 93, No. 3, pp. 278-283, ISSN 0260-8774.

Baeza, R.I., Carp, D.J., Pérez, O.E. & Pilosof, A.M.R. (2002). κ-Carrageenan - protein interactions: effect of proteins on polysaccharide gelling and textural properties. *LWT – Food Science and Technology,* Vol. 35, No. 8, pp. 741-747, ISSN 0022-1155.

Bagley, E. B. & Christianson, D.D. (1982). Swelling capacity of starch and its relationship to suspension viscosity-effect of cooking time, temperature and concentration.*Journal of Texture Studies,* Vol. 13, No. 1, pp. 115-126, ISSN 0022-4901.

Bayley, S.T. (1955). Hydrated K^+ ions and sulphate groups X-ray and infrared studies on carrageenan. Biochimica et Biophysica Acta, Vol. 17, No. 1, pp. 194-205, ISSN 0304-4165, cited in Rochas, C. (1982). Étude de la transition sol-gel du kappa-carraghénane. Thèse Docteur ès Sciences Physiques. Université Scientifique et Médicale et Institute National Polytechnique de Grenoble. Grenoble, France.

Bonferoni, M.C., Chetoni, P., Giunchedi, P., Rossi, S., Ferrari, F., Burgalassi, S. & Caramella, C. Carrageenan-gelatin mucoadhesive system for ion-exchange based ophthalmic delivery: in vitro and preliminary in vivo studies. *European Journal of Pharmaceutics and Biopharmaceutics,* Vol. 57, No. 3, pp. 465-472, ISSN 0939-6411.

Cayot, N. (2006) Preliminary tests on a flavoured model system: elaboration process and rheological characterization of a custard dessert. *Flavour and Fragrance Journal,* Vol. 21, No. 1, pp. 25–29, ISSN 0882-5734.

Cha, D. S., Choi, J.H., Chinnan, M.S. & Park, H.J. (2002). Antimicrobial films based on Na-alginate and κ-carrageenan. LWT – *Food Science and Technology,* Vol. 35, No. 8, pp. 715-719, ISSN 0022-1155.

Cha, D.S., Cooksey, K., Chinnan, M.S. & Park, H.J. (2003). Release of nisin from various heat-pressed and cast films. *LWT – Food Science and Technology,* Vol. 36, No. 2, pp. 209-213, ISSN 0022-1155.

Chandrasekaran, R. (1998). X-ray diffraction of food polysaccharides, in *Advances in food and nutrition research,* Vol 42, Taylor, S. pp. 131-210, Academic Press, ISBN 10: 0-12-374120-3, USA.

Chen, Y., Liao, M.-L., Boger, D.V. & Dunstan, D.E. (2001). Rheological characterisation of κ-carrageenan/locust bean gum mixtures. *Carbohydrate Polymers,* Vol. 46, No. 2, pp. 117-124, ISSN 0144-8617.

Chen, Y., Liao, M.-L., & Dustan, D.E. (2002). The rheology of K^+ - κ-carrageenan as a weak gel. *Carbohydrate Polymers,* Vol. 50, No. 2, pp. 109-116, ISSN 0144-8617.

Choi, J.H., Choi, W.Y., Cha, D.S., Chinnan, M.J., Park, H.J., Lee, D.S. & Park, J.M. (2005). Diffusivity of potassium sorbate in κ-carrageenan based antimicrobial film. *LWT – Food Science and Technology,* Vol. 38, No. 4, pp. 417-423, ISSN 0022-1155.

Chronakis, I.S., Piculell, L. & Borgström, J. (1996). Rheology of kappa-carrageenan in mixtures of sodium and cesium iodide: two types of gels. *Carbohydrate Polymers,* Vol. 31, No. 4, pp. 215-225, ISSN 0144-8617.

Chronakis, I.S., Doublier, J.L. & Piculell, L. (2000). Viscoelastic properties of kappa- and iota-carrageenan in aqueous NaI from the liquid-like to the solid-like behavior. *International Journal of Biological Macromolecules,* Vol. 28, No. 1, pp.1-14, ISSN 0141-8130.

Damasio, M.H., Fiszman, S.M., Costell, E. & Duran, L. (1990). Influence of composition on the resistance to compression of kappa carrageenan-locust bean gum-guar gum mixed gels: relationship between instrumental and sensorial measurements. *Food Hydrocolloids,* Vol.3, No. 6, pp. 457-464, ISSN 0268-005X.

Depypere, F., Verbeken, D., Torres, J.D. & Dewettinck, K. (2009). Rheological properties of dairy desserts prepared in an indirect UHT pilot plant. *Journal of Food Engineering,* Vol. 91, No. 1, pp140-145, ISSN 0260-8774.

Descamps, O. Langevin, P. & Combs, D.H. (1986). Physical effect of starch/carrageenan interactions in water and milk. *Food Technology,* Vol. 40, No. 4, pp 81-90, ISSN 0015-6639.

Doublier, J.L. (1981). Rheological studies on starch flow behaviour of wheat starch pastes. *Starch/Starke,* Vol. 33, No. 12, pp. 416-420, ISSN 0038-9056.

Doublier, J.L. (1987). A rheological comparison of wheat, maize, faba bean and smooth pea starches. *Journal of Cereal Science,* Vol. 5, No. 3, pp. 247-262, ISSN 0733-5210.

Dunstan, D.E., Chen, Y., Liao, M.L., Salvatore, R., Boger, D.V. & Prica, M. (2001). Structure and rheology of the κ-carrageenan/locust bean gum gels. *Food Hydrocolloids,* Vol. 15, No. 4, pp. 475-484, ISSN 0268-005X.

El-Mohdy, H.L.A. & Rehim, H.A.A. (2008). Radiation-induced kappa carrageenan/acrylic graft-copolymers and their application as catalytic reagent for sucrose hydrolysis. *Chemical Engineering Journal,* Vol. 145, No.1, pp. 154-159, ISSN 1385-8947.

Ellis, A., Keppeler, S. & Jacquier, J.C. (2009). Responsiveness of κ-carrageenan microgels to cation surfactants and neutral salt. *Carbohydrate Polymers,* Vol. 78, No. 3,pp. 384-388, ISSN 0144-8617.

Fennema, O. R., (2002). *Food Chemistry,* (3rd Ed), CRC Press, ISBN 0824796918, USA.

Ferry, J. D. (1980). *Viscoelastic properties of polymers,* (3rd edition). John Wiley & Sons, ISBN 0471048941, NY USA.

Hambleton, A., Voilley, A. & Debeaufort, F. (2011). Transport parameters for aroma compounds through ι-carrageenan and sodium alginate-based edible films. *Food Hydrocolloids,* Vol. 25, No. 5, pp. 1128-1133, ISSN 0268-005X.

Harding, S.E., Day, K., Dhami. R. & Lowe, P.M. (1996). Further observations on the size, shape and hydration of kappa-carrageenan in dilute solution. *Carbohydrate Polymers,* Vol. 32, No. 2, (November 1996), pp. 81-87, ISSN 0144-8617.

Heertje, I. & Pâkes, M. (1995). Advances in Electron Microscopy, in *Newphysico-chemical techniques for the characterization of complex food systems,* Dickinson E. (ed), pp. 1-52, Blackie Academic & Professional, ISBN 0751402524, London.

Hu, Y., Tang, T., Yang, W. & Zhou, H. (2009). Bioconversion of phenylpyruvic acid to L-phenylalanine by mixed-gel immobilization of Escherichia coli EP8-10.*Process Biochemistry,* Vol. 44, No. 2,pp. 142-145, ISSN 1359-5113.

Huang, M., Kennedy, J.F., Li, B., Xu, X. & Xie, B.J. (2007). Characters of rice starch gel modified by gellan, carrageenan and glucomannan: A texture profile analysis study. *Carbohydrate Polymers,* Vol. 69, No. 3,pp. 411-418, ISSN 0144-8617.

Ikeda, S., Morris, V. & Nishinari, K. (2001). Microstructure of aggregated and nonaggregated κ-Carrageenan helices visualized by atomic force microscopy. *Biomacromolecules,* Vol. 2, No. 4, pp. 1331-1337, ISSN 1525-7797.

Ikeda, S. & Nishinari, K. (2001). "Weak gel"-type rheological properties of aqueous dispersions of nonaggregated κ-carrageenan helices. *Journal of Agricultural and Food Chemistry,* Vol. 49, No. 9, pp.4436-4441, ISSN 0021-8561.

Kampf, N. & Nussinovitch, A. (2000). Hydrocolloid coating of cheeses. *Food Hydrocolloids,* Vol. 14, No. 6, pp. 531-537, ISSN 0268-005X.

Kara, S., Tamerler, C., Bermek, H. & Pekcan, Ö. (2003). Cation effects on sol-gel and gel-sol phase transitions of κ-carrageenan-water system. *International Journal of Biological Macromolecules,* Vol. 31, No. 4-5, pp.177-185, ISSN 0141-8130.

Karbowiak, T., Debeaufort, F., Champion, D. & Voilley, A. (2006). Wetting properties at the surface of iota-carrageenan-based edible films. *Journal of Colloid and Interface Science,* Vol. 294, No. 2, pp. 400-410, ISSN 0021-9797.

Kavanagh, G. & Ross-Murphy, S.B. (1998) Rheological characterization of polymer gels. *Progress in Polymer Science,* Vol. 23, No. 3, 533-562, ISSN 0079-6700

Kováčová, R., Štětine, J. & Čurda, L. (2010). Influence of processing and κ-carrageenan on properties of whipping cream. *Journal of Food Engineering,* Vol. 99, No. 4, pp. 471-478, ISSN 0260-8774.

Lai, V. M.-F., Huang, A.L. & Lii, C.Y. (1999). Rheological properties and phase transition of red algal polysaccharide-starch composites. *Food Hydrocolloids,* Vol. 13, No. 5, pp. 409-418, ISSN 0268-005X.

Lai, V.M.E., Wong, P.A.-L. & Lii, C.-Y. (2000). Effects of cation properties on sol-gel transition and gel properties of κ-carrageenan. *Journal of Food Science,* Vol. 65, No. 8,pp. 1332-1337, ISSN 0022-1147.

Loisel, C., Tecante, A., Cantoni, P. & Doublier, J. L. (2000). Effect of temperature on the rheological properties of starch/carrageenan mixtures, In *Gums and Stabilizers for the Food Industry 10,* Williams, P.A. & Phillips, G.O. (eds), ISBN 185573 788 4, pp. 181-187, Royal Society of Chemistry, Cambridge, UK.

Lundin, L. & Hermanson, A.-M. (1997). Rheology and microstructure of Ca- and Na–κ-carrageenan and locust bean gum gels. *Carbohydrate Polymers,* Vol. 34, No. 4, (January 1997), pp. 365-375, ISSN 0144-8617.

MacArtain, P., Jaacquier, J.C. & Dawson, K.A. (2003). Physical characteristics of calcium induced κ-carrageenan networks. *Carbohydrate Polymers,* Vol. 53, No. 4, (March 2003), pp. 395-400, ISSN 0144-8617.

Mangione, M.R., Giacomazza, D., Bulone, D., Martorana, V. & Biario, P.L. (2003). Thermoreversible gelation of κ-carrageenan: relation between conformational transition and aggregation. *Biophysical Chemistry,* Vol. 104, No. 1, pp. 95-105, ISSN 0301-4566.

Marcuzzo, E., Sensidoni, A., Debeaufort, F. & Voilley, A. (2010). Encapsulation of aroma compounds in biopolymeric emulsion based edible films to control flavour release. *Carbohydrate Polymers,* Vol. 80, No. 3, pp. 984-988, ISSN 0144-8617.

Medina-Torres, L., Brito-De La Fuente, E., Torrestiana-Sánchez, B. & Alonso, S. (2003). Mechanical properties of gels formed by mixtures of mucilage gum (*Opuntia ficus indica*) and carrageenans. *Carbohydrate Polymers,* Vol. 52, No. 2, pp. 143-150, ISSN 0144-8617.

Meunier, V., Nicolai, T. & Durand, D. (2001). Structure of aggregating κ-carrageenan fractions studied by light scattering. *International Journal of Biological Macromolecules,* Vol. 28, No 2, pp. 157-165, ISSN 0141-8130.

Meunier, V., Nicolai, T. & Durand, D. (1999). Light scattering and viscoelasticity of aggregating and gelling κ-carrageenan. *Macromolecules,* Vol.32, No. 8, pp. 2610-2616, ISSN 0024-9797.

Michel, A-S., Mestdagh, M.M. & Axelos, M.A.V. (1997). Physicochemical properties of carrageenan gels in the presence of various cations. *International Journal of Biological Macromolecules,* Vol. 21, No. 1-2, pp. 195-200, ISSN 0141-8130.

Milas, M., Shi, X. & Rinaudo, M. (1990). On the physicochemical properties of gellan gum. *Biopolymers,* Vol. 30, No. 3-4, pp. 451-464, ISSN 0006-3525.

Miyoshi, E. & Nishinari, K. (1999). Non-Newtonian flow behaviour of gellan gum aqueous solutions. *Colloid Polymer Science,* Vol. 277, No. 8, pp. 727-734, ISSN 0303-402X.

Morris, E.R. (1990). Mixed polymer gels, In *Food Gels,* Harris, P. pp. 291-359, Elsevier Science Publishing Co., Inc., ISBN 1851664416, London.

Nagano, T., Tamaki, E. & Funami, T. (2008). Influence of guar gum on granule morphologies and rheological properties of maize starch. *Carbohydrate Polymers,* Vol. 72, No 1, pp. 95-101, ISSN 0144-8617

Nayouf, M. (2003). Étude rhéologique et structurale de la qualité texturante du système amidon/kappa-carraghénane en relation avec le traitement thermomécanique. Thèse de Doctorat. École Nationale des Ingénieurs des Techniques des Industries Agricoles et Alimentaires. Nantes, France.

Nessem, D.I., Eid, S.F. & El-Houseny, S.S. (2011). Development of novel transdermal self-adhesive films for tenoxicam, an anti-inflammatory drug. *Life Sciences,* Vol. 89, No. 13-14, pp. 430-438, ISSN 0024-3205.

Nishinari, K. & Takahashi, R. (2003). Interaction in polysaccharide solutions and gels.*Current Opinion in Colloid and Interface Science,* Vol. 8, No. 4-5, pp. 396-400, ISSN 1359-0294.

Núñez-Santiago, M.C. & Tecante, A. (2007). Rheological and calorimetric study of the sol-gel transition of κ-carrageenan. *Carbohydrate Polymers,* Vol. 69, No. 4 , pp. 763-773, ISSN 0144-8617.

Núñez-Santiago, M.C., Tecante, A., Garnier, C. & Doublier, J.L. (2011). Rheology and microstructure of κ-carrageenan under different conformations induced by several concentrations of potassium ion. *Food Hydrocolloids,* Vol. 25, No. 1, pp. 32-41, ISSN 0268-005X.

Paterson, J.L., Hardacre, A., Li, P. & Rao, M.A. (2001). Rheology and granule size distributions of corn starch dispersions from two genotypes and grown in four regions. *Food Hydrocolloids,* Vol. 15, No. 4-6, pp. 453-459, ISSN 0268-005X.

Rao, M.A. & Tattiyakul, J. (1999). Granule size and rheological behaviour of heated tapioca starch dispersions. *Carbohydrate Polymers,* Vol. 38, No. 2, pp. 123-132, ISSN 0144-8617.

Ribeiro, C., Vicente, A.A., Texeira, J.A. & Miranda, C. (2007). Optimization of edible coating composition to retard strawberry fruit senescence. *Postharvest Biology and Technology,* Vol. 44, No. 1, pp. 63-70, ISSN 0925-5214.

Rinaudo, M. (2001). Relation between the molecular structure of some polysaccharides and original properties in sol and gel states. *Food Hydrocolloids,* Vol. 15, No. 4-6, pp. 422-440, ISSN 0268-005X.

Rochas, C. (1982). Étude de la transition sol-gel du kappa-carraghénane. Thèse Docteur ès Sciences Physiques. Université Scientifique et Médicale et Institute National Polytechnique de Grenoble. Grenoble, France.

Rochas, C. & Rinaudo, M. (1982). Calorimetric determination of the conformational transition of kappa carrageenan. *Carbohydrate Research,* Vol. 105, No. 2, pp. 227-236, ISSN 0008-6215.

Roh, Y.H. & Shin, C.S. (2006). Preparation and characterization of alginate-carrageenan complex films. *Journal of Applied Polymer Science,* Vol. 99, No. 6, pp. 3483-3490, ISSN 0021-8995.

Savary, G., Lafarge, C., Doublier, J.L. & Cayot, N. (2007). Distribution of aroma in a starch-polysaccharide composite gel. *Food Research International,* Vol. 40, No. 6, pp. 709-716, ISSN 0963-9969.

Seuvre, A.-M., Turci, C. & Voilley, A. (2008). Effect of the temperature on the release of aroma compounds and on the rheological behaviour of model dairy custard. *Food Chemistry,* Vol. 108, No. 4, pp. 1176-1182, ISSN 0308-8146.

Slootmaekers, D., De Jonghe & Raynaers, H. (1988). Static light scattering from κ-carrageenan solutions. *International Journal of Biological Macromolecules,* Vol. 10, No. 3, pp. 160-168, ISSN 0141-8130.

Snoeren, T.H.M. (1976). κ-Carrageenan. A study on its physico-chemical properties, sol-gel transition and interaction with milk proteins. Ph.D. Thesis. Wageningen, Holland.

Soukoulis, C., Chandrinos, I. & Tzia, C. (2008). Study of the functionality of selected hydrocolloids and their blends with κ-carrageenan on storage quality of vanilla ice cream. *LWT-Food Science and Technology,* Vol. 41, No. 10, pp. 1816-1827, ISSN 0022-1155.

Stading, M. & Hermansson, A.M. (1993). Rheological behaviour of mixed gels of κ-carrageenan-locust bean gum. *Carbohydrate Polymers,* Vol. 22, No. 1, pp. 49-56, ISSN 0144-8617.

Stanley, N.F. (1987). Carrageenans, In *Production and utilization of products from commercial seaweeds,* McHugh, Dennis J. (ed), pp. FAO Fisheries Technical Papers, ISBN 9251026122, Australia.

Takemesa, M. & Chiba, A. (2001). Gelatin mechanism of κ and ι-Carrageenan investigated by correlation between the strain-optical coefficient and the dynamic shear modulus. *Macromolecules,* Vol. 34, No. 21, pp. 7427-7434, ISSN 0024-9797.

Takemesa, M. & Nishinari, K. (2004). The effect of the linear change density of carrageenan on the ion binding investigated by differential scanning calorimetry, dc conductivity, and kHz dielectric relaxation. *Colloids and Surface B: Biointerfaces,* Vol. 38,No. 3, pp. 231-240, ISSN 0927-7765.

Tecante, A. & Doublier, J.L. (1999). Steady flow and viscoelastic behaviour of crosslinked waxy corn starch-κ-carrageenan pastes and gels. *Carbohydrate Polymers,* Vol. 40, No. 3, pp. 221-231, ISSN 0144-8617.

Techawipharat, J., Suphantharika, M. & BeMiller, J. (2008). Effects of cellulose derivatives and carrageenans on the pasting, paste, and gel properties of rice starches. *Carbohydrate Polymers,* Vol. 73, No. 3, pp. 417-426, ISSN 0144-8617.

van de Velde, Rollema, H.S., Grinberg, N.V., Burova, T.V., Grinberg, V.Y. & Tromp, R.H. (2002). Coil-helix transition of ι-Carrageenan as a function of chain regularity, *Biopolymers,* Vol. 65, No. 4, pp. 299-323, ISSN 0006-3525.

Verbeken, D., Thas, O. & Dewettinck, K. (2004). Textural properties of gelled dairy desserts containing κ-carrageenan and starch. *Food Hydrocolloids,* Vol. 18, No. 5, pp. 817-823, ISSN 0268-005X.

Viebke, C., Borgström, J. & Piculell, L. (1995). Characterisation of kappa- and iota-carrageenan coils and helices by MALLS/GPC. *Carbohydrate Polymers,* Vol. 27, No. 2, pp. 145-154, ISSN0144-8617.

Vreeman, H.J., Snoeren, T.H.M. & Payens, T.A.J. (1980). Physicochemical investigation of κ-carrageenan in the random state. *Biopolymers,* Vol. 19, No. 7, pp. 1357-1374, ISSN 0006-3525.

Whistler, R.L. & BeMiller, J.N. (1997). *Carbohydrate Chemistry for Food Scientists* (2nd ed), Egan Press., ISBN 1891127535, St. Paul. MN.

Yuguchi, Y., Thuy, T.T., Urakawa, H., & Kajiwara, K. (2002). Structural characteristics of carrageenan gels: temperature and concentrations dependence. *Food Hydrocolloids,* Vol. 16, No. 6, pp. 515-522, ISSN 0268-005X.

Yuguchi, Y., Urakawa, H. & Kajiwara, K. (2003). Structural characteristics of carrageenan gels: various types of counter ions. *Food Hydrocolloids,* Vol. 17, No. 4, pp. 481-485, ISSN 0268-005X.

11

Influence of Acidification on Dough Rheological Properties

Daliborka Koceva Komlenić, Vedran Slačanac and Marko Jukić
Faculty of Food Technology, Josip Juraj Strossmayer University of Osijek, Croatia

1. Introduction

Due to increasing consumer demands for more natural, tasty and healthy food, the traditional process of sourdough bread production has been enjoyed renewed success in recent years, (Brümmer & Lorenz, 1991; Thiele *et al.*, 2002; Lopez *et al.*, 2003). Sourdough was traditionally used as leavening agents until it was replaced by baker's yeast in the 19th century (Corsetti & Setanni, 2007). Today, sourdough is employed in the manufacture of a number of products, such as breads, cakes and crackers (De Vuyst & Gänzle, 2005). Some of these products have strictly regional and artisanal character; well some of these are widely distributed on the world market (De Vuyst & Neysens, 2005). Many wheat breads and cakes are characteristic for the Mediterranean countries, the San Francisco bay and Southern America, whereas a number of bakery preparations made with rye, wheat, barley, or mixtures of these flours are typical for Central Europe, Eastern Europe and Scandinavia (Stephan & Neumann, 1999).

In general, sourdough is mixture of flour and water that is than fermented with lactic acid bacteria (LAB) (Hammes & Gänzle, 1998). Ordinarily, these LAB are heterofermentative strains which produce lactic and acetic acid in the mixture, and resulting in a sour taste of the dough (Vogel et. al., 1999). The acidification process affected by the application of sourdoughs is mainly used to improve quality, taste and flavour of wheat breads (Brümmer & Lorenz, 1991; Katina *et al.*, 2006a; Arendt *et al.*, 2007), and the slow staling (Katina *et al.*, 2006b; Plessas *et al.*, 2007).

Sourdough can be freshly produced in bakeries, or can be obtained from the commercial suppliers in the some of the following applying forms: living, liquid sourdough or dried, non-fermenting sourdough (Bőcker et. al., 1990). Considerably, doughs could be acidified biologically, chemically and naturally preferment.

The addition of sourdough during production of wheat bread causes major changes in the dough characteristics (Clarke *et al.*, 2002; Clarke *et al.*, 2004; Ketabi *et al.*, 2008), especially in the flavour and structure (Clarke et. al., 2002). The effect of the fermentation process of wheat doughs containing lactic acid bacteria are complex and depending on variations between sourdoughs regard to the type of starter culture, dough yield and fermentation regime used (Wehrle et. al., 1997). This propounds requirements for the detail microbiological investigations on sourdough microflora. Study of sourdough from microbiological point of view barely started a hundred years ago (Salovaara, 1998). Today, microbial population of different types of sourdough is rather known. Many inherent properties of sourdough rely on the metabolic activities of its resident LAB: lactic fermentation, acetic fermentation, proteolysis, synthesis of volatile compounds, anti-mould activity and antiropiness are among the most important activities during sourdough fermentation (Hammess and Gänzle, 1998). The most of the effects of sourdough have been considered by pH value decrease which has been caused with organic acid production.

Between many of important effects, the drop in pH value caused by the organic acids produced influences the viscoelastic behaviour of doughs, respectively to sourdoughs rheological properties (Wehrle & Arendt, 1998). A correct description of the changes in dough behaviour is necessary to maintain handling and machinability in industrial production (Wehrle et. al., 1997). Furthermore, following the identification and classification of LAB from cereal fermentation, basic and applied today face the challenge of identifying functional characters of these bacteria to completely exploit their microbial metabolic potential from the production of baked goods (Vogel et. al., 2002). In European countries, production of sourdoughs from wheat flour has been predominant. Most fundamental studies on wheat sourdough have been conducted in Western Europe, and the results have been published in German-language journals and books (Brűmmer & Lorenz, 1991).

Rheological properties, acidification and flavour development are the most important parameters in fermentation process control (Hames et. al., 2005). Rheological properties of dough have been determined by a number of methods, such as dynamic rheological measurements, extensigraphs, alveographs, lubricated uniaxial compression, oscillatory probe rheometers etc. (Hoseney, 1994). The rheological characteristics of dough have been considerably changed with fermentation. Types of microorganisms, metabolic activity and time-dependent development pH value effect on rheological properties (Wehrle & Arendt, 1998). Acids strongly influence on the mixing properties of dough. Doughs with lower pH requires a slightly shorter mixing time and have less stability than dough with normal pH level (Hoseney, 1994). Changes in pH values caused by production of lactic acid also influence on the rheological properties of dough (Wehrle *et al.*, 1997). Dough with containing acid has been characterized by increased phase angle and reduced complex modulus indicative of overmixing (Wehrle et al 1997). Small physical and chemical changes in the gluten network can result in significant changes in rheological properties. Clarke et al. (2002) were concluded that addition of sourdough prepared either from a single strain starter culture or a mixed strain starter culture had significant impact on the rheological properties of wheat flour dough.

Koceva Komlenić et al (2010) investigated the influence of chemical and biological acidification on dough rheological properties. According to their experimental results, rheological properties strongly depend on acidification type. Dough with lower pH value showed less stability during mixing, decreased extensibility and gelatinization maximum. In general, the rheological properties of dough greatly improved when the sourdough was added.

Regard to all facts mentioned above, fermentation of dough with LAB greatly effects on the properties of many bakery products. Structure properties are one of the most important. From that reason, rheological measurements play important role in definition of quality of products from sourdough. In this paper, influence of sourdough fermentation on the structure of sourdough was observed.

2. Microbiology of sourdough

It is well known that the type of bacterial flora developed in each fermented food depends on water activity, pH (acidity), minerals concentration, gas concentration, incubation temperature and composition of food matrix (Font de Valdez et. al. 2010). The microflora of raw cereals is composed of bacteria, yeast and fungi (10^4 - 10^7 CFU/g), while flour usually contains 2 x 10^4 - 6 x 10^6 CFU/g (Stolz, 1999). In sourdough fermentation major role play heterofermentative species of LAB (Salovaara, 1998; Corsetti & Settani, 2007), especially when sourdoughs are prepared in a traditional manner (Corsetti et. al., 2003).

Type Ia	Type Ib	Type Ic	Type II	Type III
Obligate heterofermentative	Obligate heterofermentative	Obligate heterofermentative	Obligate heterofermentative	Obligate heterofermentative
Lb. sanfranciscensis	*Lb. brevis*	*Lb. pontis*	*Lb. brevis*	*Lb. brevis*
	Lb. brevis	*Lb. fermentum*	*Lb. fermentum*	
	Lb. buchneri	*Lb. reuteri*	*Lb. frumenti*	
	Lb. fermentum		*Lb. pontis*	
	Lb. fructivorans		*Lb. panis*	
	Lb. pontis		*Lb. reuteri*	
	Lb. reuteri		*Lb. sanfranciscensis*	
	Lb. sanfranciscensis		*W. confusa*	
	W. cibara			
	Facultative heterofermentative			Facultative heterofermentative
	Lb. alimentarius			*Lb. plantarum*
	Lb. casei			*P. pentosaceus*
	Lb. paralimentarius			
	Lb. plantarum			
	Obligate homofermentative	Obligate homofermentative	Obligate homofermentative	
	Lb. acidophilus	*Lb. amylovorus*	*Lb. acidophilus*	
	Lb. delbrueckii		*Lb. delbrueckii*	
	Lb. farciminis		*Lb. amylovorus* (rye)	
	Lb. mindensis		*Lb. farciminis*	
			Lb. johnsonii	

Table 1. Classification of sourdoughs and the corresponding chracteristic microflora (De Vuyst & Neysens, 2005)

Sourdough LAB generally belongs to the genus *Lactobacillus, Leuconostoc, Pediococcus* or *Weisella* (Ehrmann & Vogel, 2005; De Vuyst & Neysens, 2005). A few less than 50 different species of LAB from sourdough have been reported by De Vuyst & Neysens (2005). Although species belonging *Leuconostoc, Pediococcus, Weisella, Enterococcus* and *Streptococcus* genera have been isolated from sourdough, *Lactobacillus* strains are the most frequently observed bacteria in sourdoughs (Corsetti & Settanni, 2007). Characteristic microflora of sourdough is presented in Table 1. Undesirable bacteria such as *Staphylococus aureus* and *Bacillus cereus*, as well as other bacteria, may be present in minor concentration (De Vuyst and Neysens, 2005). Corsetti & Settani (2007) reported that the recently described species *Lactobacillus frumenti, Lactobacillus mindensis, Lactobacillus paralimentarius, Lactobacillus spicheri, Lactobacillus rossiae, Lactobacillus acidifarinae, Lactobacilus zymae, Lactobacillus hammesii, Lactobacillus nantensis* and *Lactobacillus siliginis* were first isolated from the sourdough matrices. 15 lactobacilli species known to occur in sourdough are also known to live in human and animal intestines. *Lactobacillus reuteri, Lactobacillus acidophilus* and *Lactobacillus plantarum* are some of these intestinal lactobacilli (Hammes & Gänzle, 1998).

Kumar et. al. (2004) reported a phylogram based on 16S rRNA gene sequences of sourdough lactobacilli. A total of 41 species are included in phylogram tree, some of them, such as *Lactobacillus rhamnosus,* have been rarely or only once reported to be found in sourdough (Spicher & Löner, 1985). Hammes et. al. (2005) reported that about 30 species are considered to be typical of sourdough environments. *Lactobacillus sanfranciscensis, Lactobacillus brevis* and *Lactobacillus plantarum* are the most frequently lactobacilli isolated from sourdough (Gobbetti, 1998; Corsetti et. al. 2001; Valmorri et. al., 2006; Corsetti & Settanni, 2007). *Lb. sanfranciscensis* was first reported in the San Francisco sourdough French bread process, to be responsible for acid production (Corsetti et. al., 2001). *Lb. sanfranciscensis* has been predominant bacteria in traditional production by various stages of continuous production and production by commercial starter cultures (Gobbetti & Corsetti, 1997). Gobbetti et. al. (1998) reported on the *Lb. sanfranciscensis* – *Lb plantarum* association in Italian wheat sourdough. *Lb. alimentarius,* which probably belong to *Lb. paralimentarius,* were first isolated from Japanese sourdough (Cai et. al., 1999). *Lb. brevis* and *Lb. plantarum* have been associated with *Lb. fermentum* in Russian sourdoughs (Kazayanska et. al., 1983). *Lb. fermentum* dominates Swedish sourdoughs and German type II sourdoughs (Meroth et al., 2004.) Gobbetti et. al. (1994b) reported that *Lb. acidophilus* is found in Umbrian (Italian region), while Corsetti et. al. (2003) described a new sourdough associated species, *Lb. rossae,* in sourdoughs of central and southern Italy (Settanni et. al., 2005a). African sorghum sourdoughs, which are produced at higher temperature (> 35 °C) contain obligate heterofermentative *Lb. fermentum, Lb. pontis* and *Lb. reuteri* species, as well as the obligate homofermentative *Lb. amylovorus* (De Vuyst & Neysens, 2005).

According to all facts mentioned above, it is so clear that large biodiversity in lactoballi composition of the sourdough exists, regarding to type of sourdough (Table 2), as well as to some cultural, geographical and traditional reasons.

Country	Product/method of isolation and identification	Lactic acid bacteria
Belgium	Wheat/rye sourdoughs polyphasic approach	*Lb. brevis, Lb. plantarum, Lb. sanfranciscensis, Lb. paralimentarius, P. pentosaceus, Lb. helveticus*
Finiand	Rye sourdough phenotypical	*Lb. acidophilus, Lb. plantarum, Lb. casei*
Denmark	Sour rye dough phenotypical	*Lb. reuteri, Lb. panis, Lb. amylovorus*
France	Wheat bread phenotypical	*Lb. plantarum, Lb. casei, Lb. delbrueckii* subsp. *delbrueckii, Lb. acidophilus, Lb. brevis, Leuc. mesenteroides* subsp. *mesenteroides, Leuc. mesenteroides* subsp. *dextranicum, P. pentosaceus, Lb. Curvatus*
Germany	Wheat sourdough phenotypical	*Lb. delbrueckii, Lb. planlarum, Lb. casei, Lb. fermentum, Lb. buchneri, Lb. brevis*
	Rye bread phenotypical	*Lb. acidophilus, Lb. farciminis, Lb. alimentarius, Lb. casei, Lb. plantarum, Lb. brevis, Lb. sanfranciscensis, Lb. fructivorans, Lb. fermentum, Lb. buchneri*

Country	Product/method of isolation and identification	Lactic acid bacteria
	Rye sourdough phenotypical	*Lb. acidophilus, Lb. casei, Lb. plantarum, Lb. farciminis, Lb. alimentarius, Lb. brevis, Lb. buchneri, Lb. fermentum, Lb. fructivorans, Lb. sanfranciscensis, Pediococcus* spp.
	Wheat sourdoughs (Panettone, wheat bread) phenotypical	*Lb. plantarum, Lb. casei, Lb. farciminis, Lb. homohiochii, Lb. brevis, Lb. hilgardii* (spontaneous); *Lb. sanfranciscensis, Lb. brevis, Lb. hilgardii, W. viridescens* (masa madre)
	Rye sourdough RAPD-PCR	*Lb. amylovorus, Lb. pontis, Lb. frumenti, Lb. reuteri*
	Rye bran PCR-DGGE	*Lb. sanfranciscensis, Lb. mindensis* (type I rye sourdough); *Lb. crispatus, Lb. pontis, Lb. panis, Lb. fermentum, Lb. frumenti* (type II rye sourdough); *Lb. johnsonii, Lb. reuteri* (type II rye bran sourdough)
Greece	Wheat sourdoughs polyphasic approach	*Lb. sanfranciscensis, Lb. brevis, Lactobacillus* spp.[a], *Lb. paralimentarius, W. cibaria*
Italy	Panettone phenotypical	*Lb. brevis, Lb. plantarum*
	Panettone, Brioche phenotypical	*Lb. sanfranciscensis, Lb. fermentum, Lb. plantarum, Leuc. mesenteroides, Pediococcus* spp.
	Umbrian wheat sourdoughs phenotypical	*Lb. sanfranciscensis, Lb. plantarum, Lb. farciminis*
	Pizza (Naples) phenotypical	*Lb. sakei, Lb. plantarum, Leuc. gelidum, Leuc. mesenteroides*
	Verona sourdoughs RAPD-PCR	*Lb. sanfranciscensis*
	Lombardian mother sponges speciessspecific PCR	*Lb. sanfranciscensis*
	Apulian wheat sourdoughs 16S rDNA sequencing 165/23S rRNA spacer region PCR	*Lb. sanfranciscensis, Lb. alimentarius, Lb. brevis, Leuc. citreum, Lb. plantarum, Lb. lactis* subsp. *lactis, Lb. fennentum, Lb. acidophilus, W. confusa, Lb. delbmeckii* subsp. *bulgaricus*
Iran	Sangak phenotypical	*Leuc. mesenteroides, Lb. plantarum, Lb. brevis. P. cerevisiae*
Mexico	Pozol (maize) 16S rDNA sequencing	*Lb. lactis, S. suis. Lb. plantarum, Lb. casei. Lb. alimentarius. Lb. delbrueckii*
Morocco	Sourdough ferments traditional starter sponges phenotypical	*Lb. plantarun, Lb. brevis, Lb. buchneri, Lb. casei. Leuc. mesenteroides, Pediococcus* sp.
	Soft wheat flour phenotypical	*Lb. plantarum, Lb. delbrueckii, Lb. buchneri, Lb. casei, Lb. sanfranciscensis, Leuc.mesenteroides, P. pentosaceus*
Portugal	Broa phenotypical	*Leuconostoc* spp., *Lb. brevis, Lb. curvatus, Lb. delbrueckii, Lb. lactis* subsp. *lactis, E. casseliflavus, E. durans, E. faecium, S. constellantus, S. equinus*
Russia	Rye sourdough phenotypical	*Lb. plantarum, Lb. brevis, Lb. fermentum*
Spain	Wheat sourdough phenotypical	*Lb. brevis, Lb. plantarum*
	Wheat sourdough phenotypical	*Lb. brevis, Lb. plantarum, Lb. cellobiosus. Leuc. mesenteroides*
Sudan	Kisra (sorghum sourdough)	*Lb. fermentum, Lb. reuteri, Lb. amylovorus*
	Kisra RAPD	*E. faecalis, Lb. laclis, Lb. fermentum, Lb. reuteri, Lb. vaginalis, Lb. helveticus*
Sweden	Rye/wheat phenotypical	*Lb. fermentum. Lb. delbrueckii, Lb. acidophillus, Lb. plantarum, Lb. rhamnosus, Lb. farciminis. Lb. fermentum, Lb. sanfranciscensis, Lb. brevis. W. viridescens*

Country	Product/method of isolation and identification	Lactic acid bacteria
	Rye sourdough phenotypical	*Laclobacillus* sp., *P. pentosaceus*
USA	San Francisco sourdough French bread phenotypical	*Lb. sanfranciscensis*

Table 2. Biodiversity of sourdough lactic acid bacteria in sourdough of different origin (De Vuyst & Neysens, 2005)

The following yeasts have been detected in cereals (9 x 10^4 CFU/g) and flour (2 x 10^3 CFU/g): *Candida, Cryptococcus, Pichia, Rodothorula, Torulaspora, Trychoporon, Saccharomyces* and *Sporobolomyces*. *Saccharomyces cerevisiae* is not found in the raw materials. Its occurance in sourdough has been explained by the application of baker's yeast in most daily bakery practice (Corsetti et. al., 2001). Among fungi (ca. 3 x 10^4), *Alternaria, Cladosporium, Drechslera, Fusarium; Helminthosporium, Ulocladium; Aspergilus* and *Penicillium* were found in raw cereals and flour. Possibility to dough be contaminated with these species exist, but in the most of cases it could be avoided (De Vuyst & Neysens, 2005).

3. Role of LAB in sourdough fermentation

The metabolic activities of lactobacilli during sourdough fermentation improve dough properties, bread texture and flavour; retard the stalling process of bread; and prevent bread from mould and bacterial spoilage (Gerez et. al., 2009; de Valdez et. al., 2010). Additionally, LAB could contribute to nutritive value and healthiness of bread (Gobbetti et. a., 2007). The sourdough microflora is usually composed of stable associations of yeasts and lactobacilli.

3.1 Carbohydrate metabolism

In dough fermentation carbonhydrate metabolism varies depending on involved LAB strains, type of sugars, presence of yeasts and processing conditions (Gobetti et. al., 1994a). The importance of antagonistic and synergistic interactions between lactobacilli and yeasts are based on the metabolism of carbon hydrates and amino acids and the production of carbon dioxide (Gobetti & Corsetti 1997). Lactic and acetic acid are predominant products of sourdough fermentation (Figure 1).

Diversity in metabolic pathways between different sourdoughs is large. Recipe and geographical origin make the difference in the aroma of breads consumed in the different areas (Hansen & Schieberle, 2005). All these parameters influence on characteristics and give originality to different breads from sourdoughs. It is the basic reason why bread from biologically fermented dough posses a superior sensory quality than bread produced from chemically acidified dough (Kirchhoff & Schieberle, 2002). The ratio between lactic and acetic acid is an important factor that might affect the aroma profile and structure of final product. Acetic acid, produced by heterofermentative LAB, is responsible for a shorter and harder gluten, while lactic acid can gradually account for a more elastic gluten structure (Lorenz, 1983; Corsetti & Settani, 2007). *Lb. sanfrancienscis*, isolated from traditional Italian sourdoughs ferment only glucose and maltose. However, *Lb. sanfrancienscis* strains isolated from some other sourdoughs use other sugars, such as sucrose, raffinose, galactose, melobiose, ribose and fructose (Tieking et al., 2003). Ginés et. al. (1997) presented metabolic potential of *Lb. reuteri* CRL1100, a strain isolated from homemade sourdough. This microorganism metabolizes glucose and galactose, but not fructose and cellobise. The most of lactobacilli strain in sourdough are unable to ferment sucrose, a disaccharide whose metabolism by heterofermentative lactobacilli in sourdough appears to be attributable to glycosil-transferases rather than to invertase activities (Font de Valdez et. al., 2010). Recently, it has been reported that certain lactic acid bacteria are able to produce exopolysacharides (EPS), which might have a positive affect on bread volume and shelf life (Font de Valdez et. al., 2010). Following EPS have been produced during fermentation of sourdough: fructans (levan, inulin) and glucans (reuteran, dextran, xanthan). EPS influence significantly to the texture improvement of sourdough and bread (Tieking et. al., 2003). The production of EPS in sufficient amounts during sourdough fermentation would create the possibility to replace hydrocolloids in baking (Tieking et. al., 2003). Hydrocolloids have been reported to improve bread quality through stabilization of water-flour association in sourdoughs matrix (Tieking et. al., 2003; Font de Valdez et. al., 2010).

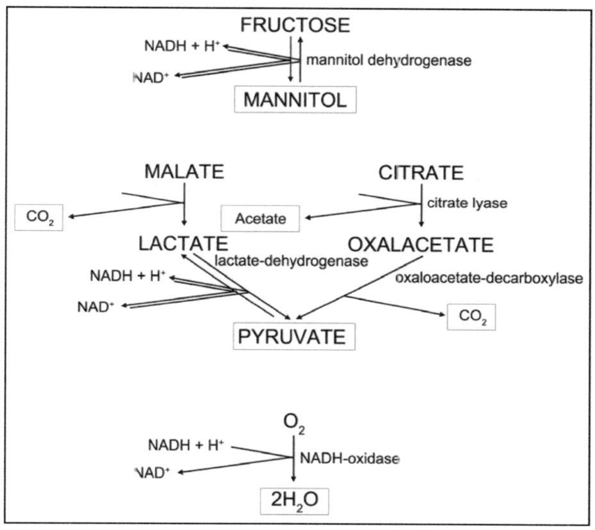

Fig. 1. Fate of potential electron acceptors upon carbohydrates fermentation by *Lb. sanfranciscensis* (Corsetti et al, 2007)

3.2 Metabolism of proteins

During sourdough fermentations protein modification and protein degradation have been occurred. Protein degradation that occurs during sourdough fermentations is among the key metabolic route that directly effects on dough texture and flavour of sourdough. During dough fermentation, LAB releases small peptides and free amino acids important for microbial growth and acidification of dough (Rollán & Font de Valdez, 2001). Also, small peptides and free amino acids are important as precursors for flavour development of the leavened baked products (Thiele et. al., 2002). According to the results of studies performed by Gerez et. al. (2006) 13 nine lactobacilli and four pediococci were able to use gluten as a nitrogen source. Gerez et. al. (2006) also reported an increase in essential amino acids (treonine, valine, lysine and phenylalanine) in a gluten based medium fermented by LAB strains. Thiele et al. (2004) reported that the degree of protein degradation in wheat sourdough is usually similar to that observed in chemically acidified dough. However, bacterial proteolysis during sourdough fermentation was shown to contribute much more to the development of typical sourdough flavours of baked breads compared to breads produced from acidified or yeasted dough (Hansen et. al., 1989a, Hansen et. al., 1989b). Gluten proteins determine, to a great extent, the rheological properties of wheat dough. Substantial hydrolysis of gliadinin and glutenin proteins occurs during sourdough fermentation. Proteolityc activity in sourdough originates not only from LAB enzymes, than derives also from the cereal materials present in sourdough (Thiele, 2002; Thiele, 2004). Except activity of own enzymes, LAB contribute to overall proteolysis during sourdough fermentation by creating optimum (acidic) conditions for activity of cereal proteinases (Vermeulen et al. 2006). The partial hydrolysis of glutenins during sourdough fermentation results in depolymerisation and solubilisation of the gluten macro peptide (GMP). After 24 hours of fermentation with defined lactobacill strains, all gluten proteins were SDS-soluble (Thiele et. al., 2003). Glutathione (GSH) is the most relevant reducing agent in wheat doughs (Grosh & Wieser, 1999). Heterofermentative lactobacilli express glutathione reductase during growth in dough and reduce extracellular oxidized glutathione (GSSG) (Jänsch et. al., 2007). The continuous transformation of GSSG to GSH by LAB metabolism maintains high SH levels in wheat doughs, and increase the amount of SH-groups in gluten proteins (Vermeulen et. al., 2006)

Increased proteolysis during sourdough fermentation leads to the liberation of amino acids in wheat dough. Importance of amino acid conversion to typical flavour volatile compounds has been reported by a number of researchers (Juillard et. al., 1998; Tamman et. al., 2000; Zotta et al. 2006). Sourdough fermentation with LAB results in an increase of amino acids concentration during fermentation, whereas dough fermentation with yeast reduces the concentration of free amino acids (Thiele et. al., 2002). Furthermore Gassenmayer & Schieberle (1995) reported that addition of amino acids (e.g. ornithine, leucine and phenylalanine) to dough resulted in an enhanced conversion to flavour compounds. The level of individual amino acids in wheat dough depends on the pH level of dough, fermentation time and the consumption of amino acids by the fermentative microflora (Thiele et. al., 2002). In wheat sourdoughs, *Lb. brevis linderi, Lb safransciensis, Lb. brevis* and *Lb. plantarum* have been reported to increase the levels of aliphatic, dicarboxylic and hydroxyl amino acids (Gobbetti et. al., 1994a, Gobbetti et. al., 1994b). The yeasts, *S. cerevisiae* and *S. exiguous* decrease the total level of amino acids.

3.3 Development of aroma compounds

The key degradation reaction of amino acids during fermentation of sourdough is the Erlich pathway, leading to aldehydes or the corresponding alcohol, while during baking takes place the Strecker reaction which also lead to aldehydes, but also to the corresponding acids (Hoffmann & Schieberle, 2000). Differences other than acetic acid production, in the overall aroma profile of final bread depend to the type of dominating lactobacilli, as well as to the type of flour. E. g., firm wheat fermented with heterofermentative strains had higher contest of ethyl acetate and hexyl acetate compared to cases when the homofermentative strains were used (Lund et. al., 1989). On the other hand, higher content of aldehydes was higher in rye sourdough fermented with homofermentative LAB (Lund et. al., 1989).

The arginine metabolism by *Lb pontis* and *Lb. sanfransciensis* has been demonstrated to have an impact on bread flavour (Thiele et. al., 2002; De Angelis et. al., 2002). Phenylalanine metabolism was studied in *Lb. plantarum* and *Lb. sanfransciensis*, besides producing phenyllactic acid and 4-hydroxy compounds, which have documented antifungal activity (Vermeulen et. al., 2006). Gerez et. al. (2006) demonstrated that gluten-breakdown with lactobacilli and pediococci, beside reducing gluten-allergen compounds. Furthermore increased the basic amino acid concentration in broth cultures, mainly due to an increase in the ornithine amount, which is considered to be key flavour precursor in wheat bread, generating 2-acetyl-pyrolin (Gassenmeyer & Schieberle, 1995; Thiele et. a., 2002). Czerny & Schieberle (2002) reported that during fermentation of dough, LAB did not generate new aroma compounds than those present in raw materials. However, they demonstrated that many compounds as acetic acid and 3–methylbutanal were increased, whereas aldehydes were decreased.

From all referred above, LAB play very important role in overall sourdough fermentation process. Together with yeasts is responsible for unique quality of the baking bread. More research, however, is needed to clarify all metabolic pathways during fermentation of sourdough.

4. Classification of sourdoughs

Sourdoughs have been classified on the base of procedures during their production (Böcker et. al., 1995). Sourdoughs have been classified into three types:

1. type I sourdoughs or traditional sourdoughs
2. type II sourdoughs or accelerated sourdoughs
3. type III sourdoughs or dried sourdoughs

Each type of sourdough is characterised by a specific sourdough LAB microflora (Table 1).

4.1 Type I sourdough

Type I sourdoughs are produced with traditional techniques, and are characterized by continuous (daily) refreshments to keep the microorganisms in an active state. Type I sourdough is indicated by high metabolic activity, above all regard to leavening, i. e. gas production. The process is conducted at room temperature (20 - 30 °C) and the pH is approximately 4.0. Examples of baked breads with type I sourdough are Francisco sourdough French bread, Panettone and other brioches, Toscanon and Altamura bread, Pugliese, and three-stage sourdough rye bread. Traditional, type I sourdough encompass pure culture, pasty sourdough starter preparations from different origin (type Ia), spontaneously developed, mixed culture sourdoughs made from wheat and rye or mixture thereof and prepared through multiple stage fermentation processes (type Ib), and sourdough made in tropical regions fermented at high temperatures (Stolz, 1999). Pure culture sourdoughs (type Ia) are derived from natural sourdough fermentations.

4.2 Type II sourdough

Type II of sourdough has been prepared in semi-fluid silo condition. Those bakery pre-products serve mainly as dough acidifiers. Several modified, accelerated processes with continuous propagation and long-term one-step fermentations are common now. They guarantee more production reliability and flexibility. A recent trend of industrial bakeries exists in the instalment of continuous sourdough fermentation plants (Stolz & Böcker, 1996).

Typical type II process last for 2-5 days and are often carried out at increased fermentation temperature (usually > 30 °C) to speed up the process. Those sourdoughs exhibit a high acid content at a pH of <3.5 after 24 hours of fermentation. The microorganisms are commonly in the late stationary phase and therefore exhibit restricted metabolic activity. The high dough yields of these preparations permit pumping of the dough. They are frequently used in local bakeries. Those sourdoughs are stored fresh until use (up to one week), they can be produced in large quantities. In industry, they are applied for the production of dried sourdough products as well (De Vuyst & Neysens, 2005).

4.3 Type III sourdough

Type III sourdoughs are dried doughs in powder form, which are initiated by defined starter cultures. They are used as acidifier supplements and aroma carries during bread-making. They mostly contain LAB that are resistant to drying and are able to survive in that form, e.g. heterofermentative *Lb. brevis*, facultative heterofermentative *P. pentosaceus* and *Lb. plantarum* strains. The drying process (spray-drying or drum-drying) also leads to an increased shelf-life of the sourdough and turns it into a stock product until further use. Dried sourdoughs are convenient, simple in use, and result in standardized end products. They can be distinguished in colour, aroma and acid content (De Vuyst & Neysens, 2005).

5. Rheology of the sourdough: Influence of LAB action

5.1 Effects of LAB to dough structure

Cereal grains contain starch and non-starch polysaccharides, the latter composed of glucose (β-glucans), fructose (polyfructan), xylose and arabinose (arabinoxylan) (Belitz & Grosh, 1999). Starch is partially digestible, while some other polysaccharides are not, and would represent dietary fibres. Wheat flours contain, in addition to polyfructans, also nystose, kestose and other fructooligosaccharides of the inulin type (Campbell et al., 1999). Their prebiotic potential has been well examined and documented (Van Loo et al., 1999; Corsetti & Settani, 2007). The structural effects of sourdough in wheat-based system may first be due to the direct influence of low pH on structure-forming dough components, such as gluten, starch, arabinoxylan etc. (Angioloni et. al., 2006). Dough is very sensitive to changes in ionic strength and pH and such changes could have direct impact on the constituents of dough (Clarke et al., 2002). The drop in pH value caused by the produced organic acids influences the viscoelastic behaviour of dough. A correct description of the changes in dough behaviour is necessary to maintain handling and machinability in industrialized production (Wehrle et. al., 1997). A number of earlier studies have examined influence of acids and different pH values on the dough properties. All of these confirmed that changes in the absolute pH value of sourdough significantly influence sourdough components. It has been well documented (Tsen, 1966; Wehrle et. al.,1997; Takeda et. al., 2001; Tieking et. al., 2003) and supported by the findings of the many rheological studies which indicated differences between the nonacidified and chemically or biologically acidified doughs (Clarke et. al., 2002). Osborne (1907) reported a century ago that the presence of acids increased the solubility of the glutenin fraction extracted from the wheat flour. Barber et. al. (1992) reported that there could be mild acid hydrolysis on starch in sourdough system.

The pH profile may affect the time frame during which the acid influences the constituent ingredients of the dough. The changing pH values during sourdough fermentation period may also afford passage through a range of pH values close to the optimum for various enzymes present in the dough system. It is so-called secondary (indirect) effect of sourdough acidification (Clarke et al., 2004). The activity of proteolytic and amylolytic enzyme present may be influenced to a greater degree by the pH profile of the biological acidification fermentation period in contrast to the rather instantaneous nature of the chemically acidified regime. Optimum activity of these enzymes, which play significant role in changes of dough constituents, achieve optimum activity at pH 4-5 for the proteolytic and pH 3.6 - 6.2 for the amylolytic enzymes (Belitz & Grosh, 1992). Other enzymes that might affect the structural components of the dough the activity of which is pH dependent include peroxidases, catalases, lipoxigenases and polyphenol oxydases (Belitz & Grosh, 1992; Clarke et. al., 2002). Results obtained by the the fundamental rheological tests, baking tests, and farinograms show that activity of some enzymes in the biologically acidified dough led to structural changes in the dough (Corsetti et. al., 2000; Clarke et. al., 2002; Clarke et. al., 2004). Corsetti et. al. (2000) also reported that even limited photolytic degradation of wheat proteins affects the physical properties of gluten, which in turn can have a major effect on bread firmness and staling.

Gas production during fermentation of sourdough has been marked as a one of the most important parameters to affect on sourdough structure and rheology. Hammes and Gänzle (1998) have noted that the contribution of yeasts and LAB to the overall gas volume differs with the type of starters and the dough technology applied. These authors have also reported that gas formation by the sourdough microflora is only of minor importance if baker's yeast added. Clarke et. al. (2002) proved that the amount of gas produced by the sourdough microorganisms does not contribute remarkably to the increase in loaf specific volume. However, Hammes & Vogel (1995) indicated certain differences connected to the type of used LAB starter. Thus, obligately heterofermentative *Lb. pontis* significantly higher influenced to the gas formation in sourdough than facultatively hetrofermentative *Lb. plantarum*.

Exopolysaccharides (EPS) in sourdough are products of LAB metabolism. EPS have been recognized as the one of the important structure stability factors (Tieking et. al., 2003). Two classes of EPS from LAB can be distinguished, extracellulary synthetized homopalysacharides (HoPS), composed of only one type of monosaccharide and are synthesised by extracellular glucan and fructosyltransferases (glycosyltransferases) using sucrose as the glycosyl donor, and heteropolysaccharides (HePS) with irregular repeating units (Corsetti & Settani, 2007). HoPS are today generally applied to improve the texture of baked goods, while HePS usually have been used only in fermented dairy industry (Laws & Marshall, 2001). Sourdough lactobacilli have not been found to produce HePS.

One of the key sourdough LAB starters, *Lb. sanfranciscensis*, has been well known and characterized for its contribution to the enhancement of polysaccharide content due to the production of EPS (Korakli et. al., 2001; Korakli et. al., 2003; Tieking et. al., 2003). Formation of EPS is a well accepted characteristic of sourdough LAB, since this feature influencing on the viscosity of sourdough (Vogel et. al., 2002).

Such as presented, through the many microbiological and biochemical reaction starter microorganisms in sourdough affect on its important structural changes. From that reason, choose of adequate starters and control of structural changes during process of fermentation and baking is essential.

5.2 Dough structure

The main constituents are those derived from the flour, most importantly the proteins, both soluble and insoluble. The carbohydrates, which include starch, sugars, soluble and insoluble polysaccharides, are the most abundant. The lipids form a small, but significant part of the flour. The molecules of some minor constituents are composed of the lipid part (glycolipids) or are protein part (lipoproteins). These molecules are interesting because of their possible role in the interaction between more abundant constituents. Finally, flour enzymes may affect the dough properties (Chavan & Chavan, 2011). Water, one of the basic constituents, plays a key role in the formation of dough. During dough mixing, some air is occluded. The air forms the nuclei of the gas cells, which expand during fermentation and oven rise (Gan et. al., 1995). The remaining constituents include all other ingredients that are added to dough, including yeasts, salt, malt, enzymes, flour improvements, sugar, fats, emulsifiers, milk or soy solids, mould inhibitors, or, in the case of sourdough LAB starters in the one of the mentioned forms.

Dough is composed of a continuous phase in which gas cells are dispersed. This continuous phase is called "dough phase" (Bloksma & Bushuk 1988). The following constituents can be distinguished:

1. Starch granules, occupying about 60% of the volume of the dough phase. Large elliptical granules occur side by side with small spherical granules. Size of this granule is significantly differing between normal wheat dough and sourdough (Brummer & Lorenz, 1991);
2. Swollen protein
3. Yeast cells, with the diameter of about 2 μm
4. LAB cells (in the case of sourdough)
5. Lipids
6. Irregularly shaped remnants of cell organelles and wheat grain tissues (Pomeranz, 1988). Immediately after mixing, the gas cells form spherical holes with diameters between 10 and 100 μm. Their number at that stage is estimated to be between 10^{11} and 10^{13} m^{-3}. Clarke et al. (2002) reported that LAB starters do not significantly influencing on gas transformation in dough, but certain differences between wheat dough and sourdough has been revealed by Hammes & Vogel (1995).

Wheat flour dough, same like sourdough, is viscoelastic, exhibits both flow and elastic recovery. When a piece of dough is placed on a flat surface, if the humidity is high enough so it will flow. The amount of flow depends on the balance of viscous and elastic properties (Faridi & Faubion, 1986). In fact, dough is not truly elastic such as rubber or similar artificial materials. If a piece of dough be stretched rapidly and the forse be released immediately, it will only partially recover its original shape (Hoseney, 1994). In dough, the cross-links between molecules are secondary (noncovalent) bonds that are constantly breaking and reforming other units (Rao, 1984). This appears that starch is a not an inert ingredient in flour-water systems. Instead starch appears to act as filler in the gluten polymer. Filled polymers are known to have a larger modulus than their unfilled counterparts.

In general, sourdough structure could be described similar to the model of the only yeast-fermented dough. However, according to all facts mentioned in previous text, it is so obvious that sourdough have different structure, due to added LAB starters (Koceva Komlenić et. al., 2010).

5.3 Fundamentals in dough rheology

Rheology is the study of how materials deform, flow, or fail when force is applied. If a material viscosity is constant regardless of stirring or flowing through pipe, these materials are called Newtonian. However, in many food systems, including flour-water systems, the viscosity changes (decreases) as the shear rate is increased (Bagley, 1992). These systems show more complicated non-Newtonian behaviour, and we cannot define the system by a single viscosity value, but must give viscosity at each shear rate. Additionally, viscosity can also be affected by the time involved in making the measurement (Hoseney, 1994). Many different kind of rheological moduli exist in applied rheology. In general, modulus refer to the stiffness of the material and is proportionally constant relating stress to strain. It tells how much force is required to produce a specific deformation of the material under test (Rao, 1986).

In cereal technology, farinograph, mixograph, extensograph and amylograph have been referred to as rheological measurements (Honesey, 1994). These instruments measure how doughs deform and flow. The problem with the use of these instruments for rheological studies is that cannot define the stress at any moment of time during the test. This is not to say that above mentioned instruments are not useful. They have stood the test of time and can give much useful information. They are particularly used when used to characterize flour. It has been important to know whether the mixing properties of the flour used today are similar to or different from those of the flour used yesterday. The mixograph or farinograph can easily and rapidly detect this (Faridi & Faubion, 1986). Unlike, from fundamental measurements we can learn much more about rheological behaviour of dough. We can find effect of various interactions and how the properties of the dough change as a function of time or temperature. Additionally, the measurements can often be made on the complete dough system so the results are easy to interpret. The second reason is, that it has now become much easier to obtained good rheological data with the advent of minicomputers incorporated to the measurements and their related equipments (Hoseney, 1994).

Rheological measurements recently used for the characterization of dough rheology are described below.

5.4 Rheological tests on dough

The physical characteristics of doughs are important in relation to the uses of flours. In the case of sourdough, LAB fermentation is another factor. Pseudo-rheological characteristics are investigated mainly with the following methods:

1. The Brabender Farinograph measures and records the resistance of dough to mixing as it is formed from flour and water, developed and broken down. This resistance is called consistency. The maximum consistency of the dough is adjusted to a fixed value by

altering the quantity of water added. This quantity, the water absorption may be used to determine a complete mixing curve, the various features of which are guide to the strength of the flour (Figure 2)

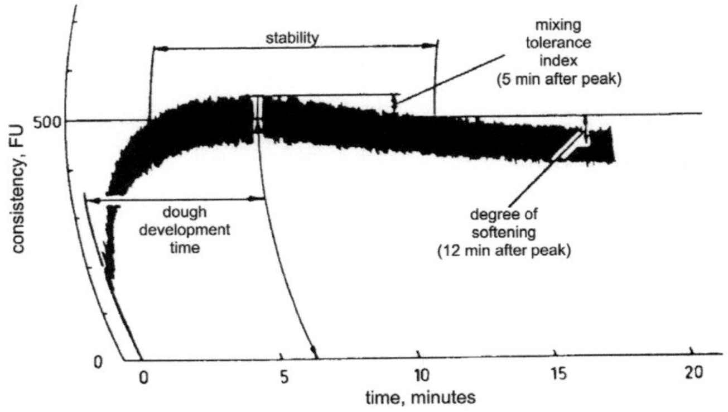

Fig. 2. Representative farinogram showing some commonly measured indexes. A consistency of 500 farinograph units (FU) corresponds to a power of 68 and 81 W per kilogram of dough in mixtures for 300 and 50 g of flour, respectively (Bloksma &Bushuk, 1988).

2. The Brabender Extensograph records the resistance of dough to stretching and the distance the dough stretches before breaking. A flour-salt-water dough is prepared under standard conditions in the Brabender Farinograph and moulded on the Extensograph into a standard shape. After a fixed period the dough is stretched and a curve drawn, recording the extensibility of the dough and its resistance to stretching (Fig. 3). The dough is removed and subjected to a further two stretches. The Extensograph has replaced the Extensometer in the Brabender instrument range.

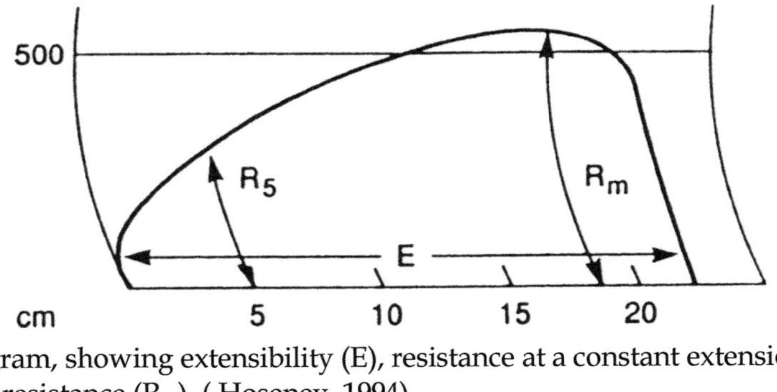

Fig. 3. Extensigram, showing extensibility (E), resistance at a constant extension of 5 cm (R_5) and maximum resistance (R_m). (Hoseney, 1994).

3. The Chopin Alveograph uses air pressure to inflate a bubble of dough until it bursts. The instrument continuously records the air pressure and the time that elapses before the dough breaks (Figure 4).
4. The Brabender Amylograph continuously measures the resistance to stirring of a 10% suspension of flour in water while the temperature of the suspension is raised at a constant rate of 1.5 °C/min from 20° - 95 °C. It is of use in testing flour for soups (etc.),

for which purpose the viscosity of the product after gelatinization is an important characteristic, as well as for adjusting the malt addition to flours for breadmaking

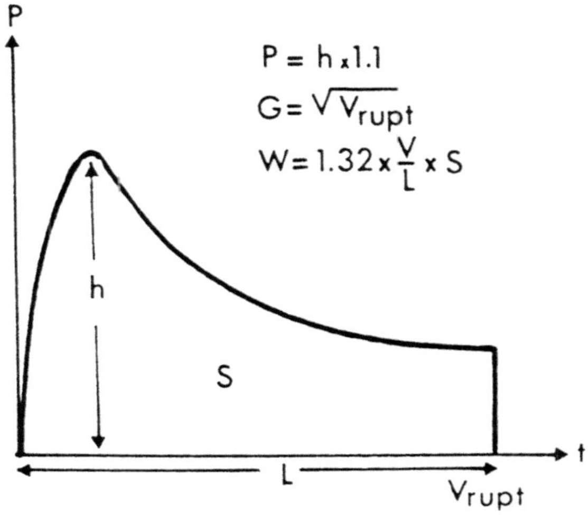

Fig. 4. A representative alveogram. P = overpressure (mm), L = abscissa at rupture (mm), G = swelling index (ml), V = volume of air (ml), W = deformation energy (10^{-4} J), h = maximum height (mm) and S = area under the curve (cm^2). (Hoseney, 1994).

5. The Rapid Visco Analyzer (RVA), may be regarded as a derivative of the Amylograph. Measurements of viscosity are made using small samples, containing 3-4 g of starch, in periods which may be as short as 2 min. Use of disposable containers and mixer paddles eliminates the needs for careful washing of the parts between tests. As with the Amylograph, the characteristics of starch pastes and the effects of enzyme of them can be recorded on charts. They can also be transferred in digital form direct to a data-handling computer.
6. True dynamic rheological instruments. In recent years frustration with instrument-dependent units obtained with some of the above methods, together with the poor reproducibility, from one instrument to another of the same type, has led cereal chemists to turn to true rheological measurements. Suitable instruments for use with doughs, slurries and gels, derived from flours, include Bohlin VOR (viscometric, oscillation and relaxation), the Carri Med CSL Rheometer and the Rheometrics RDA2. In addition to providing excellent reproducibility, these instruments, which are also used in many non-foods, and many non-cereal food materials, allow comparisons to be made across a wide range of substances. Although they are expensive, they will undoubtedly enable the development of tests that can be performed on simpler, dedicated instruments (Faridi & Faubion, 1990; Kent & Evers, 1994).

5.5 Documented rheological results on the sourdough: Recapitulation

In a number of previous studies differences between simple yeast-fermented doughs, chemically acidified (using lactic or acetic acid) doughs and biologically acidified (using a starter culture) dough were investigated by the use of empirical and dynamic rheology measurements (Zeng et. al., 1997; Wehrle et. al., 1997; Wehrle & Arendt, 1998; Lee et. al.,

2001; Clarke et. al. 2002; Clarke et. al., 2004; Angioloni et. al., 2006). Hoseney (1994) reported that acid developed during fermentation with LAB strongly influence the mixing behaviour of doughs, whereby doughs with lower pH values require a slightly shorter mixing time and have less stability than normal doughs. The water absorption of flour is an important factor influencing the handling properties and machinability of dough in large mechanized bakeries and is related to the quality of the finished baked product (Caterall, 1998). Incorporation of sourdough changed the mixing behaviour, resulting in a significant decrease in water absorption relative to control (Maher Galal et. al., 1978; Wehrle et. al., 1997; Clarke et. al., 2002; Koceva Komlenić et. al., 2010). The addition of sourdough prepared with starter culture significantly reduced the stability of the dough relative to both the control (yeast fermented) and chemically acidified dough (Wehrle et. al., 1997; Clarke et. al., 2002; Clarke et. al., 2004). A large difference was found for the degree of softening. Addition of sourdough or acid significantly increased the degree of softening (Maher Galal et. al., 1978; Clarke et. al., 2002). Clark et. al. (2002) reported that biologically acidified doughs (sourdoughs) showed a significantly greater degree of softening than the chemically acidified (with lactic or acetic acid) doughs (Table 3. and Figure 5.).

Sample**	Water absorption (g/100g)	Dough development time (min)	Stability (min)	Degree of softening (BU)
Control I	**62.15 ± 0.06 g**	**2.25 ± 0.29 e**	**2.03 ± 0.05 a**	**61.25 ± 2.50 e**
LAB I	60.23 ± 0.05 h	3.25 ± 0.29 c	0.90 ± 0.23 e	82.50 ± 2.89 b
DS I	64.28 ± 0.05 e	2.80 ± 0.23 d	1.70 ± 0.12 b	66.25 ± 2.50 cd
LA I	62.35 ± 0.06 f	2.15 ± 0.17 e	1.40 ± 0.01 c	62.50 ± 2.89 de
Control II	**67.75 ± 0.06 c**	**3.98 ± 0.05 b**	**1.28 ± 0.05 cd**	**65.00 ± 5.77 cde**
LAB II	65.58 ± 0.05 d	4.63 ± 0.10 a	0.65 ± 0.06 f	122.50 ± 2.89 a
DS II	69.23 ± 0.05 a	4.20 ± 0.14 b	1.05 ± 0.29 e	81.25 ± 2.50 b
LA II	68.10 ± 0.12 b	4.03 ± 0.05 b	0.90 ± 0.23 e	67.50 ± 2.89 c

*Values are mean ± SD of four independent determinations; Mean values followed by common letter within the same column are not significantly different ($p < 0.05$)
**Control, control dough; I and II, flour types: T-550 and T-110, respectively; LAB, biologically acidified dough by addition of sourdough prepared with *Lactobacillus brevis* L-75; DS, biologically acidified dough by addition of dry sourdough; LA, chemically acidified dough by addition of lactic acid

Table 3. Farinogram results* for different wheat doughs (control, biologically and chemically acidified doughs). (Koceva Komlenić et al. 2010).

The effect of sourdough addition on extensibility of dough were also investigated (Tsen, 1966; Corsetti et. al., 2000; Clarke et. al., 2002; Katina et. al. 2006a; Katina et. al. 2006b, Arendt et. al., 2007; Koceva Komlenić et. al., 2010). The Extensograph gives information about about a dough extensibility and resistance to extension (Walker & Haazelton, 1996). Generally, the addition of sourdough or acid significantly reduced extensibility of the dough. Clarke et. al. (2002) demonstrated the higher reduction of extensibility when the mixed-strain starter culture was used compared to single strain used (Figure 6). It is in agreement with the results obtained by Tsen (1966).

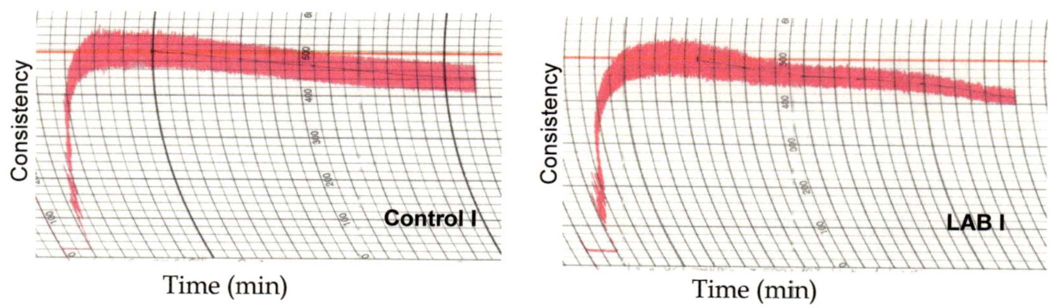

Fig. 5. Farinograms of doughs: Control I (prepared with flour type I) and LAB I (prepared with flour type I and *L. brevis* preferment). (Koceva Komlenić et al. 2010).

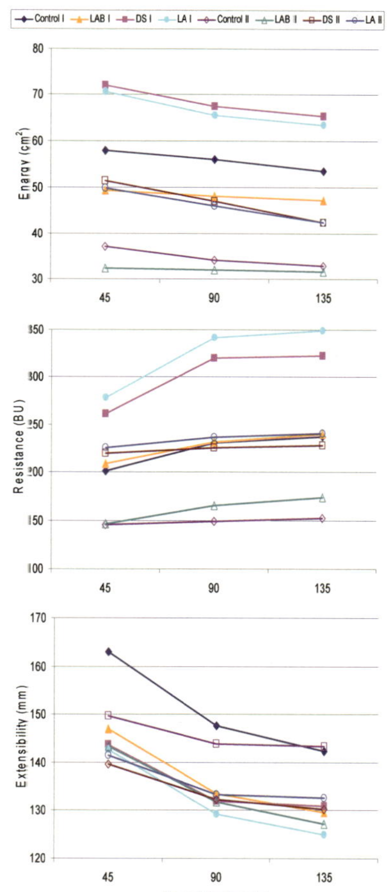

Fig. 6. Extensograph analysis of doughs with sourdough and lactic acid: I and II (different flour type), LAB (biologically acidified dough by addition of sourdough prepared using a starter culture of *L. brevis* L-75), DS (biologically acidified dough by addition of dry sourdough), LA (chemically acidified dough by addition of lactic acid).
(Koceva Komlenić et al. 2010)

Dynamic oscillatory measurements examined the effects of sourdough or acid addition on the viscoelastic properties of the dough. The rheological characteristics of fermented dough are determined by many factors. At the beginning of the mixing process, physical actions, such as hydration, take place. The gluten network is formed by proteins, and starch granules absorb water. Enzyme activity of amylases, proteases and hemicellulases causes breakdown of several flour components. Changes of pH level also alter the rheological behaviour of the dough Wehrle et al. 1997). Even small chemical and physical changes in gluten network cal lead to significant changes in rheological characteristics. Formation of gas during fermentation of sourdough leads to an increased volume and decreased density.

According to the most of the authors, main variables of viscoelastic properties of the dough are phase angle (δ) and complex modulus G* (Figure 7 and 8.). The results of fundamental oscillatory measurements gave the key attributes of rheological behaviour of the dough. The addition of sourdough prepared by LAB starters did increase the phase angle values relative to the control or the chemically acidified dough significantly (Clarke et. al., 2002). It means that addition of sourdough reduced the elasticity of the dough in contrast to the chemical acidification, which did not same effect (Hoseney, 1994). Some authors observed that was no significant differences between G* values for the control and chemically acidified doughs over all frequencies, while the values for the control were significantly higher than those obtained from sourdough over all frequencies. The same was true for the chemically acidified doughs which yielded greater G* values than the sourdoughs. An increase in phase angle and decrease in G* due to addition of sourdough indicated that that the dough was less elastic and became simultaneously less firm at low rate of strain applied (0.1%). These results also clearly suggest that chemical acidification was not directly comparable with biological acidification. Thiele et. al. (2002) reported that, with regard to the levels of amino acids in wheat dough, may influence on sourdough rheology. Clarke et al. (2004) also reported a dramatic loss firmness and elasticity in preferments during 24-hours fermentation of sourdough. These results suggest that cereal proteases with acidic optima play a central role in the rheological changes taking place during sourdough fermentation.

Wehrle & Arend (1998) analyzed rheological characteristic of controlled and spontaneous fermented sourdough. During the first 5 hours of fermentation, phase angles did not change significantly. After 5 hours phase angles of sourdough increased sharply. Maximum phase angles were achieved after fermentation time of 20 hours. A high phase angle indicates viscous behaviour of material. Fermented sourdough after 20 hours of fermentation is an entirely viscous manner, with phase angles close to 90°, which indicates an ideal fluid. Phase angle was closely related to CO_2 production by heterofermentative microorganisms in the dough. Results of Wehrle & Arendt indicate the beginning of gas production in the sourdough after 5th hour of fermentation (Table 4). larke et. al. (2004) also observed a dramatic loss of firmness and elasticity during 24-hr period irrespective of the presence acid or lactic acid bacteria. The presence of acid, however, enhanced the effects. Wehrle et. al. (1997) reported that doughs with acid initially were firmer and more viscous than doughs without acids, but showed less stability during mixing. Addition of acids intensified decrease in G* and increase in δ with longer mixing times when measured immediately after mixing.

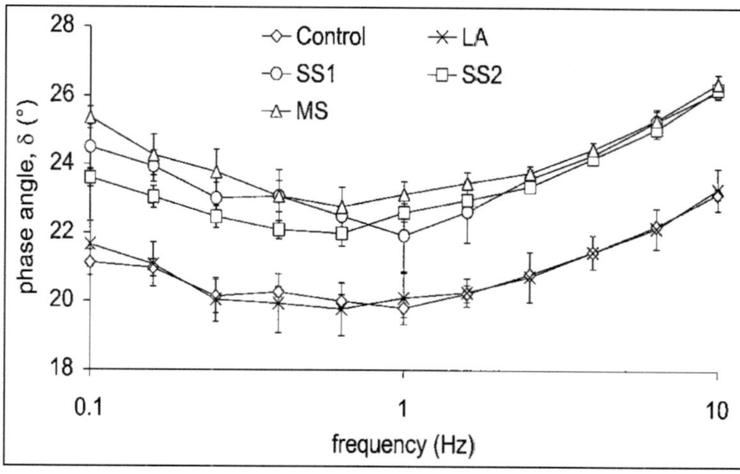

Fig. 7. Phase angle (δ) as a function of frequency for wheat dough formulations: control (nonacidified), LA (chemically acidified with lactic acid), SS1 (added sourdough prepared using a single strain starter culture of *Lactobacillus brevis* L-62), SS2 (added sourdough prepared using a single strain starter culture of *L. plantarum* L2-1), MS (added sourdough prepared using a mixed strain starter culture, Böcker Reinzucht-Sauerteig Weizen). Mean value ± standard deviation of three replicates (Clarke et al. 2002).

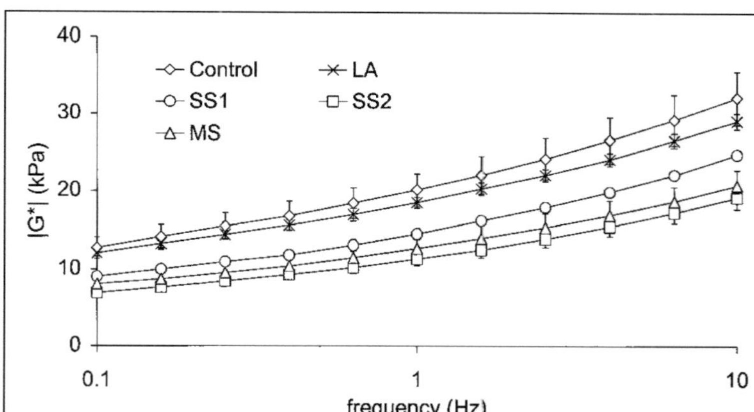

Fig. 8. Absolute value of the complex dynamic modulus (|G*|) as a function of frequency for wheat dough formulations: control (nonacidified), LA (chemically acidified with lactic acid), SS1 (added sourdough prepared using a single strain starter culture of *Lactobacillus brevis* L-62), SS2 (added sourdough prepared using a single strain starter culture of *L. plantarum* L2-1), MS (added sourdough prepared using a mixed strain starter culture, Böcker Reinzucht-Sauerteig Weizen). Mean value ± standard deviation of three replicates. (Clarke et al. 2002).

Rheological properties, as well as acidification and flavour development, are important parameters in controlling fermentation process. Types of microorganisms, metabolic activity, and time-dependent development of pH levels have an effect on the final rheological properties.

	Control	LA[b]	SS1[c]	SS2[d]	MS[e]
Dough development curve parameters					
Hm[f] (mm)	87.0 ± 1.0b	79.0 ± 2.6ab	75.2 ± 1.2a	76.0 ± 3.3a	87.8 ± 6.3b
h[g] (mm)	86.3 ± 0.4b	78.3 ± 3.8ab	71.8 ± 2.3a	72.7 ± 3.6a	87.3±6.9b
(Hm – h)/Hm[h] (%)	0.7±0.8ab	0.9 ± 1.6abc	4.5 ± 2.1c	4.3 ± 0.8bc	0.6 ± 0.7a
T1[i] (min)	177 ± 3b	179 ± 2b	140 ± 22a	132 ± 6a	179 ± 1b
Gaseous release curve parameters					
T´1[j] (min)	114.0 ± 9.1b	117.0 ± 4.0b	76.5 ± 10.4a	96.0 ± 21.3ab	67.0 ± 5.7a
Vt[k] (mL)	1,942 ± 50ab	2,019 ± 56b	1,986 ± 52ab	2,021 ± 16b	1,870 ± 75a

[a] Mean value ± standard deviation of two replicates. Mean values followed by a common letter within the same row are not significantly different (P < 0.05).
[b] Chemically acidified by the addition of lactic acid.
[c] Added sourdough prepared using a single strain starter culture of *Lactobacillus brevis* L-62.
[d] Added sourdough prepared using a single strain starter culture of *L. plantarum* L2-1.
[e] Added sourdough prepared using a mixed strain starter culture, Böcker Reinzucht-Sauerteig Weizen.
[f] Maximum height of dough development curve.
[g] Dough development height at the end of 3-hr test period.
[h] % Reduction in dough development height at the end of test period relative to T1.
[i] Time of maximum height of dough development curve.
[j] Time of maximum height of gaseous release curve.
[k] Total volume of carbon dioxide released by dough.

Table 4. Development and Gaseous Release Characteristics of Doughs from Rheofermentometer Testinga (Clarke et al. 2002)[a]

6. Perspectives

The composition and processing of cereals grains, substrate formulation, growth capability and productivity of starter culture, stability of probitic strains during storage, organoleptic properties and nutritional value of the final product are key parameters to be considered (Charalampopoulos, 2002). Probiotics bacteria from the sourdough starters must meet the criteria not only for good survival during fermentation process but also for fermentation and symbiosis with other starter cultures used (Kedia et. al., 2007). Nowadays, the use of probiotic bacteria as starter cultures in baked goods is at its early stage and merits further investigation (Font de Valdes et. al., 2010). Since sourdough systems are so complex, and the determination of rheological properties of dough should be comprehensive to reveal influence of sourdough composition and all components of flour (gluten, starch, lipids, water soluble proteins and pentosans). Regardless of whether the empirical or fundamental rheological testing is used and despite the decades of bread dough rheology testing there is still no universal equipment or method that can provide the data that are 100% accurate in evaluating performance during dough processing. Over time, numerous models of dough rheological behavior have been developed to simplify describing the rheological changes in bread dough system and they are still emerging.

Recently, Tanner *et al.* (2008) introduced the use of a simple Lodge-type model (Lodge, 1964), including a power law memory function, and a damage function (assumed to be a function of strain) to represent the breakdown of the molecular structure within the dough, for a description of the rheological behavior of wheat dough. The model produced a satisfactory reconstruction of stress data in shear start up and elongation flow at low deformation rates. The model has relatively few parameters, all of which can easily be found from simple experiments. The sequence of step strains, with both reversing and steadily increasing steps, is reasonably well described by the model. The model can therefore describe small-strain motions, steady shearing, steady uniaxial and biaxial elongations, recoil, stress relaxation and step strain motions. Since then, several modifications of this model have been developed. Hicks *et al.* (2011) find that the damage term is explicitly a function of strain, a concept that may carry over to elongation flow and that rapid decrease of the damage function post material fracture is a representation of the transition of dough from solid- to liquid-like rheological behavior. Tanner *et al.* (2011) described an improved damage function model for bread dough rheology that enables describing uniaxial and biaxial stretching with the same damage function derived from shear data.

Further investigations of dough rheological behavior is required to obtain quality results and to develop models that accurately simulate processing situations and parameters and provides a platform for comprehensive process control.

7. References

Angioloni, A., Romani, S. Pinnavaia, G. G., Dalla Rosa, M. (2006). Characteristics of bread making doughs: infuence of sourdough fermentation on the fundamental rheological properties. *European Food Research and Technologie,* 222, pp. 54-57

Arendt, E.K., Ryan, L.A. M. & Dal Bello, F. (2007). Impact of sourdough on the texture of bread. *Food Microbiology,* 24, pp. 165-174.

Bagley, E. B. (1992). Mechanistic basicof rheological behaviour of foods. In: *Physical Chemistry of Foods.* Schwartzberg H. G. & Hartel R. W. (eds.), Marcel Dekker, New York.

Barber, B., Ortolá, C., Barber, S., & Fernández, F. (1992). Storage of packaged white bread. III. Effects of sour dough and addition of acids on bread characteristics. *Zeitschrift für Lebensmittel Untersuchung und Forschung,* 194, pp. 442-449.

Belitz, H. D., & Grosh, W. (1999). *Food chemistry.* Springer-Verlag: Berlin, Germany

Belitz, H.-D., & Grosch, W. (1992). *Lehrbuch der Lebensmittelchemie.* 4th ed. Springer-Verlag: Berlin, Germany.

Bloksma, A. H., & Bushuk, W. (1988). Rheology and chemistry of dough. In: *Wheat chemistry and technology,* Vol. 2, Pomeranz, Y. (Edt). American Association of Cereal Chemists, St. Paul, Minnesota. pp. 131-217.

Böcker, G., Stolz, P., & Hammes, W. P. (1995). Neue Erkenntnisse zum Ökosystem Sauerteig und zur Physiologie des Sauerteig-Typischen Stämme *Lactobacillus sanfrancisco* und *Lactobacillus pontis. Getreide Mehl und Brot,* 49, pp. 370-374.

Böcker, G., Vogel, R. F., & Hammes,W. P. (1990). *Lactobacillus sanfrancisco* als stabiles Element in einem Reinzucht-Sauerteig- Präparat. Getreide Mehl und Brot, 44, pp. 269-274.

Brümmer, J.-M. & Lorenz, K. (1991). European developments in wheat sourdoughs. *Cereal Foods World,* 36, pp. 310-314.

Cai, Y., Okada, H., Mori, H., Benno, Y., & Nakase, T. (1999). *Lactobacillus paralimentarius* sp. nov., isolated from sourdough. *International Journal of Systematic and Evolutionary Microbiology*, 49, pp. 1451-1455.

Campbell, J. M., Bauer, L. L., Fahey, G. C., Hogarth, A. J. C. L., Wolf, B. W., & Hunter, D. E. (1997). Selected fructooligosaccharide (1-kestose, nystose, and 1F-b-fructofuranosylnystose) composition of foods and feeds. *Journal of Agricultural and Food Chemistry*, 45, pp. 3076-3082.

Catterall, P. (1998). Flour milling. In: *Technology of Breadmaking*. S. P. Cauvain and L. S. Young, eds. Blackie Academic and Professional: London. pp. 296-329.

Charalampopoulos , D. , Pandiella , S.S. , and Webb , C. (2002) Growth studies of potentially probiotic lactic acid bacteria in cereal - based substrates. *Journal of Applied Microbiology*, 92 , pp. 851 - 859 .

Chavan, R. S. & Chavan, S. R. (2011). Sourdough technology - A traditional way for wholesome foods: A review. *Comprehensive Reviews in Food Science and Food Safety*. 11, pp. 170-183

Clarke, C. I., Schober, T. J., Dockery, P., O'Sullivan, K. & Arendt, E. K. (2004). Wheat sourdough fermentation: effects of time and acidification on fundamental rheological properties. *Cereal Chemistry*, 81, pp. 409-417.

Clarke, C.I., Schober, T.J. & Arendt, E.K. (2002). Effect of single strain and traditional mixed strain starter cultures on rheological properties of wheat dough and on bread quality. *Cereal Chemistry*, 79, pp. 640-647.

Corsetti, A. & Settanni, L. (2007). *Lactobacilli* in sourdough fermentation. *Food Research International* 40, pp. 539-558

Corsetti, A., De Angelis, M., Dellaglio, F., Paparella, A., Fox, P. F., Settanni, L., & Gobbetti, M. (2003). Characterization of sourdough lactic acid bacteria based on genotypic and cell-wall protein analyses. *Journal of Applied Microbiology*, 94, pp. 641-654.

Corsetti, A., Gobbetti, M., De Marco, B., Balestrieri, F., Paoletti, F., Russi, L., and Rossi, J. (2000). Combined effect of sourdough lactic acid bacteria and additives on bread firmness and staling. *Journal of Agricultural and Food Chemistry*, 48, pp. 3044-3051.

Corsetti, A., Lavermicocca, P., Morea, M., Baruzzi, F., Tosti, N., & Gobbetti, M. (2001). Phenotypic and molecular identification and clustering of lactic acid bacteria and yeasts from wheat (species *Triticum durum* and *Triticum aestivum*) sourdoughs of southern Italy. *International Journal of Food Microbiology*, 64, pp. 95-104.

Czerny, M., & Schieberle, P. (2002). Important aroma compounds in freshly ground wholemeal and white wheat flour-identification and quantitative changes during fermentation. *Journal of Agricultural and Food Chemistry*, 50, pp. 6835-6840.

De Angelis, M., Mariotti, L., Rossi, J., Servili, M., Fox, P. F., Rollán, G. C., et al. (2002). Arginine catabolism by sourdough lactic acid bacteria: purification and characterization of the arginine deiminase pathway enzymes from Lactobacillus sanfranciscensis CB1. *Applied and Environmental Microbiology*, 68, pp. 6193-6201.

De Vuyst, L., & Gänzle, M. (2005). Second international symposium on sourdough: from fundamentals to applications. *Trends in Food Science and Technology*, 16, pp. 2-3.

De Vuyst, L., & Neysen, P. (2005). The sourdough microflora: biodiversity and metabolic interactions. Trends *in Food Science and Technology*, 16, pp. 43-56.

Ehrmann, M.A. , and Vogel , R.F. (2005) Molecular taxonomy and genetics of sourdough lactic acid bacteria. *Trends in Food Science and Technology* 16 , pp. 31 - 42 .

Faridhi, H. & Faubion, J. M. (1990) *Dough Rheology and Baked Product Texture*. Van Norstrand Rheinhold. New York.

Faridi H. & Faubion J. M. (1986). *Fundamentals of dough Rheology*. America Association of Cereal Chemists, St. Paul, Minnesota

Font de Valdez, G., Gerez, C. L., Torino, M. I. & Rollán, G. (2010). New Trends in Cereal - based Products Using Lactic Acid Bacteria. In F. Mozzi, R. R. Raya, G. M. Vignolo (Eds.), *Biotechnology of lactic acid bacteria: novel applications*. Wiley-Blackwell, Iowa, pp. 273-287

Gan Z., Ellis P. R., Schofield J. D. (1995). Mini review: Gas cell stabilisation and gas retention in wheat bread dough. *Journal Cereal Science*, 21, pp. 215-230

Gassenmeier, K., & Schieberle, P. (1995). Potent aromatic compounds in the crumb of wheat bread (French-type)-influence of pre-ferments and studies on the formation of key odorants during dough processing. *Zeitschrift fur Lebensmittel Untersuchung und Forschung*, 201, pp. 241-248.

Gerez , C.L., Torino, M.I., Rollán , G. & Font de Valdez , G. (2009). Prevention of bread mould spoilage by using lactic acid bacteria with antifungal properties. *Food Control* 20, pp. 144 - 148 .

Gerez, C.L., Rollán, G. & Font de Valdez, G. (2006). Gluten breakdown by lactobacilli and pediococci strains isolated from sourdough. *Letters in Applied Microbiology*, 42, pp. 459 - 464.

Ginés , S. , Bárcena , J.M., Ragout , A., & Font de Valdez, G. (1997) Effect of the carbon source on the fermentation balance of *Lactobacillus reuteri*. *Microbiologie, Aliments, Nutrition*, 15 , pp. 23 - 27 .

Gobbetti, M. (1998). The sourdough microflora: interactions of lactic acid bacteria and yeasts. *Trends in Food Science and Technology*, 9, pp. 267-274.

Gobbetti, M., & Corsetti, A. (1997). *Lactobacillus sanfrancisco* a key sourdough lactic acid bacterium: a review. *Food Microbiology*, 14, pp. 175-187.

Gobbetti, M., Corsetti, A., Rossi, J., La Rosa, F., & De Vincenzi, S. (1994a). Identification and clustering of lactic acid bacteria and yeasts from wheat sourdoughs of central Italy. *Italian Journal of Food Science*, 6, pp. 85-94.

Gobbetti, M., Corsetti, A., & Rossi, J. (1994b). The sourdough microflora. Interactions between lactic acid bacteria and yeasts: metabolism ofcarbohydrates. *Applied Microbiology and Biotechnology*, 41, pp. 456-460.

Gobbetti, M., Rizzello, J.C., Di Cagno, R. & De Angelis, M. (2007) Sourdough lactobacilli and celiac disease. *Food Microbiology*, 24, pp. 187-196.

Grosh, W., and Wieser, H. (1999). Redox reactions in wheat dough as affected by ascorbic acid. *Journal of Cereal Science*, 29, pp. 1-16.

Hammes, W. P. & Vogel, R. F. (1995). The genus Lactobacillus. Pp. 19-54 in: *The Lactic Acid Bacteria*. Vol. 2. B. J. B. Wood and W. H. Holzapfel, eds. Chapman and Hall: London.

Hammes, W. P., Brandt, M. J., Francis, K. L., Rosenheim, M., Seitter, F. H., & Vogelmann, S. (2005). Microbial ecology of cereal fermentations. *Trends in Food Science and Technology*, 16, pp. 4-11.

Hammes, W.P. & Gänzle, M.G. (1998). Sourdough breads and related products. In: *Microbiology of fermented foods*. Vol 1, 2nd Ed, (edited by B.J.B. Wood). Pp. 199-216. London, UK: Blackie Academic & Professional

Hansen, A., & Schieberle, P. (2005). Generation of aroma compounds during sourdough fermentation: applied and fundamental aspects. *Trends in Food Science and Technology*, 16, pp. 85-94.

Hansen, A., Lund, B., & Lewis, M. J. (1989a). Flavour production and acidification of sour doughs in relation to starter culture and fermentation temperature. *Lebensmittel Wissenschaft und Technologie*, 22, pp. 145-149.

Hansen, A., Lund, B., & Lewis, M. J. (1989b). Flavour of sourdough rye bread crumb. *Lebensmittel Wissenschaft und Technologie*, 22, pp. 141-144.

Hicks, C.I., See, H., Ekwebelam, C. (2011). The shear rheology of bread dough: modeling. *Rheologica Acta*. 50, pp. 701–710.

Hoffmann, T., & Schieberle, P. (2000). Formation of aroma-active Strecker aldehydes by a direct oxidation of Amadori compounds. *Journal of Agricultural and Food Chemistry*, 48, pp. 4301-4305.

Hoseney, R.C. (1994). *Principles of cereal science and tehnology*. American Association of Cereal Chemists, Inc. St. Paul Minesota. USA .

Jänsch, A., Korakli, M., Vogel, R. F. & Gänzle, M. G. (2007). Glutathione Reductase from Lactobacillus sanfranciscensis DSM20451: Contribution to Oxygen Tolerance and Thiol Exchange Reactions in Wheat Sourdoughs. *Applied And Environmental Microbiology*, 73. pp. 4469-4476

Juillard, V., Guillot, A., Le Bars, D., & Gripon, J. C. (1998). Specificity of milk peptide utilization by Lactococcus lactis. *Applied and Environmental Microbiology*, 64, pp. 1230-1236.

Katina, K., Heiniö, R.L., Autio, K. & Poutanen, K. (2006a). Optimization of sourdough process for improved sensory profile and texture of wheat bread. *LWT - Food Science and Technology*, 39, pp. 1189-1202.

Katina, K., Salmenkallio-Marttila, M., Partanen, R., Forssell, P. & Autio, K. (2006b). Effects of sourdough and enzymes on staling of high-fibre wheat bread. *LWT - Food Science and Technology*, 39, pp. 479-491.

Kazanskaya, L. N., Afanasyeva, O. V., & Patt, V. A. (1983). Microflora of rye sours and some specific features of its accumulation in bread baking plants of the USSR. In J. Holas & F. Kratochvil (Eds.), *Developments in food science. Progress in cereal chemistry and technology* (pp. 759-763). London: Elsevier.

Kedia , G. , Wang , R. , Patel , H. , and Pandiella , S.S. (2007) Use of mixed cultures for the fermentation of cereal - based substrates with potential probiotic properties. *Process Biochemistry*, 42 , pp. 65 - 70 .

Kent, N. L. & Evers, A. D. (1994). *Technology of Cereals*. Elsevier Science, Oxford.

Ketabi, A., Soleimanian-Zad, S., Kadivar, M., Sheikh-Zeinoddin, M. (2008). Production of microbial exopolysaccharides in the sourdough and its effects on the rheological properties of dough. *Food Research International*, 41, 10, pp. 948-951

Kirchhoff, E., & Schieberle, P. (2002). Quantitation of odor-active compounds in rye flour and rye sourdough using a stable isotope dilution assay. *Journal of Agricultural and Food Chemistry*, 50, pp. 5311-5378.

Koceva Komlenić, D., Ugarčić-Hardi, Ž., Jukić, M., Planinić, M., Bucić-Kojić, A., Strelec, I. (2010). Wheat dough rheology and bread quality effected by Lactobacillus brevis preferment, dry sourdough and lactic acid addition. *International journal of food science & technology*. 45, pp. 1417-1425

Korakli, M., Pavlovic, M., Gänzle, M. G., & Vogel, R. F. (2003). Exopolysaccharide and kestose production by Lactobacillus sanfranciscensis LTH2590. *Applied and Environmental Microbiology*, 69, pp. 2073-2079.

Korakli, M., Rossman, A., Gänzle, M. G., & Vogel, R. F. (2001). Sucrose metabolism and exopolysaccharide production in wheat and rye sourdough by *Lactobacillus sanfranciscensis*. *Journal of Agricultural and Food Chemistry*, 49, pp. 5194-5200.

Kumar, S., Tamura, K., & Nei, M. (2004). MEGA3: Integrated software for molecular evolutionary genetics analysis and sequence alignment. *Briefings in Bioinformatics*, 5, pp. 150-163.

Laws, A., & Marshall, V. M. (2001). The relevance of exopolysaccharides to the rheological properties in milk fermented with ropy strains of lactic acid bacteria. *International Dairy Journal*, 11, pp. 709-721.

Lee, L., Ng, P. K. W., Whallon, J. H. & Steffe, J. F. (2001). Relationship between rheological properties and microstructural characteristics of nondeveloped, partially developed, and developed doughs. *Chereal Chemistry*. 78, pp. 447-452

Lodge AS (1964) Elastic liquids. Academic, London

Lopez, H.W., Duclos, V., Coudray, C., Krespine, V., Feillet-Coudray, C., Messager, A., Demigné, C. & Rémésy, C. (2003). Making bread with sourdough improves mineral bioavailability from reconstituted whole wheat flour in rats. *Nutrition*, 19, pp. 524-530.

Lorenz, K. (1983). Sourdough processes. Methodology and biochemistry. *Baker's Digest*, 55, pp. 85-91.

Lund, B., Hansen, A., & Lewis, M. J. (1989). The influence of dough yield on acidification and production of volatiles in sour doughs. *Lebensmittel Wissenschaft und Technologie*, 22, pp. 150-153.

Maher Galal, A., Varriano-Marston, E. & Johnson, J. A. (1978). Rheological dough properties as affected by organic acids and salt. *Cereal Chemistry*. 55, pp. 683-691.

Meroth, C. B., Hammes, W. P., & Hertel, C. (2004). Characterisation of the microbiota of rice sourdoughs and description of Lactobacillus spicheri sp. nov. *Systematic and Applied Microbiology*, 27, pp. 151-159.

Osborne, T. B. (1907). The proteins of the wheat kernel. Carnegie Institute of Washington publication 84. Judd and Detweiler: Washington, DC.

Plessas, S., Trantallidi, M., Bekatorou, A., Kanellaki, M., Nigam, P. & Koutinas, A.A. (2007). Immobilization of kefir and *Lactobacillus casei* on brewery spent grains for use in sourdough wheat bread making. *Food Chemistry*, 105, pp. 187-194.

Rao, V. N. M. (1984). Dynamic force deformation properties of foods. *Food Technology* 38, pp. 103-109

Rollán , G. & Font de Valdez, G. (2001). The peptide hydrolase system of *Lactobacillus reuteri*. *International Journal of Food Microbiology*, 70, pp. 303 - 307.

Salovaara, H. (1998). Lactic acid bacteria in cereal-based products. In S. Salminen & A. von Wright (Eds.), *Lactic acid bacteria microbiology and functional aspects* (pp. 115-138). New York: Marcel Dekker.

Settanni, L., Van Sinderen, D., Rossi, J., & Corsetti, A. (2005a). Rapid differentiation and in situ detection of 16 sourdough Lactobacillus species by multiplex PCR. *Applied and Environmental Microbiology*, 71, pp. 3049-3059.

Spicher, G., & Lönner, C. (1985). Die mikroflora des sauerteiges. XXI. Mitteilung: die in sauerteigen schwedischer bäckereien vorkommenden lactobacillen. *Zeitschrift für Lebensmittel Untersuchung und Forschung*, 181, pp. 9-13.

Stephan, H., & Neumann, H. (1999a). Technik der Roggen-Sauerteigführung. In G. Spicher, & H. Stephan (Eds.), Handbuch Sauerteig: *Biologie, Biochemie, Technologie* (5th ed.) (pp. 161-245). Hamburg: Behr's Verlag.

Stephan, H., & Neumann, H. (1999b). Technik der Weizenvorteigund Weizensauerteigführung. In G. Spicher, & H. Stephan (Eds.), Handbuch Sauerteig: *Biologie, Biochemie, Technologie* (5th ed.) (pp. 247-275). Hamburg: Behr's Verlag.

Stolz, P. (1999). Mikrobiologie des Sauerteiges. In G. Spicher & H. Stephan (Eds.), Handbuch Sauerteig: *Biologie, Biochemie, Technologie* (pp. 35-60). Hamburg: Behr's Verlag.

Stolz, P. (1999). Mikrobiologie des Sauerteiges. In G. Spicher, & H. Stephan (Eds.), Handbuch Sauerteig: *Biologie, Biochemie, Technologie* 5th ed. (pp. 35-60). Hamburg: Behr's Verlag.

Stolz, P., Hammes, W. P., & Vogel, R. F. (1996). Maltosephosphorylase and hexokinase activity in lactobacilli from traditionally prepared sourdoughs. *Advances in Food Science*, 18, pp. 1-6.

Takeda, K., Matsumura, Y. & Shimizu, M. (2001). Emulsifying and surface properties of wheat gluten under acidic conditions. *Journal of Food Science*, 66, pp. 393-399.

Tamman, J. D., Williams, A. G., Novle, J., & Lloyd, D. (2000). Amino acid fermentation in non-starter Lactobacillus spp. isolated from cheddar cheese. *Letters in Applied Microbiology*, 2000, pp. 370-374.

Tanner, R.I., Qi, F., Dai, S. (2011). Bread dough rheology: an improved damage function model. *Rheologica Acta*. 50, pp. 75–86.

Tanner, R.I., Qi, F., Dai, S. (2008). Bread dough rheology and recoil: I. Rheology. *Journal of Non-Newtonian Fluid Mechanics*. 148, pp. 33-40.

Thiele, C., Ga¨nzle, M. G., & Vogel, R. F. (2003). Fluorescence labeling of wheat proteins for determination of gluten hydrolysis and depolymerisation during dough processing and sourdough fermentation. *Journal of Agricultural and Food Chemistry*, 51, pp. 2745-2752.

Thiele, C., Gänzle, M.G., & Vogel, R.F. (2002). Contribution of sourdough lactobacilli, yeast and cereal enzymes to the generation amino acids in dough relevant for bread flavour. *Cereal Chemistry*, 79, pp. 45-51.

Thiele, C., Grassl, S. & Gänzle, M. G. (2004). Gluten hydrolysis and depolymerization during sourdough fermentation. *Journal of Agricultural and Food Chemistry*, 52, pp. 1307-1314.

Tieking, M., Korakli, M., Ehrmann, M. A., Gänzle, M. G., & Vogel, R. F. (2003). In situ production of exopolysaccharides during sourdough fermentation by cereal and intestinal isolates of lactic acid bacteria. *Applied and Environmental Microbiology*, 69, pp. 945-952.

Tieking, M., Korakli, M., Ehrmann, M. A., Gänzle, M. G., & Vogel, R. F. (2003). In situ production of exopolysaccharides during sourdough fermentation by cereal and intestinal isolates of lactic acid bacteria. *Applied and Environmental Microbiology*, 69, pp. 945-952.

Tsen, C. C. (1966). A note on effects of pH on sulfhydryl groups and rheological properties of dough and its implication with the sulfhydryl-disulfide interchange. *Cereal Chemistry*. 43, pp. 456-460.

Valmorri, S., Settanni, L., Suzzi, G., Gardini, F., Vernocchi, P., & Corsetti, A. (2006). Application of a novel polyphasic approach to study the lactobacilli composition of sourdoughs from the Abruzzo region (central Italy). *Letters in Applied Microbiology*, 43, pp. 343-349

Van Loo, J.,Cummings, J., Delzenne, N., Englyst, H., Franck, A., Hopkins, M., et al. (1999). Functional Food Prcperties of Non-digestible Oligosaccharides: a Consensus Report from theENDOProject (DGXII AIRII-CT-1095). *British Journal of Nutrition*, 81, pp. 121-132.

Vermeulen, N., Gänzle, M. G., & Vogel, R. F. (2006). Influence of peptide supply and cosubstrates on phenylalanine metabolism of Lactobacillus sanfranciscensis DSM20451T and Lactobacillus plantarum TMW1.468. Journal of *Agricultural and Food Chemistry*, 54, pp. 3832-3839.

Vogel, R. F., Ehrmann, M. A., & Ga¨nzle, M. G. (2002). Development and potential of starter lactobacilli resulting from exploration of the sourdough ecosystem. *Antonie van Leeuwenhoek*, 81, pp. 631-638.

Vogel, R. F., Knorr, R., Müller, M. R. A., Steudel, U., Gänzle, M. G., & Ehrmann, M. A. (1999). Non-dairy lactic fermentations: the cereal world. *Antonie van Leeuwenhoek*, 76, pp. 403-411.

Walker, C. E., & Hazelton, J. L. (1996). Dough rheological testing. Cereal Foods World, 41, pp. 23-28.

Wehrle, K. & Arendt, E.K. (1998). Rheological changes in wheat sourdough during controlled and spontaneous fermentation. *Cereal Chemistry*, 75, 882-886.

Wehrle, K., Grau, H., & Arendt, E.K. (1997). Effects of lactic acid, acetic acid and table salt on fundamental rheological properties of wheat dough. *Cereal Chemistry*, 74, pp. 739-744.

Zeng, M., Morris, C. F., Batey, I. L., & Wrigley, C. W. (1997). Sources of variation for starch gelatinization, pasting, and gelation properties on wheat. *Cereal Chemistry*, 74, pp. 63-71.

Zotta, T., Piraino, P., Ricciardi, A., McSweeney, P. L., & Parente, E. (2006). Proteolysis in model sourdough fermentations. *Journal of Agricultural and Food Chemistry*, 54, pp. 2567-2574.

Part 5

Other Materials and Applications

12

Influence of Electric Fields and Boundary Conditions on the Flow Properties of Nematic-Filled Cells and Capillaries

Carlos I. Mendoza[1], Adalberto Corella-Madueño[2] and J. Adrián Reyes[3]
[1]*Institute of Materials Research, National Autonomous University of Mexico,*
[2]*Department of Physics, University of Sonora,*
[3]*Institute of Physics, National Autonomous University of Mexico,*
Mexico

1. Introduction

From a rheological point of view, nematic liquid crystals are interesting because they exhibit unique flow properties. Although some of these properties have been known for a long time, they continue to attract the attention and interest of the scientists. As a result, a large amount of theoretical, numerical, and experimental work has been produced in recent years. In particular, a number of publications treat the behavior of nematic liquid crystals in shear and Poiseuille flow fields (Denniston, Orlandini, and Yeomans 2001, Vicente Alonso, Wheeler, and Sluckin 2003, Marenduzzo, Orlandini, and Yeomans 2003, Marenduzzo, Orlandini, and Yeomans 2004, Guillen and Mendoza 2007, Medina and Mendoza 2008, Mendoza, Corella-Madueño, and Reyes 2008, Reyes, Corella-Madueño, and Mendoza 2008, Zakharov and Vakulenko, 2010).

On the other hand, it has been shown that the influence of an electric field strongly modifies the rheology of liquid crystals. This has considerable interest due to its possible application in microsystems since homogeneous fluids, like liquid crystals, present some advantages over conventional electrorheological fluids. This is mainly due to the fact that liquid crystals, in contrast to other active fluids, do not contain suspended particles, which is of particular importance for microsystems since small channels are easily obstructed by suspended particles. Also, they prevent agglomeration, sedimentation and abrasion problems (de Volder, Yoshida, Yokota, and Reynaerts 2006).

In this chapter we review recent theoretical results on the rheology of systems consisting of a flow-aligning nematic contained in cells and capillaries under a variety of different flow conditions and under the action of applied electric fields. In particular, we revise steady-state flows and the behavior of viscometric quantities like the local and apparent viscosities and the first normal stress differences. Among the important issues that were recently studied by us and by others is the possibility of multiple steady state solutions due to the competition between shear flow and electric field that give rise to a complex non-Newtonian response with regions of shear thickening and thinning. From these results one can construct a phase diagram in the electric field vs. shear flow space that displays regions for which the system may have different steady-state configurations of the director's field. The selection of a given steady-state configuration depends on the history of the sample. Interestingly, as a consequence of the hysteresis of the system, this response may be asymmetric with respect to the direction of the shear flow. Possible applications of these phenomena are also discussed together with future research.

2. Fundamentals

Liquid crystal systems (De Gennes P.G. and Prost J. 1993) are well defined and specific phases of matter (mesophases) characterized by a noticeable anisotropy in many of their physical properties as solid crystals do, although they are able to flow. Liquid crystal phases that undergo a phase transition as a function of temperature (thermotropics), exist in relatively small intervals of temperature lying between solid crystals and isotropic liquids.

Liquid crystals are synthesized from organic molecules, some of which are elongated and uniaxial, so they can be represented as rigid rods; others are formed by disc-like molecules (Chandrasekhar S. 1992). This molecular anisotropy in shape is manifested macroscopically through the anisotropy of the mechanical, optical and transport properties of these substances.

Liquid crystals are classified by symmetry. As it is well known, isotropic liquids with spherically symmetric molecules are invariant under rotational, O(3), and translational, T(3), transformations. Thus, the group of symmetries of an isotropic liquid is O(3)×T(3). However, by decreasing the temperature of these liquids, the translational symmetry T(3) is usually broken corresponding to the isotropic liquid-solid transition. In contrast, for a liquid formed by anisotropic molecules, by diminishing the temperature the rotational symmetry is broken O(3) instead, which leads to the appearance of a liquid crystal. The mesophase for which only the rotational invariance has been broken is called nematic. The centers of mass of the molecules of a nematic have arbitrary positions whereas the principal axes of their molecules are spontaneously oriented along a preferred direction **n**, as shown in Fig. 1. If the temperature decreases even more, the symmetry T(3) is also partially broken. The mesophases exhibiting the translational symmetry T(2) are called smectics (see Fig. 1), and those having the symmetry T(1) are called columnar phases (not shown).

The elastic properties of liquid crystals determine their behavior in the presence of external fields and play an essential role in characterizing many of the electro-optical and magneto-optical effects occurring in them. In this work we shall adopt a phenomenological approach to describe these elastic and viscous properties. A liquid crystal will be considered as a continuum, so that its detailed molecular structure will be ignored. This approach is feasible because all the deformations observed experimentally have a minimum spatial extent that greatly exceed the dimensions of a nematic molecule. The macroscopic description of the Van der Waals forces between the liquid crystal molecules is given in terms of the following formula (Frank F. C. 1958) for the elastic contribution to the free-energy density:

$$F_{el} = \frac{1}{2}\int_V dV \left[K_{11}(\nabla \cdot \hat{n})^2 + K_{22}(\hat{n} \cdot \nabla \times \hat{n})^2 + K_{33}(\hat{n} \cdot \nabla \times \hat{n})^2 \right]. \quad (1)$$

Here the unit vector **n** is the director, the elastic moduli K_{11}, K_{22}, and K_{33} describe, respectively, transverse bending (splay), torsion (twist), and longitudinal bending (bend) deformations. The free energy of the LC cylinder has, in addition to the above elastic part, also an electromagnetic part due to the applied electrostatic field. As we have already discussed, the first contribution is given by Eq. (1). The electromagnetic free energy density, in MKS units,

$$F_{em} = \frac{1}{2}\int_{v} dV\left[\vec{\mathrm{E}}\cdot\vec{D}^{*} + \vec{B}\cdot\vec{H}^{*}\right] \tag{2}$$

where the displacement field D and the magnetic flux vector B are related to the electric field E and magnetic field H by means of the constitutive relations

$$\vec{D} = \varepsilon_0\varepsilon\cdot\vec{E}, \vec{B} = \mu_0\mu\cdot\vec{H}, \tag{4}$$

characterized specifically by dielectric and magnetic tensors (DeGennes P. G. and Prost. J. 1993)

$$\begin{aligned} \varepsilon_{ij} &= \varepsilon_{\perp}\delta_{ij} + \varepsilon_a n_i n_j \\ \mu_{ij} &= \delta_{ij} \end{aligned} \tag{5}$$

Here δ_{ij} is the Kronecker delta, $\varepsilon_a = \varepsilon_{\|} - \varepsilon_{\perp}$ is the dielectric anisotropy of the LC, $\varepsilon_{\perp}$ and $\varepsilon_{\|}$ represent the dielectric constants perpendicular and parallel to the director. Also, ε_0 and μ_0 are the dielectric permittivity and magnetic permeability constants in vacuum.

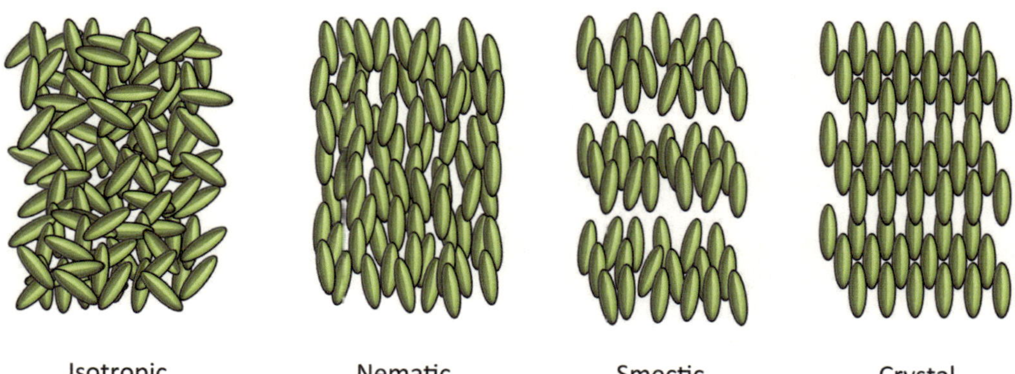

Fig. 1. Schematics of thermotropic liquid crystal phases in between the isotropic fluid and the crystal, arranged from left to right in order of increasing order and decreasing temperature.

3. Nematodynamics

The hydrodynamic description of complex condensed matter systems like superfluids, ferromagnets, polymeric solutions, etc. has been possible thanks to the deep understanding of the role played by the symmetries and thermodynamic properties of the system (Kadanoff and Martin P.C. 1963, Hohenberg and Matin P.C., Kalatnikov I. M. 1965). The extension of this linear hydrodynamic to liquid crystals has been started in the seventies, (Parodi O. 1970, Forster D. 1975), and in recent years it has been generalized to the nonlinear case and to more complex liquid crystal phases (Brand H. R. & Pleiner H. J. 1980).

The key idea of the hydrodynamic formalism is based on the observation that for most complex condensed matter systems in the limit of very large temporal and spatial scales, only a very small number of slow processes, compared with the enormous number of microscopic degrees of freedom, survives. The evolution of these processes is described by the evolution of the corresponding hydrodynamic variables that describe cooperative phenomena that are not to be relaxed in a finite time for a spatially homogeneous system. That is to say, the hydrodynamic variables are such that their Fourier transform satisfy the relation: ω(k→0)→0. Moreover the hydrodynamic variables can be identified uniquely by utilizing conservation laws (global symmetries) and symmetry breaking assumptions, for spatio-temporal scales such that the microscopic degrees of freedom have already been relaxed. For these scales, the description of the systems is exact. When the microscopic degrees of freedom reach thermodynamic equilibrium (local equilibrium) one can use thermodynamics to follow the evolution of the slow variables. Thus, one has to consider a thermodynamic potential, for instance, the internal energy as a function of the system variables (Pleiner H. 1986, Pleiner H. 1988). In a second step we obtain the dynamics of the system by expressing the currents or thermodynamic fluxes in terms of their corresponding thermodynamic forces, which are the gradients of the conjugated thermodynamic variables, and performing a series expansion of the fluxes in powers of the forces. This expansion will be expressed in terms of dynamical phenomenological coefficients (transport coefficients) which can be determined only from an experiment or a microscopic theory. Then, we separate the fluxes in those for which the entropy is conserved (reversible part) and those that make the entropy to increase (irreversible part) and use classical thermodynamic laws to find the evolution equations for the hydrodynamic variables. After obtaining these equations for the liquid crystal, it is possible to include the effects of external fields like electromagnetic fields, stresses, thermal gradients, etc.

In what follows we sketch the steps of this theoretical formalism for nematics. The first class of hydrodynamic variables is associated with local conservation laws which express the fact that quantities like mass, momentum or energy cannot be locally destroyed or created and can only be transported. If ρ(r,t), g=ρv(r,t) and ε(r,t), where v is the hydrodynamic velocity, denote respectively, the density of these quantities, the corresponding conservation equations are ((Landau L.D. and Lifshitz E. 1964).

$$d\rho / dt + \rho \nabla_i v_i = 0 \tag{6}$$

$$dv_i / dt + \nabla_j \sigma_{ij} / \rho = 0 \tag{7}$$

$$d(e / \rho) / dt + \nabla_i j^e{}_i / \rho = 0 \,. \tag{8}$$

Here $d / dt = \partial / \partial t + v_i \nabla_i$ denotes the hydrodynamic velocity, σ_{ij} is the nematic's stress and j_i^e is the energy flow.

When a phase transition to the liquid crystal state occurs after reducing the temperature, the rotational symmetry O(3) is broken spontaneously and the number of hydrodynamic variables increase. Any rotation around an axis different from $\hat{n}$ transforms the system to a different and distinguishable state form that without rotation. This rotational symmetry broken is called spontaneous since the energy is a rotational invariant and there is no energy that favors one orientation of $\hat{n}$ with respect to any other. This is equivalent to say that the state of the system becomes infinitely degenerate. Under these conditions, one soft variation of the degeneracy parameter is related to a slow relaxation of the system that increases as q→0. This type of behavior is the basic content of the Goldstone Theorem (Forster D. 1975). Therefore, the degeneracy parameter is related to the order parameter of the liquid crystal and adopts different structures for different mesophases. For a nematic phase the order parameter has the following form $Q_{ij} = S(n_i n_j - \delta_{ij}/3)$, where S is the degree of order, i. e., S=0 for the isotropic phase and S=1 for a nematic phase having the molecules completely aligned. In agreement with this statement the dynamics of Q_{ij} is determined by that of **n**. In summary, the macroscopic state of a nematic can be described by means of two scalar variables that can be chosen as ρ(r,t), $e(r,t)$, one vectorial variable, g=ρv(r,t) and one tensorial variable Q_{ij}, that can be selected, for instance, as the anisotropic part of the dielectric tensor.

Since $\hat{n}$ is related to a conservation law, its balance equation is a dynamical equation of the form

$$\left[\partial/\partial t + v_j \nabla_j\right] n_i + Y_i = 0 \tag{9}$$

where Y_i is not a current, since its surface integral is not a flux, but a quasi-current. This quantity must be orthogonal to **n** to fulfill the nematic symmetry **n**→- **n**; however, there are other contributions to Y_i which does not come from the symmetries but from thermodynamic requirements.

If a specific physical situation is given, the state of the system can be described in terms of an appropriate thermodynamic potential. This can be chosen, for example, as the total free energy E, (Callen H. B. 1985)

$$E = \epsilon V = E(V, \rho V, gV, \rho V \nabla_j n_i, \rho V n_i, V\sigma) \tag{10}$$

where V denotes the volume of the system and σ is the entropy per unit of volume. From this assumption and using Euler's relation, we can derive the Gibbs' expression

$$de = \mu d\rho + T d\sigma + \vec{v}\cdot d\vec{g} + \Phi_{ij} d\nabla_j n_i + h_i dn_i, \tag{11}$$

and the Gibbs-Duhem's relation

$$p = -e + \mu\rho + T\sigma + \vec{v}\cdot\vec{g}. \tag{12}$$

Here μ is the chemical potential, Φ_{ij} y h_i are called the molecular fields, which are defined as the partial derivatives of the thermodynamic potential with respect to the corresponding conjugated variable. Since in equilibrium the state variables are constants, any inhomogeneous distribution of these variables takes the system out of equilibrium. For this reason the gradients of these quantities are taken as thermodynamic forces. Hence, the presence of ∇μ, ∇T, $\nabla_j n_i$ and $\nabla_j \Phi_{ij}$ give rise to irreversible processes in the system. The dynamical part of the hydrodynamic equations is obtained by expressing the currents σ_{ij}, j_i^e, and Y_i in terms of the thermodynamic variables T, μ, v_i, and Φ_{ij}. If additionally we separate in these expressions the reversible part, which does not generate entropy increase and it is invariant under temporal inversion, from the irreversible part, which increases the entropy and is not invariant under the transformation t→-t, we obtain the following expressions for the fluxes (Landau L. D. and Lifshitz E. 1986, Plainer H. 1988)

$$\sigma_{ij} = \sigma_{ij}^{R} + \sigma_{ij}^{D} = p\delta_{ij} + \Phi_{kj}\nabla_i n_k - \lambda_{kji} h_k / 2 - \nu_{ijkm}\nabla_m v_k \tag{13}$$

$$Y_i = Y_i^{R} + Y_i^{D} = -\lambda_{kji}\nabla_j v_k / 2 + \delta_{ik}^{\perp} h_k / \gamma_1 \tag{14}$$

$$j_i^{e} = T j_i^{\sigma D} + v_j \sigma_{ij}^{D} . \tag{15}$$

In these equations the superscript indexes R and D denote, respectively, the reversible and irreversible or dissipative parts, and

$$\Phi_{il} = K_{ilmn}\nabla_n n_m \tag{16}$$

where

$$K_{ilmn} = K_1 \delta_{il}^{\perp} \delta_{mn}^{\perp} + K_2 n_p n_q \varepsilon_{pli} + K_3 n_i n_m \delta_{lm}^{\perp} . \tag{17}$$

Here K_1, K_2 y K_3 are the elastic constants of the nematic and ε_{ijk} is the totally antisymmetric tensor of Levy-Civitta. The projector tensor is $\delta_{lm}^{\perp} = \delta_{ik} - n_i n_k$ and λ_{kji} can be expressed as

$$\lambda_{kji} = (\lambda - 1)\delta_{kj}^{\perp} n_i + (\lambda + 1)\delta_{ki}^{\perp} n_j . \tag{18}$$

In this expression $\lambda = \nu_1/\nu_2$, is the reversible parameter, also called flux alignment parameter, being ν_1 and ν_2 two of the five independent viscosities of the nematic. The molecular field h_k, that we have already defined as $h_k \equiv h_i' - \nabla_j \Phi_{ij}$, turns out to be explicitly

$$h_k = K_{kjnl}\nabla_j \nabla_l n_n + \delta_{kq}^{\perp}\left(\frac{1}{2}\partial / \partial n_q K_{pjkl} - \partial / \partial n_q K_{qjkl}\right)\nabla_l n_k \nabla_j n_p . \tag{19}$$

Finally, the viscous stress tensor ν_{ijkl} contains five independent viscosities for the nematic, ν_i, i=1,2,...,5

$$\begin{aligned} \nu_{ijkl} &= \nu_2(\delta_{jl}\delta_{ik} + \delta_{il}\delta_{jk}) + 2(\nu_1 + \nu_2 - 2\nu_3) n_i n_j n_k n_l + \\ &(\nu_3 - \nu_2)(n_j n_l \delta_{ik} + n_j n_k \delta_{il} + n_i n_k \delta_{jl} + n_i n_l \delta_{jk}) + (\nu_4 - \nu_2)\delta_{ij}\delta_{kl} . \\ &(\nu_5 + \nu_4 + \nu_2)(\delta_{ij} n_k n_l + \delta_{kl} n_i n_j) \end{aligned} \tag{20}$$

It should be mentioned that a different choice for this tensor has been done in the ELP formulation (Ericksen, J. L. 1960, Leslie, F. M. 1966, Parodi, 0. 1970). The complete stress tensor for this formulation is given by Eq.(39) which for this case replaces Eq.(13).

The second law of thermodynamics establishes that any irreversible process that occurs in the system should increase the entropy. Thus, the entropy obeys the following balance equation

$$\partial\sigma / \partial t + \nabla_i \left(\sigma v_i + j_i^{\sigma R} + j_i^{\sigma D}\right) = R / T \tag{21}$$

where R is the dissipation function for irreversible processes. This quantity can be interpreted as the energy per unit of volume dissipated by the microscopic degrees of freedom and divided by the temperature (R/T), represents the entropy production of the nematic. If, as we did previously, we relate Eq.(21) with Eqs. (6), (7), (8) and (9), by using the Gibbs' expression (11) and the expressions (13)-(21), we obtain an explicit formula for R, that is

$$\begin{aligned} R &= -\nabla_i (j_i^D - T o_{ij}^D \nabla_j v_i + h_i \delta_{ij}^{\perp} Y_j^D) - j_i^{\sigma D} \nabla_i T - \sigma_{ij}^D \nabla_j v_i + h_i \delta_{ij}^{\perp} Y_j^D \\ &= \frac{1}{2\gamma_1} h_i \delta_{ij}^{\perp} h_j + \frac{1}{2} \nu_{ijkl} \nabla_j v_i \nabla_l v_k + \frac{1}{2} \kappa_{ij} \nabla_i T \nabla_j T, \end{aligned} \tag{22}$$

where γ_1^{-1} is the rotational viscosity and the tensor κ_{ij} describes the heat conduction (thermal conductivity). The second law of thermodynamics requires R to be a definite positive form, which in turns implies that every single coefficient of the previous expression is positive. Notice that Eq.(22) implies as well that the dissipative currents and quasi-currents are given by the partial derivatives of the dissipation function, that is

$$j_i^{\sigma D} = \partial R / (\partial \nabla_i T) = \kappa_{ij} \nabla_j T, \tag{23}$$

$$\sigma_{ij}^D = \partial R / \partial(\nabla_j v_i) = \nu_{ijkl} \nabla_l v_k, \tag{24}$$

$$Y_k^D = \partial R / \partial h_k = \delta_{ik}^{\perp} h_i / \gamma_1 . \tag{25}$$

In summary equations (6), (7), (8), (9) and (22) constitute a complete set to describe the irreversible dynamics of a low molecular weight nematic (thermotropic) in absence of external fields.

4. Constitutive equations

It is usual that applied external fields like electric and magnetic fields, gravity, temperature gradients, pressure and concentration, shear and vortex flows carry out the nematic to a new equilibrium state so that these fields must be included in the hydrodynamic equations.

It is well known that for any polarizable medium an electric field E induces a polarization P $=D-\varepsilon_0 E$, where D is the displacement electric vector. Now, in a nematic the molecular dipolar moments are oriented approximately parallel with respect to the long axis of the molecules. Thus, the induced polarization gives rise to a director orientation. In contrast the influence of the magnetic field in a nematic is much weaker and in general, the induced magnetization can be neglected. A very well known result based on conventional thermodynamic arguments establishes that the work associated to an electric field E $=-\nabla\Phi$, is given by $dw_{el} = -(1/2)\vec{E} \cdot \vec{D}$, which should be added to the Gibbs' expression (11) and to the Gibbs-Duhem's relation (12). By modifying these expressions and using a procedure completely analogous to the one we followed in the last section, it is possible to show that in the presence of an electric field Eq. (7) transforms into

$$dv_i / dt + \nabla_j \sigma_{ij} / \rho = \rho_E E_i + P_j \nabla_j E_i \tag{26}$$

where the charge density is given by $\rho_E = \varepsilon_0 \text{div } D$. To linear order in the thermodynamic forces, the expression for σ_{ij} has to include in addition the electric contributions, so that we replace Eq. (13) by the expression

$$\sigma_{ij} = \sigma_{ij}^{R} + \sigma_{ij}^{D} = \tilde{p}\delta_{ij} + \Phi_{lj}\nabla_i n_l - (1/2)\lambda_{kji}h_k - \nu_{ijkl}\nabla_l v_k \tag{27}$$

with $\tilde{p} = p - \varepsilon E^2/2$. Analogously, the currents (14), now are given by

$$Y_i = Y_i^R + Y_i^D = -(1/2)\lambda_{kji}\nabla_j v_k + (1/\gamma_1)\delta_{ik}^{\perp}h_k - \zeta_{ijk}^{e}\nabla_j E_k \tag{28}$$

$$j_i^e = \sigma_{ij}^E E_j + \kappa_{ij}^E \nabla_j T + \nabla_j\left(\zeta_{kjl}^E h_k\right), \tag{29}$$

where σ_{ij}^E is the electric conductivity, and in consequence the entropy current is

$$j_i^{\sigma} = -\kappa_{ij}\nabla_j T - \kappa_{ij}^E E_j \ . \tag{30}$$

Here the material tensors of second rank, κ_{ij}^E and σ_{ij}^E have uniaxial form and each one should be expressed in terms of two dissipative transport coefficient, that is,

$$\alpha_{ij} = \alpha_{\perp}\delta_{ij}^{\perp} + \alpha_{\parallel}n_i n_j \ . \tag{31}$$

On the other hand, the third order tensor ζ_{kji}^E is irreversible and contains a dynamical coefficient, the flexoelectric coefficient ζ^E,

$$\zeta_{ijk}^E = \zeta^E(\delta_{ij}^{\perp}n_k + \delta_{ik}^{\perp}n_j). \tag{32}$$

Following the same steps we used to obtain Eq.(22), the dissipation function R show in this case additional terms which involve the electric field,

$$\begin{aligned} 2R &= h_i\delta_{ij}^{\perp}h_j + \nu_{ijkl}\nabla_j v_i \nabla_l v_k + \kappa_{ij}\nabla_i T \nabla_j T \\ &+ \sigma_{ij}^E E_i E_j + \kappa_{ij}^E E_i \nabla_j T - \zeta_{ijk}^E h_i \nabla_j E_k. \end{aligned} \tag{33}$$

Most of the parameters involved in the hydrodynamic and electrodynamic equations for a nematic have been measured for different substances that show a uniaxial nematic phase. Among these one can mention the elastic constants (Blinov L. M. and Chigrinov V. G. 1994); specific heat, the flux alignment parameter λ and the viscosities ν_i, i=1,2...5, the inverse of the diffusion constant γ_1, the thermal conductivity (Ahlers, Cannell, Berge and Sakurai 1994), and the electric conductivity σ_{ij}^E.

Finally the dynamical equations for a nematic in an isothermal process can be obtained by inserting Eqs. (27) and (28) in Eqs. (7) and (9). This leads to

$$dv_i/dt + (1/\rho)\nabla_j\left[p\delta_{ij} + \Phi_{lj}\nabla_l n_i - \lambda_{kji}h_k - \nu_{ijkl}\nabla_l v_k\right] = 0 \tag{34}$$

$$dn_i/dt + (1/\gamma)\delta_{ik}^{\perp}h_k - \lambda_{kji}\nabla_j v_k = 0 \tag{35}$$

5. Apparent viscosity

The viscosity function or apparent viscosity connects the force per unit area and the magnitude of the local shear (Carlsson T. 1984). It depends on the orientation of the director through the expression

$$\eta(\theta) = (2\alpha_1 \sin^2\theta\cos^2\theta + (\alpha_5 - \alpha_2)\sin^2\theta + (\alpha_6 + \alpha_3)\cos^2\theta + \alpha_4) / 2, \tag{36}$$

where α_1, α_2,α_3,α_5 and α_6 are the Leslie coefficients (Parodi O. 1970). Since the orientation angle θ is given by Eq. (35), from the above equation it follows that the dependence of η on θ indicates that the system is non-Newtonian in its behavior, in the sense that η is strongly dependent on the driving force. If we integrate the result over the cross section area of the flow we obtain the averaged apparent viscosity

$$\bar{\eta} = (1 / A_t) \int_{A_t} \eta(\theta) dA \tag{37}$$

where A_t is the total area of the cross section.

6. First normal stress difference

One of the distinctive phenomena observed in the flow of liquid crystal polymers in the nematic state is that of a negative steady-state first normal stress difference, N_1, in shear flow over a range of shear rates. N_1 is zero or positive for isotropic fluids at rest over all shear rates, which means that the force developed due to the normal stresses, tends to push apart the two surfaces between which the material is sheared. In liquid crystalline solutions, positive normal stress differences are found at low and high shear rates, with negative values occurring at intermediate shear rates (Kiss G. and Porter R. S. 1978).

On the other hand, Marrucci et al (Marrucci G. and Maffettone 1989) have solved a two dimensional version of the Doi model for nematics (Doi M. and Edwards S.. F. 1986), in which the molecules are assumed to lie in the plane perpendicular to the vorticity axis, that is, in the plane parallel to both, the direction of the velocity and the direction of the velocity gradient. Despite this simplification, the predicted range of shear rates over which N_1 is negative, is in excellent agreement with observations. This result opens up the possibility that negative first normal stress differences may be predicted in a two dimensional flow.

We shall now examine the effects produced by the stresses generated during the reorientation process by calculating the viscometric functions that relate the shear and normal stress differences. For a planar geometry and using the convention in (Bird R. et al 1971) the first normal stress difference is defined by

$$N_1 = \sigma_{xx} - \sigma_{zz}, \tag{38}$$

where σ_{ij} are the components of the stress tensor of the nematic used by De Gennes (DeGennes J. P. and Prost J. 1993).

$$\begin{aligned}\sigma_{ij} &= \alpha_1 n_i n_j n_\mu n_\rho A_{\mu\rho} + \alpha_2 n_i \Omega_j + \alpha_3 n_j \Omega_i + \\ &\alpha_4 A_{ij} + \alpha_5 n_i n_\mu A_{\mu j} + \alpha_6 n_i n_\mu A_{\mu j}.\end{aligned} \tag{39}$$

Here $A_{ij}\equiv(1/2)(\partial v_i/\partial x_j+\partial v_i/\partial x_j)$ is the symmetric part of the velocity gradient and $\Omega\equiv dn/dt-(1/2)\nabla\times v\times n$ represents the rate of change of the director with respect to the background fluid. The α_i for i=1,..6, denote the Leslie coefficients of the nematic.

The integration of the first normal stress difference, Eq.(38) over the whole cell and along the velocity gradient direction renders the net force between the plates as a function of the Reynolds number, which is proportional to N

$$f=(1/A_t)\int_{A_t} N_1(\theta)dA \tag{40}$$

A positive force exerted by the fluid motion tends to push the plates apart, or otherwise, if the force is negative, the fluid tends to pull the plates close together.

7. Nematic cells under shear flow

In this section we study the flow properties of nematic-filled cells under shear flow. The cell geometry is important because many micro-fluidic devices are designed with channel-like shapes and its mathematical treatment is simpler as compared to the case of capillaries.

Liquid crystals and their electrorheological properties under flows with a constant shear rate over the height of the channel have been treated in a number of papers. However, in all of these papers the studies have focused on situations in which the anchoring was the same at all boundaries. Only recently, a systematic study of the influence of different boundary conditions on the shear flow was treated together with the influence of an applied electric field.

7.1 Hybrid-aligned nematic cell

First we are going to present the case of a hybrid-aligned nematic (HAN) cell since it is a common geometry used in devices (Guillen and Mendoza 2007). In this geometry, the director is aligned perpendicularly (also called homeotropic alignment) to one of the boundaries of the confining cell while it is parallel (also called homogeneous alignment) to the opposite boundary as shown in Fig. 2.

The separation between the plates l is small compared to the transverse dimensions L of the cell, which is under the action of a perpendicular electric field E. The director's configuration is given by

$$\hat{\mathbf{n}}=[\sin\theta(z),0,\cos\theta(z)], \tag{41}$$

where $\theta(z)$ is the angle with respect to the z axis. We assume strong anchoring conditions at the plates of the cell

$$\theta(z=-l/2)=0, \quad \theta(z=l/2)=\pi/2. \tag{42}$$

As shear flow is applied as depicted in Fig. 2

$$\mathbf{v} = [v_x(z), 0, 0], \tag{43}$$

which satisfies the nonslip boundary conditions

$$v_x(z = \pm l/2) = \pm v_0. \tag{44}$$

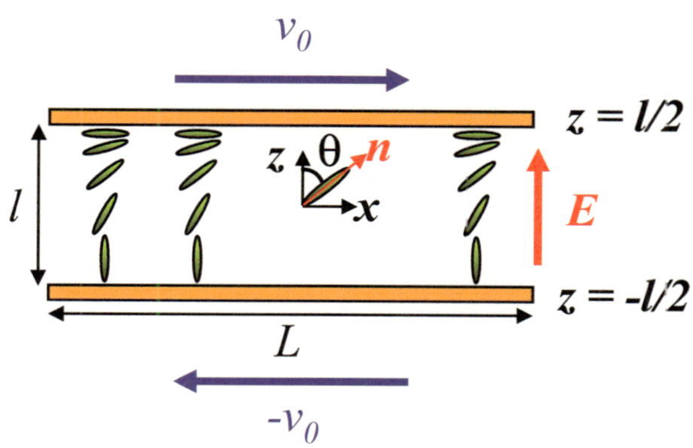

Fig. 2. Schematics of a HAN cell subjected to a normal electric field and a shear stress. Adapted from Guillen and Mendoza 2007.

Within the framework of the Ericksen, Leslie, and Parodi theory one can obtain the torque acting on a sheared molecule,

$$T_v = (\alpha_3 \sin^2 \theta - \alpha_2 \cos^2 \theta)\frac{dv_x}{dz}. \tag{45}$$

A second torque that acts on the LC molecules is due to the electric field

$$T_{el} = -\frac{\varepsilon_a \varepsilon_0}{2} E^2 \sin(2\theta), \tag{46}$$

An elastic torque can be derived form the Frank-Oseen elastic energy to give

$$\begin{aligned} T_e = (K_1 \sin^2 \theta + K_3 \cos^2 \theta)\left(\frac{d^2\theta}{dz^2}\right) \\ + (K_1 - K_3)\sin\theta\cos\theta\left(\frac{d\theta}{dz}\right)^2. \end{aligned} \tag{47}$$

Finally, the rotational inertia and viscous damping gives the following contribution to the torques

$$T_{\text{dyn}} = I\frac{d^2\theta}{dt^2} + \gamma\frac{d\theta}{dt}, \tag{48}$$

All the above contributions result in the differential equation for the director´s orientation that describes the equilibrium of torques

$$I\frac{d^2\theta}{dt^2}+\gamma\frac{d\theta}{dt}=(K_1\sin^2\theta+K_3\cos^2\theta)\left(\frac{d^2\theta}{dz^2}\right)$$
$$+(K_1-K_3)\sin\theta\cos\theta\left(\frac{d\theta}{dz}\right)^2 \tag{49}$$
$$-\frac{\varepsilon_a\varepsilon_0}{2}E^2\sin(2\theta)$$
$$+(\alpha_3\sin^2\theta-\alpha_2\cos^2\theta)\frac{dv_x}{dz}.$$

Since we are only interested in the final stationary state, the above equation reduces to (Guillen and Mendoza 2007)

$$\frac{d^2\theta}{d\varsigma^2}-q\sin(2\theta)+\frac{m}{\eta(\theta)}(\alpha_3\sin^2\theta-\alpha_2\cos^2\theta)=0, \tag{50}$$

where we have used the linear momentum conservation equation

$$\frac{d}{dz}\left[\eta(\theta)\frac{dv_x}{dz}\right]=0, \tag{51}$$

with $\eta(\theta)$ the position dependent viscosity of the liquid crystal given by Eq. (36), and we have assumed the equal elastic constant approximation. Also, we have defined a dimensionless field strength

$$q\equiv\frac{\varepsilon_a\varepsilon_0}{2K}l^2E^2 \tag{52}$$

a dimensionless shear rate

$$m\equiv\frac{lv_0c(q,m)}{K} \tag{53}$$

$$c(q,m)=\frac{2}{\int_{-1/2}^{1/2}d\varsigma'/\eta[\theta(\varsigma')]}$$

and the normalized variable $\zeta\equiv z/l$.

Equation (50) can be solved numerically using the "shooting" method to obtain the stationary configuration of the nematic's director.

Explicit numerical results are given for the particular case of the flow-aligning liquid crystal 4'-*n*-pentyl-4-cyanobiphenyl (5CB) with the following parameters $T = 10$ °C with $T_{IN} = 35$ °C, $\kappa = 1.316$, $K_1 = 1.2\_10^{-11}$ N, $\alpha_1 = -0.0060$ Pa s, $\alpha_2 = -0.0812$ Pa s, $\alpha_3 = -0.0036$ Pa s, $\alpha_4 = 0.0652$ Pa s, $\alpha_5 = 0.0640$ Pa s, $\alpha_6 = -0.0208$ Pa s, $\gamma_1 = 0.0777$ Pa s, $\gamma_2 = -0.0848$ Pa s.

In Fig. 3 we show the orientational profile for various values of q and m. We observe a tendency of the molecules to align with the direction of the electric field. In contrast, θ

increases as the value of m increases, for m>0, which means that the molecules tend to be aligned with the direction of the flow. On the other hand, for m<0, a remarkable difference is observed. In this case the cell shows two different regions, in the lower part of the cell, where the effect of the flow dominates, the molecules are tilted to the left whilst on the upper part, where the anchoring dominates, they are tilted to the right.

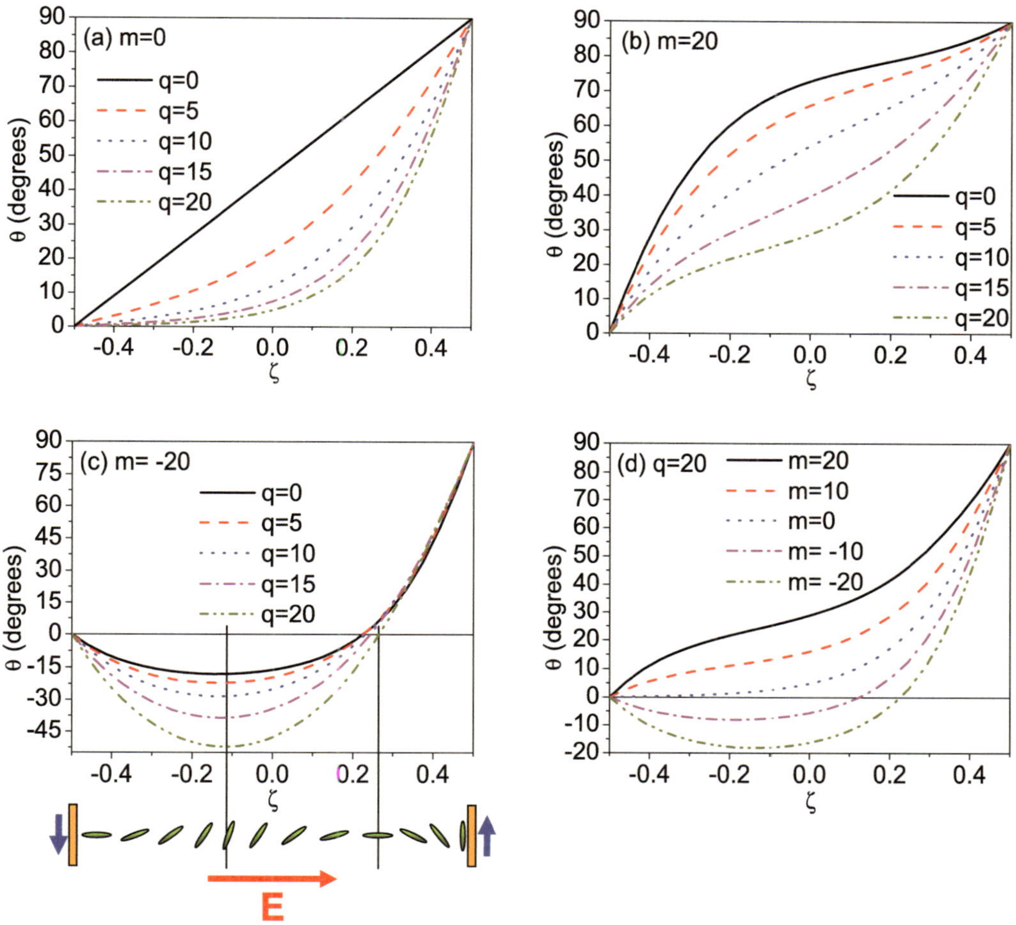

Fig. 3. Nematic's orientation θ as function of its position in the cell, ζ. Adapted from Guillen and Mendoza 2007.

The velocity profiles can be obtained from Eq. (51) and are shown in Fig. 4. Note the different behavior between a positive and a negative flow.

The position dependent viscosity can be calculated from Eq. (36) and it is shown in Fig. 5. It is maximum at the lower plate where the molecules are perpendicular to the direction of flow and minimum at the upper plate where the molecules are parallel to the direction of the flow. At intermediate positions the viscosity takes intermediate values.

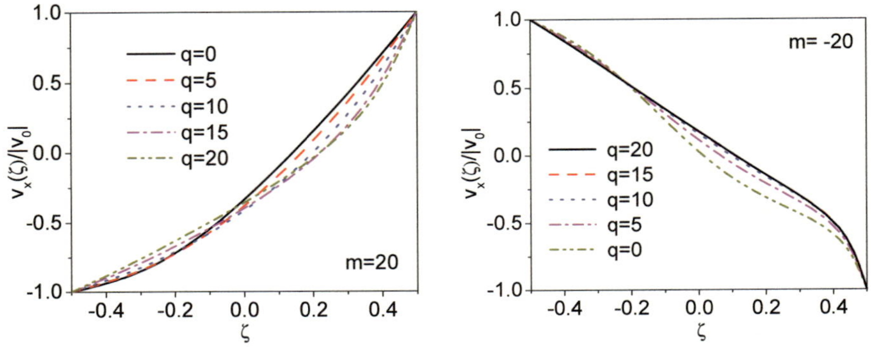

Fig. 4. Velocity profiles. Adapted from Guillen and Mendoza 2007.

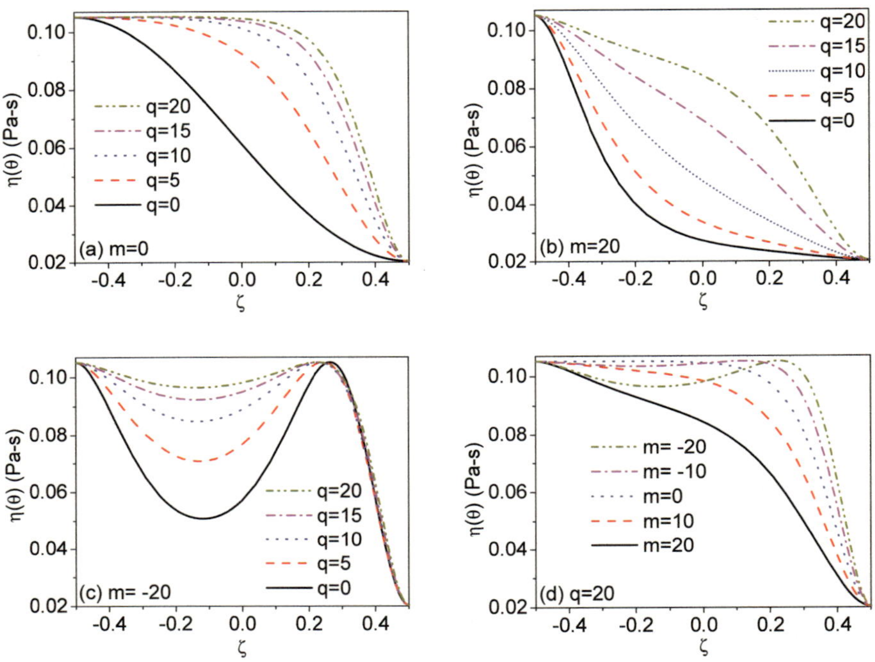

Fig. 5. Local viscosity for different values of the electric field and the shear flow. Adapted from Guillen and Mendoza 2007.

The averaged apparent viscosity [Eq. (37)] is depicted in Fig. 6. We observe a moderate electrorheological effect and an interesting non-Newtonian behavior with alternate regions of shear thickening (shaded region) and thinning.

Finally, the first normal stress difference

$$N_1[\theta(\varsigma)] = -\frac{|v_0|}{2l}\sin 2\theta(\alpha_1\cos 2\theta + \alpha_2 + \alpha_3)\frac{d\tilde{v}_x}{d\varsigma} + \frac{K}{l^2}\left(\frac{d\theta}{d\varsigma}\right)^2. \tag{54}$$

is plotted in Fig. 7 and the corresponding averaged value is shown in Fig. 8

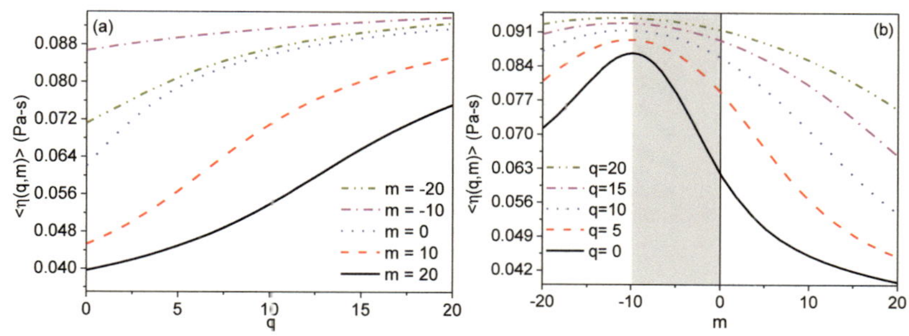

Fig. 6. Averaged apparent viscosity. Adapted from Guillen and Mendoza 2007.

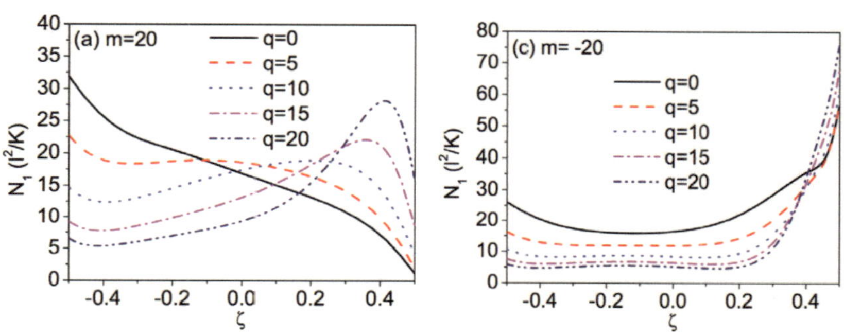

Fig. 7. First normal stress difference. Adapted from Guillen and Mendoza 2007.

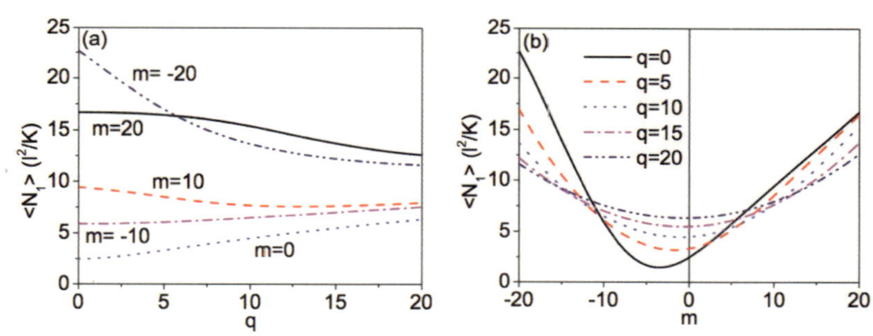

Fig. 8. Averaged first normal stress difference. Adapted from Guillen and Mendoza 2007.

7.2 Homogeneous nematic cell

In this subsection we study the flow of a homogeneous nematic cell as depicted in Fig. 9.

The only difference of this cell as compared to the HAN cell is that here the alignment is homogeneous at both plates. At first sight one may think that this small difference may only produce slight changes in the rheological behavior of the cell. However, this is not the case and a completely different behavior arises. The most striking feature is the appearance of multiple steady-state configurations for certain combinations of the applied electric field and shear flow (Medina and Mendoza 2008).

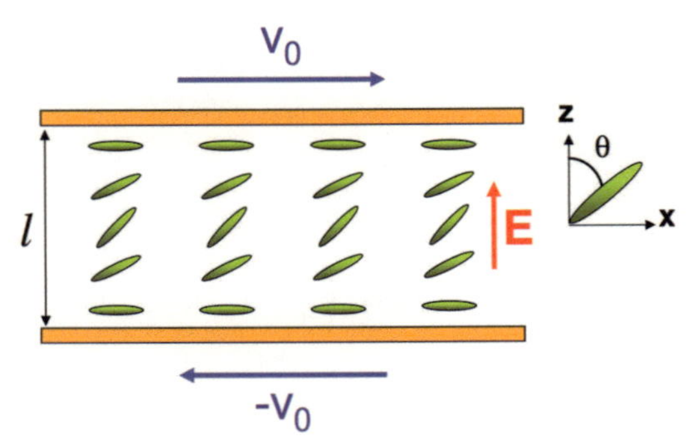

Fig. 9. Schematics of a homogeneous cell subjected to a normal electric field and a shear stress. Adapted from Medina and Mendoza 2008.

Using again the theory of Ericksen, Leslie, and Parodi together with the momentum conservation we obtain the differential equations that govern the steady state of the system

$$0 = \left(\sin^2\theta + \kappa\cos^2\theta\right)\frac{d^2\theta}{d\varsigma^2} + (1-\kappa)\sin\theta\cos\theta\left(\frac{d\theta}{d\varsigma}\right)^2 \qquad (55)$$
$$-q\sin(2\theta) + \frac{m}{\eta(\theta)}\left(\alpha_3\sin^2\theta - \alpha_2\cos^2\theta\right),$$

Here α_i are the Leslie viscosities and $\kappa \equiv K_3/K_1$, with the homogeneous

$$\theta(\varsigma = \pm 1/2) = \pi/2, \qquad (56)$$

and non-slip boundary conditions.

The stationary configuration of the nematic's director can be found by solving Eq. (55) numerically using the "shooting" method. Results are presented for 5CB as before.

In Fig. 10 we show the nematic's configuration for different values of the applied electric field, q, without flow (a) and with flow (b). In this last case two sets of solutions of Eq. (55) are shown.

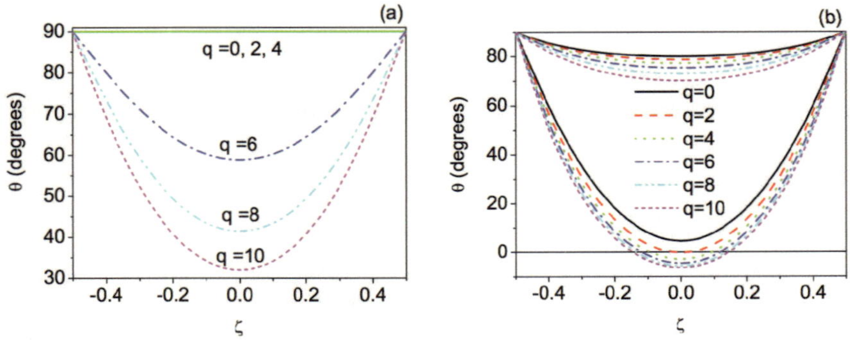

Fig. 10. Nematic's configuration for (a) m=0 and (b) m=20. In the latter case two solutions are shown, for the second set of solutions we plot 180° - θ. Adapted from Medina and Mendoza 2008.

In Fig. 11 we show the nematic's configuration for q=2 and different values of the shear flow. The case m=5 lies in a region where exist only one solution of Eq. (55) while the case m=10 lies in the region where Eq. (55) accepts multiple solutions. A phase diagram in the q-|m| space is also shown in Fig. 11 that separates the region with only one solution from the region with multiple solutions. In the lower right panel we sketch the steady-state nematic's configuration for these cases. The selection of one of the configurations over the other depends on the history of the sample as exemplified in Fig. 12. In this figure, we have recasted the phase diagram drawing the positive and negative parts of the m-axis and considered two different processes depicted by the arrows in the phase diagram. The two processes start at zero applied electric field, but with opposite starting shear flows (points A and A′ in the diagram). Then, following the processes depicted by the arrows in the phase diagram they arrive to the same final q, m pair with different configurations (point B′).

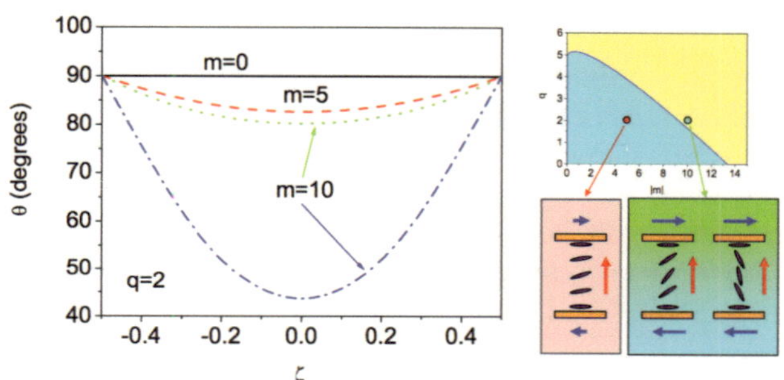

Fig. 11. Left: Nematic's configuration. Right up: Phase diagram showing the region (blue) with unique steady-state solutions and a region (yellow) with multiple solutions. Right down: Sketch of the configurations. Adapted from Medina and Mendoza 2008.

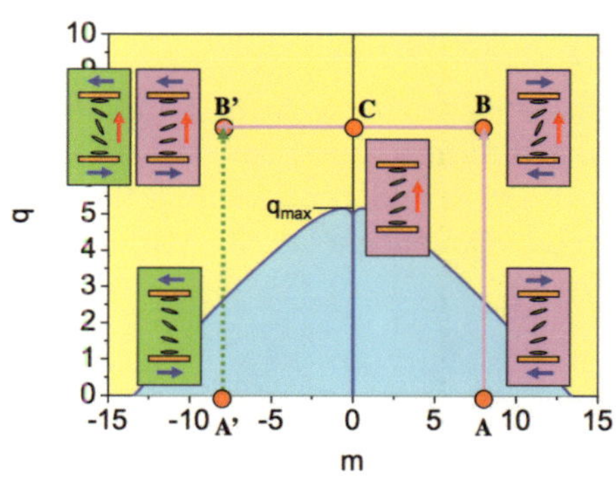

Fig. 12. Sketch of two possible trajectories in the phase diagram that gives rise to two different steady states for a given pair q and m (point B′). Adapted from Medina and Mendoza 2008.

In Fig. 13 we show the averaged viscosity as function of m for the trajectory starting at point A in Fig. 12. We observe an interesting non-Newtonian behavior with alternate regions of shear thickening and thinning. The second trajectory (the one starting at A′ in Fig. 12) would produce the same curves for the viscosity but interchanging m with -m. A moderate electrorheological effect is also evident in this figure.

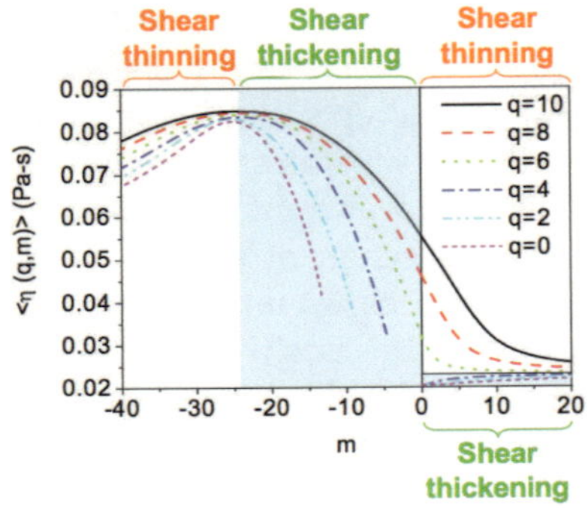

Fig. 13. Averaged apparent viscosity as a function of m showing regions of shear thickening and shear thinning. Adapted from Medina and Mendoza 2008.

In summary, we have shown that nematic cells are very sensitive to the boundary conditions at the plates of the cell. A HAN cell and a homogeneous cell behave in a completely different way, the homogeneous cell showing regions of multiple steady-state configurations that give rise to a history dependent rheological behavior that is absent in the HAN case. Both cases show complex non-Newtonian behavior with regions of shear thickening and thinning. Homeotropic cells and weak anchoring conditions remain to explore.

8. Nematic capillaries

In this section we study the flow properties of nematic-filled capillaries under the action of an electric field for two different flow conditions. In first place we treat the case of capillaries subjected to a pressure gradient and in second place we consider the case of a Couette flow.

8.1 Hybrid nematic capillary under Poiseuille flow

We consider a capillary consisting of two coaxial cylinders whose core is filled with a nematic liquid crystal subjected to the simultaneous action of both a pressure gradient applied parallel to the axis of the cylinders (Poiseuille flow) and a radial low frequency electric field as depicted in Fig. 14 (Mendoza, Corella-Madueño, and Reyes 2008).

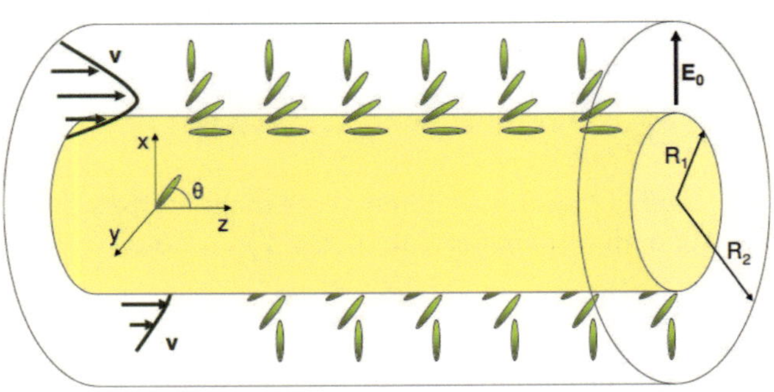

Fig. 14. Schematics of a nematic liquid crystal confined by two coaxial cylinders and subjected to a radial electric field and a pressure gradient. Adapted from Mendoza, Corella-Madueño, and Reyes 2008.

The nematic's director in cylindrical coordinates can be written as

$$\hat{\mathbf{n}} = [\sin \theta(r), 0, \cos \theta(r)], \tag{57}$$

with the hybrid hard anchoring conditions

$$\theta(r = R_1) = 0, \quad \theta(r = R_2) = \pi/2 \tag{58}$$

The constant pressure drop along the axis of the cylinders produces a flow profile given by

$$\mathbf{v} = [0, 0, v_z(r)], \tag{59}$$

with the non slip boundary conditions

$$v_z(r = R_1) = 0, \quad v_z(r = R_2) = 0. \tag{60}$$

The nematodynamic equations adopt a more involved look, as compared to the case of the cells. The reader can find the appropriate expressions in (Mendoza, Corella, Reyes 2008). Here we just present the relevant results using as in the previous sections a 5CB nematic liquid crystal.

In Fig. 15 we show the nematic's configuration as function of $x \equiv r/R_2$, parametrized with q, the ratio of the electric and elastic energies, and Λ, the ratio of the hydrodynamic and elastic energies (Mendoza, Corella-Madueño, Reyes 2008). The undistorted state corresponding to $\Lambda = 0$ and $q = 0$ is similar to the escaped configuration. For $q = 50$, $\hat{\mathbf{n}}$ is much more aligned with the radial direction than for $q = 0$. This is so because the director tends to be parallel to the electric field. For positive $\Lambda > 0$, corresponding to negative velocity, $\hat{\mathbf{n}}$ tends to be axially aligned, whereas for negative $\Lambda < 0$ the trend is the opposite. In contrast, for $q = 50$ the influence of the pressure gradient is influenced by the electric field for regions near the inner cylinder.

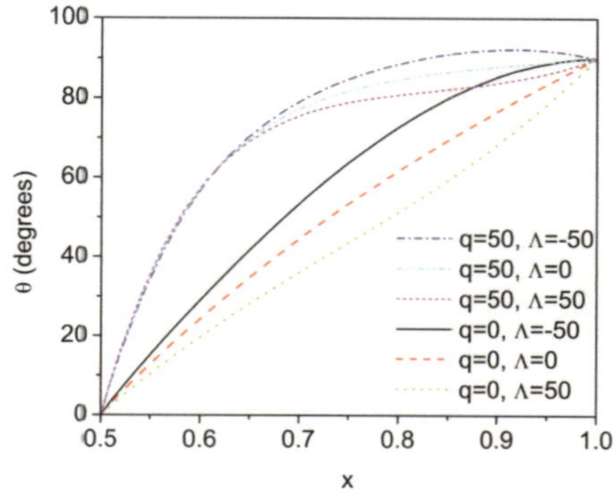

Fig. 15. Nematic's configuration θ as a function of the position $x \equiv r/R_2$ for 5CB and $R_1/R_2=0.5$. Adapted from Mendoza, Corella-Madueño, and Reyes 2008.

In Fig. 16 we show a typical velocity profile for a given value of the electric field and different values of Λ. This figure exhibits a clear difference in the magnitude of the velocity between forward and backward flows, which is a consequence of the asymmetry of the undistorted director's configuration (so called escaped configuration).

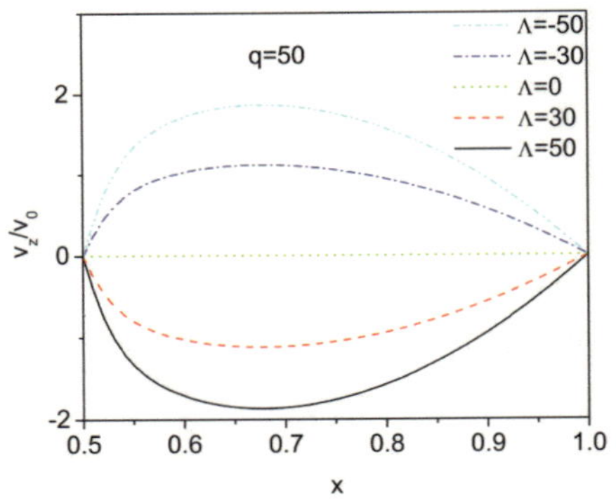

Fig. 16. Velocity profiles for a given q, and different values of Λ.

Moreover, the extreme of the curves, representing a vanishing shear stress, are closer to the inner cylinder for all the curves, with no significant dependence on the value of Λ. This behavior is different from a Newtonian fluid for which the maximum is approximately at the middle of the distance between both cylinders.

Fig. 17 presents the averaged apparent viscosity for this configuration. Note the non-Newtonian behavior of the system and the non-symmetric response with respect to the direction of the flow. In particular, in Fig. 17b we observe that for a given value of the electric field and for the range of flow considered the viscosity decreases as Λ increases. This means that for backward flow (Λ>0) the viscosity decreases as the magnitude of the flow increases whereas for forward flow (Λ<0) the viscosity increases as the magnitude of the flow increases. Therefore, we have flow thinning in one direction and flow thickening in the other. This directional response is due to the fact that the initial undistorted nematic configuration is asymmetrical. Even more, for the forward case most of the mechanical energy is elastically accumulated in distorting the nematic's configuration instead of being used to move the fluid, as compared to the backward case. In this sense the undistorted configuration is working like a biased spring inherent to the liquid, stiffer in one direction than in the other.

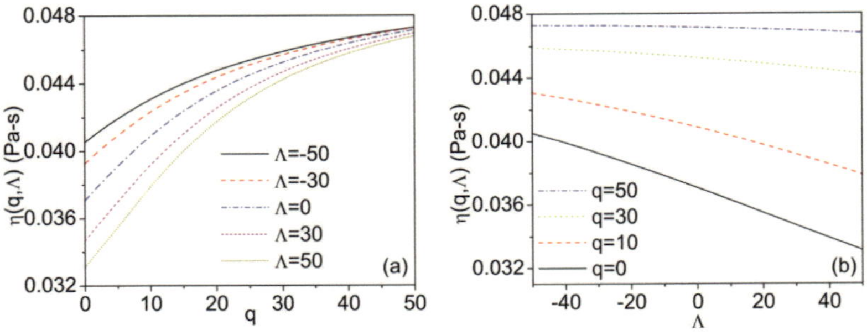

Fig. 17. Averaged apparent viscosity as a function of (a) the electric field q and (b) the pressure gradient Λ.

The averaged first normal stress difference is shown in Fig. 18. Panel (a) shows that N_1 depends almost linearly on q for backward flow, $\Lambda = -50$, whereas it has a minimum in $q =$

10 for forward flow $\Lambda = 50$. Panel (b) displays clearly the contrast between forward and backward flows for small values of q where a local minimum moves to the right as Λ increases. This shows that the directional dependence of this confined nematic can be electrically controlled.

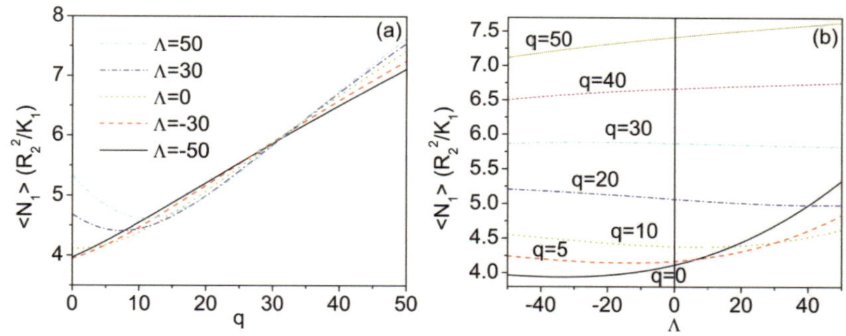

Fig. 18. Averaged first normal stress difference as function of (a) q and (b) Λ. Adapted from Mendoza, Corella-Madueño, and Reyes 2008. Adapted from Mendoza, Corella-Madueño, and Reyes 2008.

8.2 Homogeneous nematic capillary under a Couette flow

In this subsection we are going to present the case of a homogeneous nematic capillary subjected to a Couette flow and a radial electric field as shown in Fig. 19 (Reyes, Corella-Madueño, and Mendoza 2008). The inner cylinder is rotating with angular velocity Ω_1 and the outer cylinder with angular velocity Ω_2. This case resembles the one corresponding to the homogeneous cell, and in fact the phenomenology is very similar.

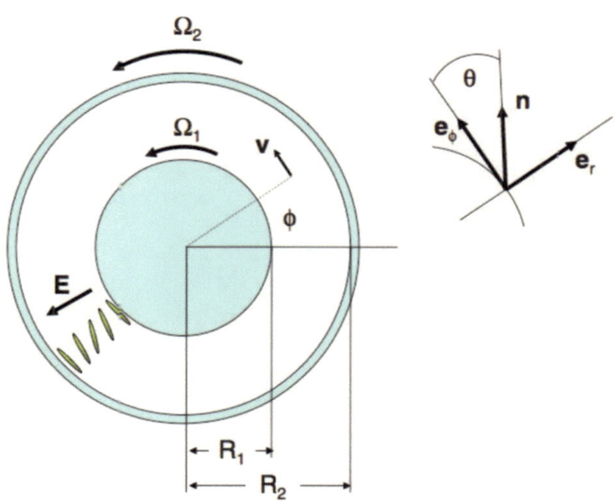

Fig. 19. Sketch of a nematic liquid crystal confined by two rotating coaxial cylinders and subjected to a radial electric field. Adapted from Reyes, Corella-Madueño, and Mendoza 2008.

According to the figure, the director can be written as

$$\hat{\mathbf{n}} = \sin\,\theta(r)\mathbf{e}_r + \cos\,\theta(r)\mathbf{e}_\phi, \tag{61}$$

and the velocity as

$$\mathbf{v} = \omega(r)r\mathbf{e}_\phi, \tag{62}$$

As in the previous situations, we are considering hard anchoring and non-slip boundary conditions at the cylinders

$$\theta(R_1) = 0, \quad \theta(R_2) = 0, \tag{63}$$

and

$$\omega(R_1) = \Omega_1, \quad \omega(R_2) = \Omega_2. \tag{64}$$

The orientational configuration for 5CB is shown in the left panel of Fig. 20. In (a) for q=20 and different values of $\Delta\Omega$. The angle grows from zero up to a maximum value, then, it decreases to zero at the outer cylinder. This maximum increases as we increase the value of $\Delta\Omega$. This simply means that the nematic's molecules tend to be more aligned with the flow

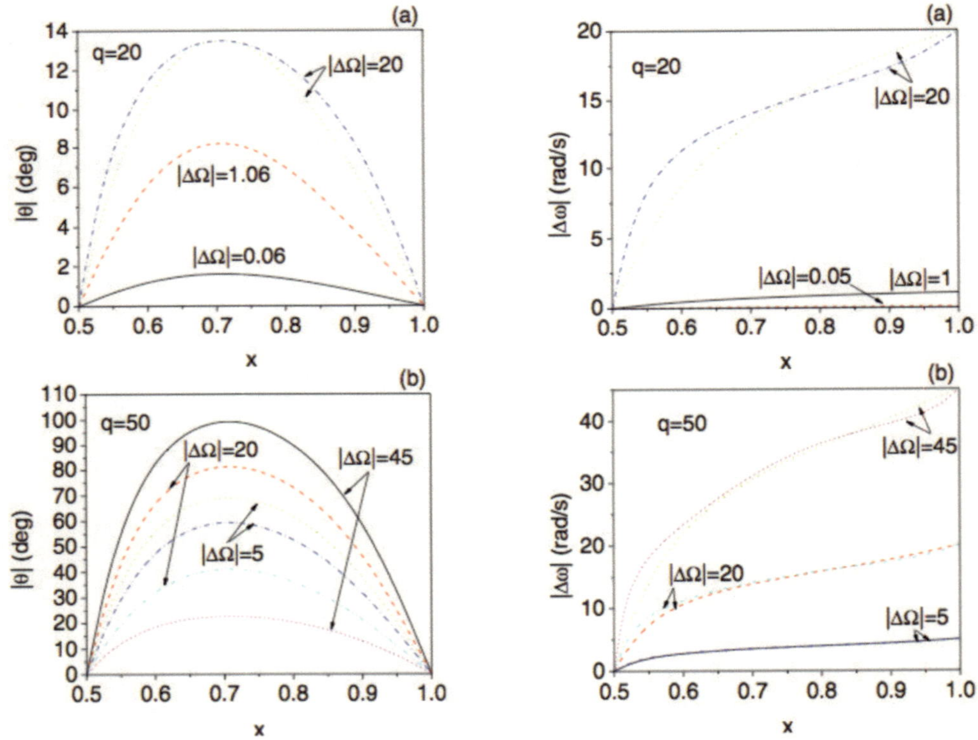

Fig. 20. Left panels: Nematic's configuration as a function of x for $R_1/R_2=0.5$ (a) $q=20$ and (b) $q=50$. Right panels: Velocity profiles as a function of x (a) $q=20$ and (b) $q=50$. The units of $\Delta\Omega$ are rad/s. Adapted from Reyes, Corella-Madueño, and Mendoza 2008.

as it increases. As we can see, for the largest value of ΔΩ shown, there are two possible stationary configurations. In (b) we plot the same as in (a) but for q=50. In this case for any value of ΔΩ the system may adopt multiple steady-state solutions. Here, we have plotted two different possible solutions. In the right panels we plot the corresponding velocity profiles.

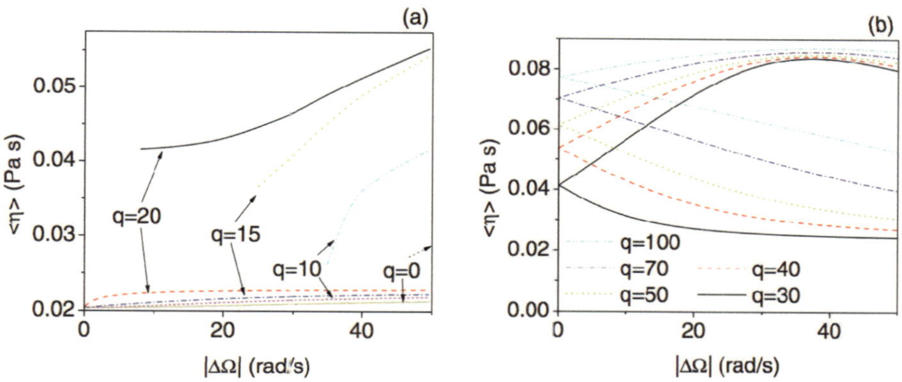

Fig. 21. Averaged apparent viscosity as a function of ΔΩ for (a) $q \leq 21$ and (b) $q \geq 21$. Adapted from Reyes, Corella-Madueño, and Mendoza 2008.

In Fig. 21 we present the average viscosity as a function of |ΔΩ|. Notice that the electrorheological effect is less pronounced for larger values of the shear flow since the cylinder's rotation turns the nematic perpendicularly to the electric field and as a consequence its influence is reduced.

9. Conclusion

We have presented a series of results that characterize the flow behavior of a flow-aligning thermotropic liquid crystal (5CB) under the action of an applied electric field in a variety of different flow geometries and boundary conditions. It is clear from these results that the influence of the boundary is enormous and may lead to completely different behaviors. Among the interesting results we can mention the existence of a rich non-Newtonian response with regions of shear thinning and thickening, a moderate electrorheological effect and a history dependent directional response.

10. Acknowledgment

CIM acknowledges partial financial support provided by DGAPA-UNAM through grant DGAPA IN-115010.

11. References

Ahlers G., Cannell D. S., Berge L. I. and Sakurai S, (1994) Thermal conductivity of the nematic liquid crystal 4-*n*-pentyl-4'-cyanobiphenyl, Phys. Rev. E Vol. 49, pp. 545

Arai, T. & Kragic, D. (1999). Variability of Wind and Wind Power, In: *Wind Power,* S.M. Muyeen, (Ed.), 289-321, Scyio, ISBN 978-953-7619-81-7, Vukovar, Croatia

Bird R. B., Armstrong R. C. and Hassager O., Dynamic of Polymeric Liquids, (Wiley, New York, 1977) Vol 1r D. (1989). Handbook of Differential Equation, Academic Press, New York Blinov L. M. and Chigrinov V. G., Electrooptic Effects in Liquid Crystal Materials (Springer, New York, 1994)

Brand H. R. and Pleiner, H., (1980). Nonlinear reversible hydrodynamics of liquid crystals and crystals, J. Phys (Paris) vol. 41, pp. 553

Brand H. R. and Pleiner H., (1982). Number of elastic coefficients in a biaxial nematic liquid crystal, Phys. Rev. A vol. 26, pp. 1783

Callen H. B., Thermodynamics and an Introduction to Thermostatistics (Wiley,New York, 1985) 2a. edición

Carlsson T., (1984). Theoretical Investigation of the Shear Flow of Nematic Liquid Crystals with the Leslie Viscosity $a_3 > 0$: Hydrodynamic Analogue of First Order Phase Transitions, Mol. Cryst. Vol. 104, pp. 307-334.

Chandrasekhar S., Liquid Crystals (Cambridge University Press, Cambridge, 1992)

de Gennes P.G., Prost J., The Physics of Liquid Crystals, (Clarendon Press, Oxford, UK, 1993)

Denniston, C., Orlandini, E. and Yeomans, J. M., (2001). C Simulations of liquid crystals in Poiseuille flow. *Comput. Theor. Polym. Sci.*, Vol. 11 pp. 389-395

de Volder, M., Yoshida, K., Yokota, S., and Reynaerts, D., (2006). J. Micromech. Microeng. Vol. 16, pp. 612.

Doi M. and Edwards S. F., The Theory of Polymer Dynamics (Oxford University Press, New York, 1986)

Ericksen, J. L., : Anisotropic fluids Arch. Ration. Mech. 4,231 (1960)

Frank F. C., (1958). Liquid crystals. On the theory of liquid crystals, Faraday Soc. Discuss. Vil. 25, pp. 19

Forster D., Hydrodynamic Fluctuations, Broken Symmetry and Correlation Functions (Benjamin, Reading, 1975)

Guillén, A.D. & Mendoza, C.I. (2007). Influence of an electric field on the non-Newtonian response of a hybrid-aligned nematic cell under shear flow. The Journal of Chemical Physics, Vol. 126, pp. 204905-1-204905-9; Guillén, A.D. & Mendoza, C.I. (2007). Erratum Influence of an electric field on the non-Newtonian response of a hybrid-aligned nematic cell under shear flow. The Journal of Chemical Physics, Vol. 127, pp. 059901-1-059901-2

Hohenberg P. and Martin P. C., (1965). Microscopic theory of superfluid helium, Ann. Phys. Vol. 34, pp. 291

Kadanoff L. P. and Martin, P. C., (1963). Hydrodynamic equations and correlation functions Ann. Phys. Vol. 24, pp. 419.

Khalatnikov I. M., Introduction to the Theory of Superfluidity (Benjamin, New York, 1965)

Kiss G. and Porter R. S., (1978). Rheology of concentrated solutions of poly(γ-benzyl-glutamate, J. Polym. Sci. Polym. Symp. Vol. 65, pp. 193

Landau L. D. and Lifshitz E., Theory of Elasticity (Pergamon, New York, 1964) 3rd. edition

Leslie F. M., (1966). Some Constitutive equations for anisotropic fluids, Quat.J. Mech. Appl. Math. Vol. 19, pp. 357.

Li, B.; Xu, Y. & Choi, J. (1996). Applying Machine Learning Techniques, *Proceedings of ASME 2010 4th International Conference on Energy Sustainability*, pp. 14-17, ISBN 842-6508-23-3, Phoenix, Arizona, USA, May 17-22, 2010.

Lima, P.; Bonarini, A. & Mataric, M. (2004). *Application of Machine Learning*, InTech, ISBN 978-953-7619-34-3, Vienna, Austria.

Marenduzzo, D., Orlandini, E. and Yeomans, J. M., (2003). Rheology of distorted nematic liquid crystals. *Europhys. Lett.*, Vol. 64, pp. 406-412

Marenduzzo, D., Orlandini, E. and Yeomans, J. M., (2004). Interplay between shear flow and elastic deformations in liquid crystals. *J. Chem. Phys.*, Vol. 121, pp. 582-591

Marrucci G. and Maffettone P. L., (1989). A description of the liquid-crystalline phase of rodlike polymers at high shear rates, Macromolecules Vol. 22, pp. 4076.

Medina, J.C. & Mendoza, C.I. (2008). Electrorheological effect and non-Newtonian behavior of a homogeneous nematic cell under shear flow: Hysteresis, bistability, and directional response. EPL Europhysics Letters, Vol. 84, pp. 16002-p1-16002-p6

Mendoza, C.I., Corella-Madueño, A., and Reyes, J.A. (2008). Electrorheological effect in a nematic capillary subjected to a pressure gradient, Physical Review E Vol. 77, pp. 011706.

Miesowicz M., (1946). The Three Coefficients of Viscosity of Anisotropic Liquids, Nature Vol. 27, pp. 158.

Negita, K. (1996). Electrorheological effect in the nematic phase of 4-n-pentyl-48-cyanobiphenyl. The Journal of Chemical Physics, Vol. 105, pp. 7837-7841

Parodi O., (1970). Stress tensor for a nematic liquid crystal, J. Phys. (Fr) Vol. 31, pp. 581.

Pleiner H., (1986). Structure of the core of a screw dislocation in smectic A liquid crystals, Liq. Cryst. Vol. 1, pp. 197.

Pleiner H., (1988). Dynamics of a disclination point in smectic-C and -C^* liquid-crystal films, Phys. Rev. A Vol. 37, pp. 3986 .

Reyes, J.A.; Manero, O. & Rodriguez, R.F. (2001). Electrorheology of nematic liquid crystals in uniform shear flow. Rheologica Acta, Vol. 40, pp. 426-433

Reyes, J.A., Corella-Madueño, A., and Mendoza, C.I. (2008). "Electrorheological response and orientational bistability for a homogeneously-aligned nematic capillary", The Journal of Chemical Physics Vol. 129, pp. 084710.

Stephen M. J. and Straley J. P., (1974). Physics of liquid crystals, Rev. Mod. Phys. Vol. 46, pp. 617.

Siegwart, R. (2001). Indirect Manipulation of a Sphere on a Flat Disk Using Force Information. *International Journal of Advanced Robotic Systems*, Vol.6, No.4, (December 2009), pp. 12-16, ISSN 1729-8806

Van der Linden, S. (June 2010). Integrating Wind Turbine Generators (WTG's) with Energy Storage, In: *Wind Power*, 17.06.2010, Available from http://sciyo.com/articles/show/title/wind-power-integrating-wind-turbine generators-wtg-s-with-energy-storage

Vicente Alonso, E., Wheeler, A. A. and Sluckin, T. J., (2003). Nonlinear dynamics of a nematic liquid crystal in the presence of a shear flow. *Proc. R. Soc.* London, Ser. A, Vol. 459, pp. 195-220

Zakharov, A.V., Vakulenko, A.A., (2010). Orientational nematodynamics of a hybrid-oriented capillary, Physics of the Solid State Vol. 52, pp. 1542.

13

Rheological Behaviors and Their Correlation with Printing Performance of Silver Paste for LTCC Tape

Rosidah Alias and Sabrina Mohd Shapee
TM Research & Development Sdn. Bhd., Malaysia

1. Introduction

Low Temperature Co-fired Ceramic (LTCC) technology has attracted much attention for high frequency applications due to the advantages in preparing 3D circuits within a ceramic block that enables burying of passive elements; resistor, inductor and capacitor (Jantunen et al., 2003). This technology also offers another advantages such as low fabrication cost due to parallel process, utilizing high conducting metal, higher interconnect density, high performance thermal management system, shrinking of circuit dimensions and high level of passive integration (Lin and Jean, 2004; Gao et al., 2010). The LTCC multilayer technology process generally consists of cutting the green tape ceramic into the required dimension, via punching, via filling, screen printing, stacking, lamination and co-firing process. All these steps are important to achieve a good quality of the final structure. Furthermore, LTCC technology requires cofiring process of conductor and ceramic-glass substrate. The co-firing mismatch should be avoided to make sure the resulting product has good appearance. The key point of LTCC technology is the screen printing process of thick-film technology which strongly depends on the optimization of the conductor paste rheological behavior. Besides, the characteristics and the quality of printed conductor lines are greatly affected by some process variables such as screen printing speed, angle and geometry of the squeegee, snap-off and screen mesh parameter including the paste characteristics (Hoornstra et al., 1997; Yin et al., 2008). The performance of the paste depends on the variety of factors including storage, how easily and accurately it can be deposited (printing) and the flow characteristics (rheology) (Nguty and Ekere, 2000). Thick-film paste must maintain good printability throughout the time on the screen and the screen parameters must allow for high throughput and repeatability results (Harper, 2001; Buzby & Donie, 2008). The successful story of making multilayer substrate is depends on this steps. Consequently, a thorough understanding of the influence of the paste rheological behavior and the combination with other parameters is essential if this objective is to be met.

Generally, the electrical connection between components and layers are given by the conductor paste such as silver, copper, gold, silver-palladium, silver-platinum and etc. Compared to the other conductor materials, silver thick film has been used as the main conductive material for LTCC technology due to their excellent electrical properties, thermal conductivity and the lower cost price of conductor material (Wu et al., 2010; Chen et al., 2010). The characteristics of the silver particles strongly affect the paste rheology and subsequently alter the densification of the printed track. Different formulations of the paste are used to produce conductors that basically consists of several major ingredients; (1) the metal powders, which provide the conductive phase; (2) glasses or oxides, which act as a permanent binder and also promote sintering of metal powders during firing and enable binding of the functional film to the substrate; (3) organic vehicle, which disperse the functional and binder components to impart the desired rheological properties to the paste; and (4) a solvent or thinner that establishes the viscosity of the vehicle phase (Taylor et al., 1981; Lin and Wang, 1996; Rane et al., 2004; Sergent, 2007). Such pastes are screen printed in the desired pattern on the substrate, dried and fired at a certain temperatures to form a conducting thick-film with a thickness ranging from 3 to 30 μm (Rane et al., 2003).

1.1 Basic information/thick-film materials

Thick-film is generally considered to consist of layers of paste deposited onto the substrate. One of the key factors that distinguish thick-film circuit is a method of thick film deposition.

In LTCC process, screen printing is one of the most dominant methods to deposit conductor paste on a substrate. The screen printing of thick film patterns on the LTCC multilayer substrate is employed for thick film systems due to a low-cost and much simpler process to produce a circuit design on substrate compared to other process. The printing process are done only at the place where required and follow the design pattern which are prepared earlier (Vasudivan & Zhiping, 2010). The printing quality and the fine line printing resolution patterns is necessary in order to achieve high component density and high frequency applications (Dubey, 1975; Kim et al., 2010). Successful transferred pattern onto the substrate depends on the rheology of the paste material. The range of most available for thick-film technology is determined by their capacity to be both printed and fired. Establish thick-film technology is based on three classes of material supplies in the form of printing paste; conductors, resistors and dielectrics. For this report we only concentrate on the conductor paste.

Thick-film circuits are created in thick-film printing patterns in paste which basically composed of organic carrier, glass-frit and active elements such as gold, silver, silver platinum and silver palladium. As mention above, silver paste is the most selected material due to the good conductivity and good characteristics which depends on the functional paste whether precious metal such silver, gold or base metal such Ni, Cr and Al. An organic carrier dispersed the binder and functional components to impart the desired rheological properties of the paste while the glass-frit act as permanent binder and help to promote metal powders during sintering and also provide binding between thick-film and the substrate (Hwang et al., 2009; Shiyong et al., 2008). The printing process is known to be controlled by a number of process parameters such as screen mesh parameters, printer setting parameters, environmental conditions and paste parameters (Durairaj et al., 2009b). Some of these parameters are fixed (eg. screen mesh). However the paste parameters generally change during the printing process cycles; before printing process, during printing and after printing process as seen in Fig. 1. So, it also noted that the viscosity is a function of the time. One of the paste parameters is the rheological properties and flow history of the paste (Parikh et al., 1991; Burnside et al., 2000). The printing behavior of the paste is characterized by its rheology through the viscosity value.

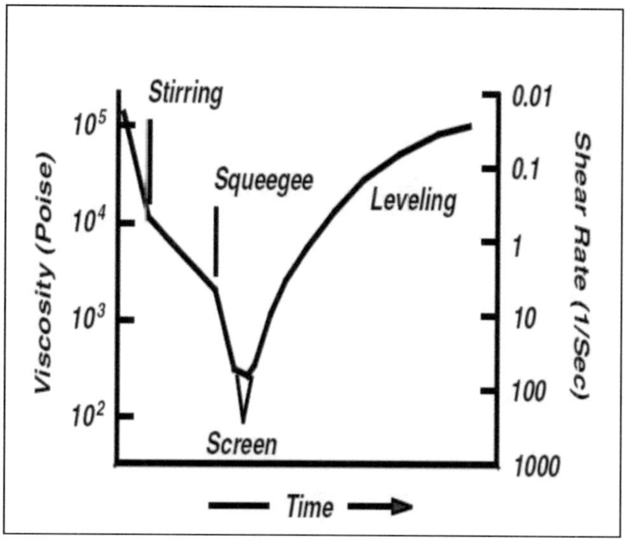

Fig. 1. Variation in the paste viscosity at different stages of the printing cycle (Barlow & Elshabini, 2007)

The viscosity is the most important rheological characteristics of the conducting paste (Gilleo, 1989). The conducting paste for screen printing process should show the pseudoplastic behavior which displays a decreasing viscosity with an increasing shear rate (Hoornstra et al., 1997). However, to be suitable for screen printing, conducting paste should be somewhat thixotropic in nature. A thixotropic fluid is one in which the shear rate/shear stress ratio is nonlinear. As the shear rate (which translates to the combination of squeegee pressure, velocity, and screen tension) is increased, the paste becomes substantially thinner, causing it to flow more readily.

The optimum operating viscosity of the paste is dependent on the parameters with the screen printing process. The variations in squeegee speed, squeegee to screen angle, squeegee pressure and snap-off distances will affects the quality of the printed film. If the printed lines have a tendency to spread on standing, it is likely that the viscosity of the paste is too low. Chiu, (2003) in his experiment also noted that the paste viscosity is a significant factor to control the line width including the thickness and roughness of the printed pattern. The paste viscosity is too high if the printed films display mesh marking, pinholes or very thin areas covered with paste.

The aim of this work is to modify the viscosity of silver paste in order to get the required thickness and fine line printing of printed material on the substrate. As well known, controlling the properties of resulting conductor thick film paste is not a simple task, so in order to comply with required properties, the conductor paste need to do some adjustment in terms of its viscosity behavior. Viscosity can be lowered (by addition of the solvent) or increased (by addition of a thixotropic nonvolatile vehicle), although the latter will require re-milling of the paste.

The investigation of rheology of the paste has been studied by several researchers especially for solder paste and conductor material for solar cell applications such as Shiyong et al., (2008), Amalu et al., (2011), Mallik et al., (2009), Durairaj et al., (2004), and Burnside et al., (2000). Their studies reported that the solder paste shows thixotropics behavior, shear thinning and yield stress. Jackson et al., (2002) noted in their paper, a good quality paste will have optimum shear thinning properties allowing flow in mesh aperture, a suitable thixotropic nature for recovery and a non tendency to slump after printing. The paste also should retain its intended printed thickness and continuity. Previous studies by Evans and Beddow, (1987) found that viscosity of solder paste increase with increasing the metal content and decrease with increase particle size distribution and temperature. However, not much literature has been found for the study of modification of wt% of thinner in order to adjust the viscosity of the paste suitable for screen printing process. The work in this study was divided into several parts; part 1 is the sample preparation of different wt% of thinner. In this part the viscosity measurement is using the Brookfield viscometer model RV-DV cone and plate with CP-52 spindle. At the same time part 2 is done; preparation of LTCC tape for the whole multilayer substrate process. The results for the part 1 will be used for part 2. After finish part 2, characterization of part 3 will be carry out. Sample analysis and detail discussion is carried out to explain the research findings.

2. Methodology

In the following sections, the various process stages will be described. The factors that have the most influence on the quality of the line resolution is the viscosity of the paste.

2.1 Sample preparation

Seven samples with different viscosities of the silver paste were used in this process. In this work, a thinner that contains texanol is used to modify the viscosity of commercial LTCC silver paste. The addition of thinner used is varies by its weight percentage from 2, 4, 6, 8, 10 to 12% of 4g silver paste. Table 1 shows a sample description that going to be used for further discussion.

Sample labels	S1	S2	S3	S4	S5	S6	S7
Weight % of Thinner	0	2	4	6	8	10	12

Table 1. Samples descriptions of silver conductor paste with varies thinner wt%.

2.2 Viscosity measurement

There is some equipment to be used for viscosity measurement which broadly classified into two categories; dynamic and kinematic viscometer. A dynamic viscometer is one of the shear rate can be controlled and measured (rotational viscometer). It is the only type of viscosity measurement that is relevant to fluids where the viscosity is related to the shear rate (non-Newtonian fluids). A kinematic viscometer is where the shear rate can neither be controlled nor measured, for example capillary viscometer.

For the purpose of this research, brief information about cone and plate geometry viscometer will be given. As noted in Brookfield catalog, cone and plate viscometer offers absolute viscosity determination with precise shear rate and shear stress information. The sample volume is extremely small amount and temperature controlled is easily accomplished. Cone and plate geometry viscometer is particularly advanced rheological analysis of non-Newtonian fluids (Phair & Kaiser, 2009).

The viscosity measurement was carried out in semi-clean room condition and the ambient temperature is maintained at 25°C. This is the key factor which might be affects the printing quality. Brookfield viscometer Model RV-DV cone & plate geometry with CP-52 spindle, using a plate 1.2 cm radius and a cone angle of 3 ° is used. Viscometer speed was set at 0.5, 1.0, 1.5, 2.0 and 2.5 RPM. The viscosity and torque readings were recorded. Viscosity of the paste for the sample 2, 4, 6, 8, 10 and 12 wt% of thinner was achieved from the measurement.

2.3 Multilayer substrate process

The quality and performance of the paste was evaluated by the screen printing process where the standard thickness is about 15-21 µm. Screen printing process was carried out for Heraeus tape HL2000 on a KEKO P-200Avf Screen Printer Machine using the TC0306 silver conductor paste from Heraeus. Heraeus HL2000 LTCC is an alumina-based ceramic glass system that has an excellent dielectric properties; the dielectric constant is 7.3 ± 0.3 and dielectric loss (tan δ) is about 0.0026. The process was started by printing the test pattern design on the standard LTCC tape dimension 21.4 cm x 21.4 cm.

Printing parameters	Value
Squeegee pressure	0.1 MPa
Squeegee speed	100 mm/s
Snap-off	1.0 mm

Table 2. Printing parameters for printing process of Heraeus HL2000 LTCC tape.

The printing process was also done in a temperature controlled room at 25 °C and using a standard 325 mesh and angle of 22.5°. The printing parameter used for the process were fixed and controlled to eliminate the variation from the printer (London, 2008). Printing parameter used for the printing process is shown in Table 2.

After the printing, the substrate was dried at 80°C for about 20 minutes to remove portion of volatile organic solvent before manual stacking process (see Fig. 2) was carried out to stack the printed tape with 8 "dummy" layers. Drying of the solvents of the conductive paste will reduce the wet thickness volume about 50%. The whole substrate was then laminate using isostatic lamination system ILS-6A at 75°C with pressure of 1500 psi (10MPa) for 10 minutes. The laminated samples were then fired up to 850°C by using an LTCC firing profile as suggested by a tape manufacturer.

The effect of the screen printing line resolution was assessed by printing a test pattern design as seen in Fig. 3. The unfired and fired samples were then measured for the line width resolution using an optical microscope OLYMPUS MX40. The line resolution is observed for the horizontal and vertical line.

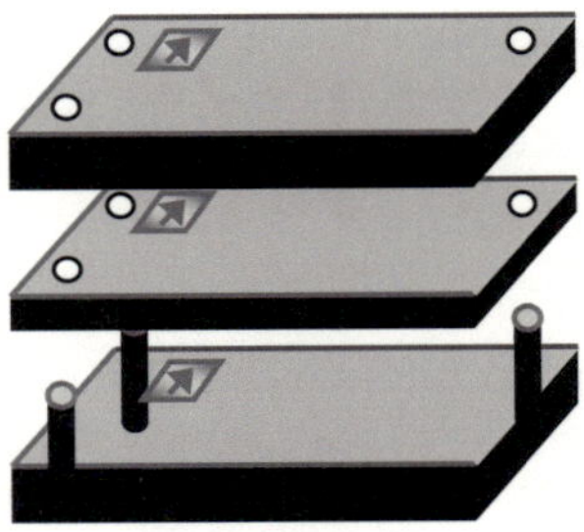

Fig. 2. The manual stacker for collating and stacking process (Alias et al., 2010)

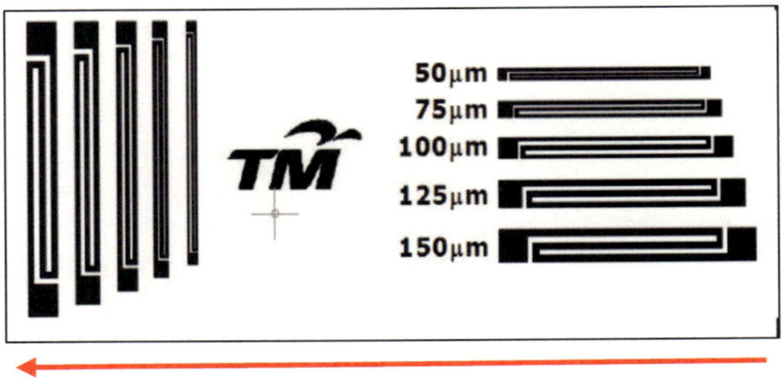

Fig. 3. Printing pattern for the line width evaluation

2.3 Analysis and measurement

2.3.1 Shrinkage and density measurement

The shrinkage of the Heraeus tape is generally calculated based on equation 1.

$$\text{Shrinkage} = \left(\frac{\text{Length}_{\text{before fired}} - \text{Length}_{\text{after fired}}}{\text{Length}_{\text{before fired}}} \right) \text{x}100\% \tag{1}$$

The properties of the final ceramic composite materials depend on the sintered density of the whole substrate. A stacked and laminated LTCC substrate before firing consists of a relatively porous compact of oxides in combination with a polymer solvent. During sintering the organic solvent evaporates and the oxides react to form crystallites, or grains of the required composition, the grains nucleating at discrete centers and growing outwards until the boundaries meet those of the neighboring crystallites. During this process, the density of the material rises; if this process were to yield perfect crystals meeting at perfect boundaries the density would rise to the theoretical maximum, i.e. the x-ray density, which is the material mass in a perfect unit crystal cell divided by the cell volume. In practice imperfections occur and the sintered mass has microscopic voids both within the grains and at the grain boundaries. The resulting density is referred to as the sintered density. The density of the sample was measured using the Archimedes principle shown in equation (2);

$$\rho = \left(\frac{W_a}{W_w}\right)\rho *_w \tag{2}$$

where W_a = weight of sample in air
W_w = weight of sample in water
$\rho*_w$ = density of water = 1 gcm^{-3}

2.3.2 Monitoring film thickness

Film thickness need to be carefully controlled and monitored. It is desirable to measure the film thickness at an early stage in the thick film process to reduce error in determination of film thickness for design requirements. For this reason, the thickness of the film should be measured before and after fired. There are some methods to determine the metal thickness using surfometer and non-destructive method. However, in this work we used SEM through cross-section view to measure the film thickness using a FEI NOVA Nano SEM 400 machine. Most of the samples were imaged several times, with at least three pictures in each case, from different areas of the sample holder.

3. Results and discussion

3.1 Viscosity analysis

The viscosity of thick-film paste is tailored to meet some requirement for screen printing process such as; 1) thixotropic behavior, 2) must have yield point and should have some degrees of hysteresis i.e; the viscosity should be higher with decreasing pressure, as the paste will be on the substrate on the time. Table 3 shows the viscosity for the entire sample for up-curve and down-curve measurement. The paste generally exhibits good shear thinning with a range of viscosity about 7000 to 65000 poise observed at various wt% of thinner. As seen in the Hereaus data sheet, suitable viscosity value for screen printing process is in the range of 2000-3000 poise.

Fig. 4 shows the variation of viscosity and shear rate of silver paste with different percentage of thinner. As we can see the trend exhibited by all the samples is generally similar to the work carried out by Bell et al., (1987) and Morissette et al., (2001) on the influence of paste rheology on print morphology and component properties. Sample 0% to 6% thinner the viscosity slightly shows decreasing trend. It also obvious that for sample 0-4wt % the trend is slightly overlapped for higher shear rate started from 3.00 to 5.00.

However, the viscosity measured for sample with higher percentage of thinner of 8%, 10% and 12% did not shows a significant change which could be due to the generation of a liquid-rich layer at the interface between the sample and cone and plate geometry causing a lower viscosity (Durairaj et al., 2009a).

Rheological characterization of the paste is to measure the relationship between shear stress and shear rate varying harmonically with the time, indicating the level of interparticle force or flocculation in the paste. The ideal paste for thick-film films should have a proper degree of pseudoplastic as well as thixotropic behavior (Wu et al., 2010). The thixotropic effect is a result of aggregation of suspended particles. Aggregation in the system caused by the attraction forces such as Van der Waals and repulsion forces due to steric and electrostatic effect on the particles. This force prevents the particles from approaching close to each other and create weak physical bond. When the suspension is sheared, this weak force are broken causing the network to break down. A paste with the excellent thixotropic could produce a good printability, help to avoid failure such as incomplete line resolution (Wu et al., 2011; Neidert et al., 2008).

	Speed (RPM)	0% thinner	2% thinner	4% thinner	6% thinner	8% thinner	10% thinner	12% thinner
Up curve	0.5	65882	64096	58540	44054	25996	21035	10716
	1	54670	52884	50801	37505	22622	17862	8938
	1.5	47758	47229	45707	34925	21101	16801	7608
	2	42962	42962	42069	32842	20042	15925	7393
	2.5	39331	39291	39251	30996	19288	15399	7174
	Speed (RPM)	0% thinner	2% thinner	4% thinner	6% thinner	8% thinner	10% thinner	12% thinner
Down curve	2.5	39609	39331	39490	31750	19526	15518	7342
	2	42863	42466	42218	33536	20489	16123	7640
	1.5	47559	46369	45906	35984	21696	17000	8004
	1	54869	52289	50999	39192	23614	18157	8434
	0.5	68660	61913	60326	45244	26988	20241	9327

Table 3. Viscosity of silver paste with varies thinner weight percentage (in Poise unit)

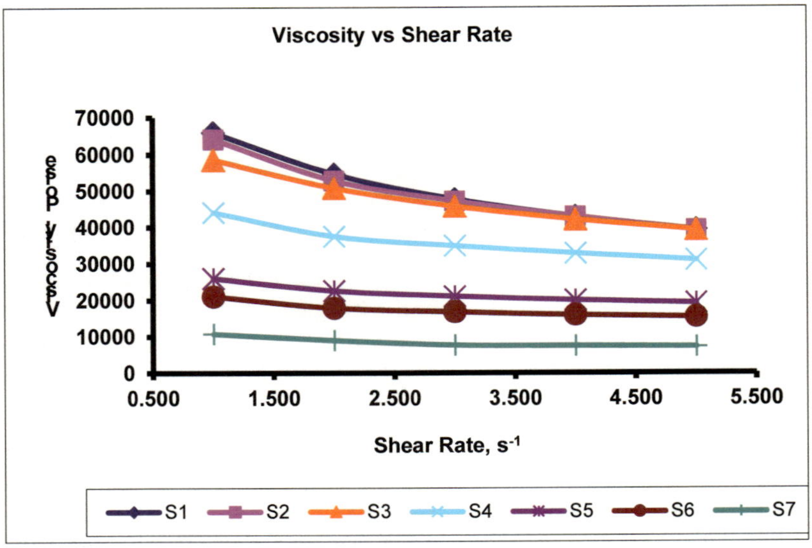

Fig. 4. Viscosity vs. shear rate of silver paste with varies weight% thinner

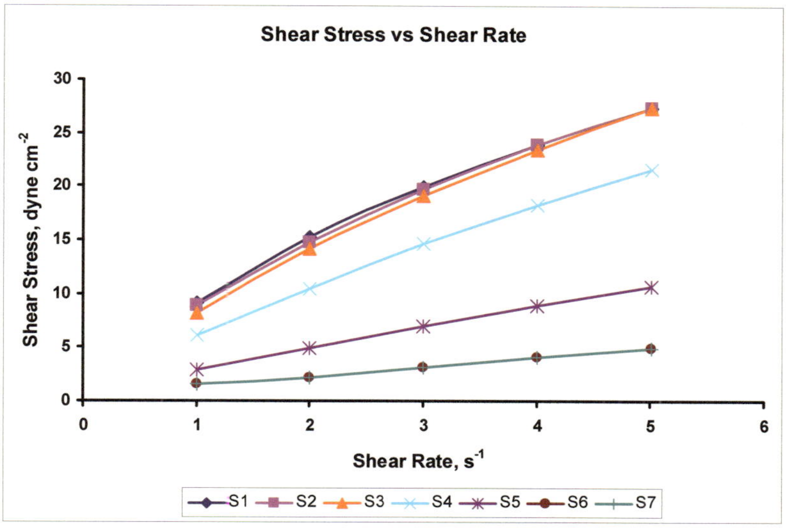

Fig. 5. Shear stress vs shear rate of silver with varies wt% of thinner

Fig. 5 shows the relation of shear stress and shear rate of silver paste with different wt % of thinner. The trend of non-Newtonian behavior is consistent with the results found by Chhabra & Richardson, (1999) for the types of time-independent flow behavior. The time-independent non-Newtonian fluid behavior observed is pseudoplasticity or shear-thinning characterized by an apparent viscosity which decreases with increasing shear rate. Evidently, these suspensions exhibit both shear-thinning and shear thickening behavior over different range of shear rate and different wt% of thinner. The viscosity and shear stress relationship with increasing percentage of thinner is plotted in Fig 6. It is clearly observed that both viscosity and shear stress decreases respectively.

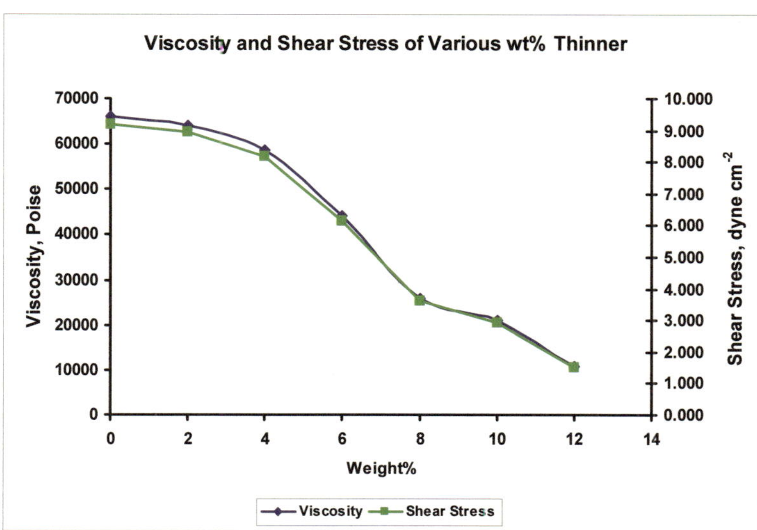

Fig. 6. Viscosity and shear stress of silver paste with varies wt% of thinner

3.2 Effects of weight percentage thinner on the line width resolution

The relationship of the line width resolution before firing process and after firing process and wt% of thinner is plotted in Fig. 7. The variation of the line width for 100 μm line resolution were found to fluctuate with increasing the wt% of thinner meaning that it did not shows a significant trend of the line width with increasing the wt % of thinner. However, due to the optimization of rheology behavior of sample S3 similar to the viscosity of the commercial pastes in the Hereaus data sheets (Hereaus design guideline) so the line width resolution using 4% thinner (S3) is acceptable compared to the rest wt % of thinner and it can be used for the printing process.

Fig. 8 shows the relationship between viscosity and shear stress with shear rate of sample S3. As we can see, the viscosity decrease and the shear stress increase as the shear rate increases. The relation of shear rate and shear stress shows that the paste has thixotropic behavior. The plot is similar to the graph by Chhabra & Richardson, (2008). Through this plot, we can see that not only the values of viscosity are seen to be different but the rate of decrease viscosity with shear rate is also seen to be varied. When shear stress increases the viscosity of the paste with good rheology decrease sharply, paste can flow rapidly through screen (Wang et al., 2002). The similar behavior of this modification paste (S3) with the commercial paste made this sample might be able to be used for fabrication of electrical connection using screen printing process.

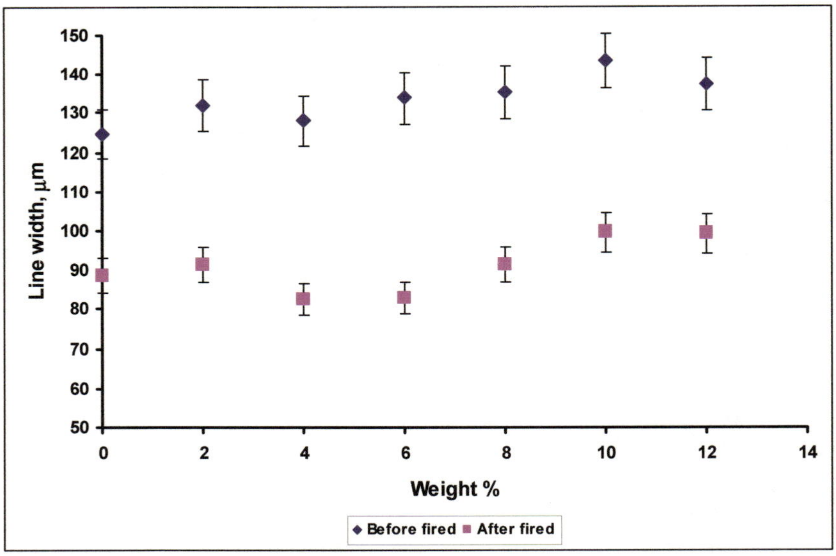

Fig. 7. Printing line width (before and after fired) of silver paste with varies wt% of thinner

The relation of printed thickness and z-shrinkage with the percentage of thinner is presented in Fig. 9. The thickness of the fired film is about 4 μm to 33 μm determined via cross section observation using SEM. It is clearly observed that increasing percentage of thinner will cause the printed thickness of the metal conductor on the substrate shows decreasing trend. When the thinner content increase in the paste the viscosity decrease and the films become thinner. It is consistent with the results found by Jabbour et al., (2001), where the thickness of the printed pattern depends on the viscosity of the paste. On the other hand, the linear shrinkage for the whole substrate did not show significant increase with increase wt % of thinner. This shrinkage can be attributed to the solvent loss. Theoretically, the thickness of metal conductor should be constant regardless of the line width since the same mesh is used (Shin et al., 2009). It could be attributed this result to the fact that the edges of the printed line should have a moderate thickness to result in uniform and continuous line width.

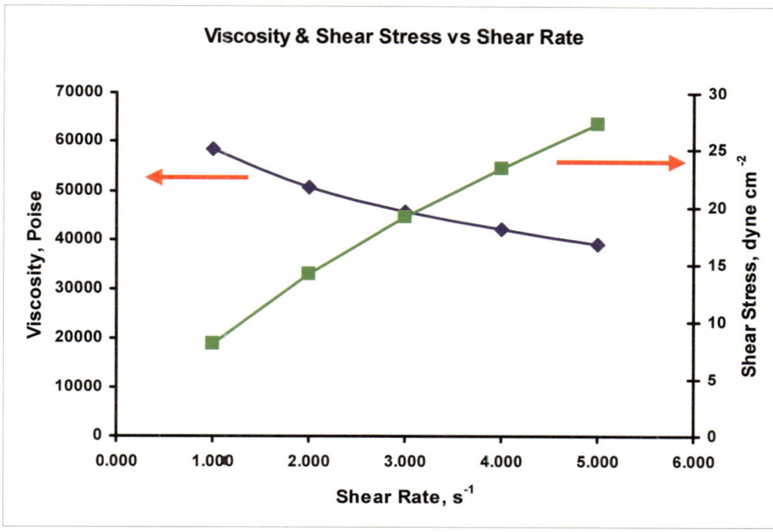

Fig. 8. Viscosity and shear stress as a function of shear rate for silver paste with 4 wt% of thinner (sample S3)

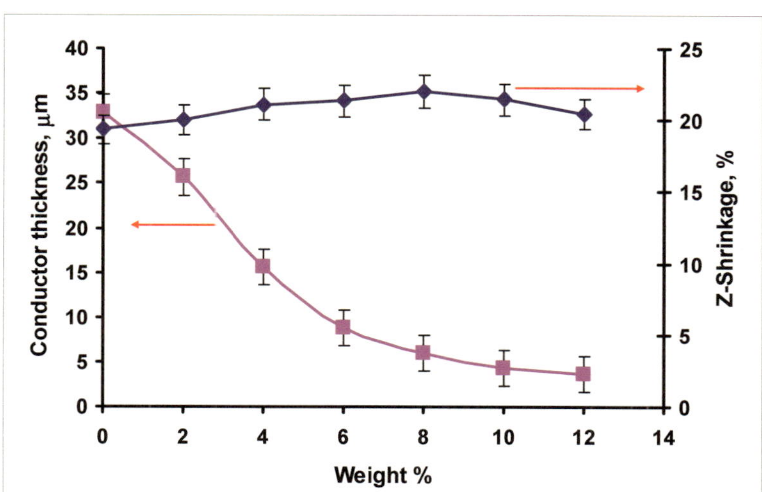

Fig. 9. Influence of printed thickness and shrinkage of silver paste with varies wt% of thinner

3.3 Printing results - Line resolution

A series of commercial pastes have been tested for the ability to easily print for various line width and spacing which designed for the horizontal and vertical line at different widths that are 50 μm, 75 μm, 100 μm and 125 μm. The different line width resolution was observed under optical microscope and was found that for sample with zero percentage of thinner (no thinner), the saw-tooth-edge line was observed as shown in Fig. 10. The line resolution is evaluated from the incomplete line and complete line for the required line width. Lines are incomplete and a portion of the lines is missing. The amount of silver seems to be insufficient to form continuous lines. It can be noted here that at lower silver content the edges of the printed lines look spread out with the paste due to its lower viscosity. This might be due to the results of insufficient adhesion of the lines to the LTCC tape. In other way, the line is broken could also be due to the mesh problem. For this study, the screen mesh 325 with the angle of 22.5 ° is used. The mesh opening and the viscosity of the paste is not compatible i.e; ink cannot pass through the mesh openings/mesh is unclean. As noted by Pudas et al., (2004) incomplete transfer pattern means the high variations in the printed results.

If it is broken line, then it shows that there are some problems regarding screen printing process (Shapee et al., 2010). It is defined has open defects that are voids on the conductor extending across more than 80% of the conductor width (Webster, 1975). Print poor resolution means that the printed film does not match the layout in terms of dimensions or shape. It may because of improper rheology or a flaw in the screen (Sergent, 2007). It is obvious that the rheology of the paste is the main factor that affects the poor print resolution. Higher viscosity is better for the good print resolution. The line is considered has a good connectivity when it is no broken line along the line. It also shows that the line resolution for the vertical direction is better than the horizontal direction. Producing optimal screen printed patterns involves consideration of many factors including printer setting, screen options, substrate preparation and paste rheology (Cao et al., 2006). Interaction between these factors must be considered (Dollen & Barnett, 2005).

If the printed pattern is too thick, the thinner should be added to the paste to adjust the rheology of the paste. As mention by Barlow & Elshabini, (2007), the thinner is useful to modify the thicker printing provided by the organic binder. The thinner can integrate together with the binder to improve paste characteristics and printing performance. The influence of paste rheology on the line resolution of as printed material has been investigated previously by Wang & Dougherty, (1994) and Vasudivan & Zhipping, (2010). As noted by Schwanke & Polhner, (2008) in their work of fine line printing enhancement, the printed pattern resolution depends on the two factors paste rheology including viscosity, particle size and its thixotropic behavior. The printed pattern for zero wt% of thinner is too thick. Its create bad surface morphology as seen in Fig. 11 through dielectric layers making it difficult to produce high resolution conductor on subsequent layer (Bender & Ferreira, 1994). Furthermore, the transferability of the paste and the uniformity of the conductors degraded when the viscosity is too high (Lahti, 2008).

The flow behavior of the paste during the movement of squeegee is an important factor for the screen printing process. During the movement of the squeegee a whirling of the paste happens which leads to a lowering the viscosity and thus, the paste is to be able to pass a screen meshes (Imanaka, 2005). It is believe that increasing wt% of thinner to the conductor paste will cause the viscosity of the paste falls so that it can be push through the mesh opening much more easily compared to the paste without thinner.

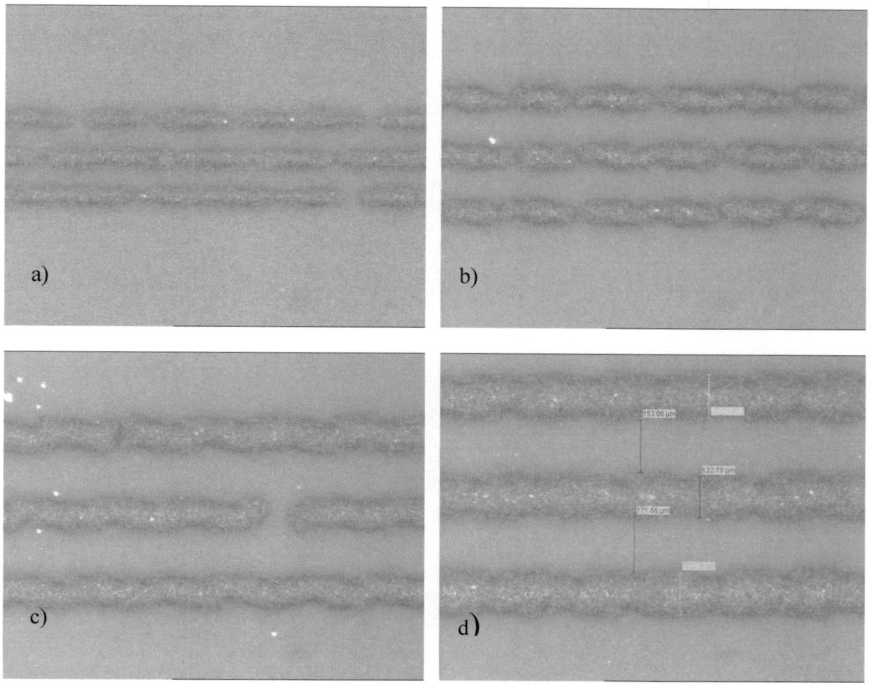

Fig. 10. Line resolution for different width a) 50 μm, b) 75 μm, c) 100 μm and d) 125 μm before firing process

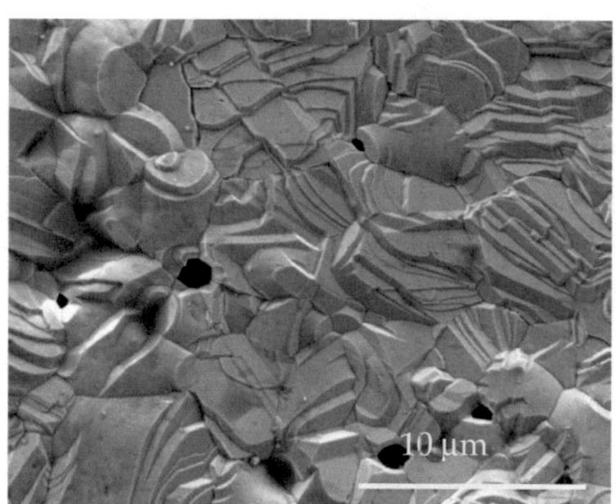

Fig. 11. Surface morphology of conductor paste without thinner

After do some observation, we decide to concentrate on the line width of 100 μm to evaluate the quality of the conductor paste with different percentage of thinner for the horizontal and vertical direction. Fig. 12 presents the representative vertical and horizontal lines of 100 μm line width for the example of 10 % thinner weight percentage. They were observed at 5x magnification. These picture shows that the screen printed line have smooth line edges and consistent line widths. No paste residue were observed between the lines, thus, these line can be used in the interconnection that require long lines (Wang et al., 2005). It also found that the vertical line resolution is better than the horizontal line resolution. It could be due to the mesh angle (Fig. 13) of the screen mesh parallel to printing in direction. This indicates that the smoothness of the edge printing for sample higher wt% of thinner is not accurate due to the screen wires interfere with the transfer of the viscous paste to the substrate, leaving voids in the printed line.

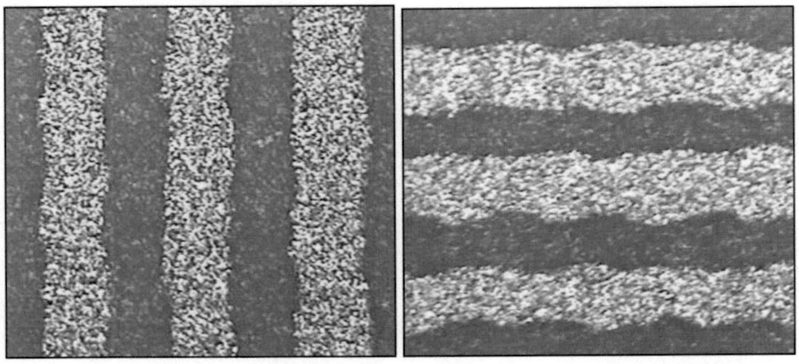

Fig. 12. 100 μm line resolution (after fired) for the silver paste with 10 % thinner weight percentage

Fig. 13. Screen mesh orientation 325 stainless steel with 22.5 °angles

The effect of screen mesh on printing quality has studied by Stalnecker Jr. (1980) and noted that to get fine line printing the selection of the screen mesh should take into account. The screens for thick film printing should be such that when viscosity and other variables are controlled it should be able to delineate reproducible thickness of the printed line (Dubey, (1974).

4. Conclusion

As a conclusion, the viscosity problem can be solved using the different wt% of thinner (2, 4, 6, 8, 10 and 12 wt %). The phenomena observed from screen-printing pastes are also identical to those reported from literature that the viscosity of the paste throughout the screen-printing process influences the efficiency of the deposition process and the film's quality. The most higher wt% of thinner pastes have a lower viscosity than those of others at a low shear rate, while their viscosities at high shear rates, which are directly affects the screen-printing quality, are excessively high, resulting in difficult snap off. The thinner may change the thixotropic of the paste and sometimes can change the paste characteristics. So, by changing the thixotropy of the paste, the printing performance can be better. It may also influence the shape of the printed pattern for the non-linearity of the viscosity versus the shear rate. However, many factors such as filler, resin, the solvent system and the solid content influenced the dispersion degree of particles in the paste, resulting in the different of rheological behaviors. The correlation of the paste rhelogical properties with the pattern transfer efficiency is important to provide better understanding of the effects and the interaction of stainless steel screen mesh process variables to optimize the required printed thickness. Such data will greatly enable the development of more rigorous rheological protocols for the characterization and optimization of the pastes for screen printing in the future.

5. Acknowledgment

The authors wish to thank Telekom Malaysia for their funding support under project IMPACT (RDTC/100745) and Assoc. Prof. Dr. Mansor Hashim for the guidance and support for this research work.

6. References

Alias, R., Ibrahim, A., Shapee, S.M., Ambak, Z., Yusoff, Z.M. & Saad, M.R. (2010). Processing Defects Observation of Multilayered Low Temperature Co-fired Ceramic Substrate. *Proceedings of International Conference on Electronic Packaging 2010 (ICEP2010)*, Hokkaido, Japan, May 2010, pp. 300-304.

Amalu, E.H., Ekere, N.N. & Mallik, S. (2011). Evaluation of Rheological Properties of Lead-Free Solder Pastes and their Relationship with Transfer Efficiency During Stencil Printing Process. *Materials and Design*, Vol. 32, (February 2011), pp. 3189-3197.

Barlow, F.D. & Elshabini, A. (2007). *Ceramic Interconnect Technology Handbook*. CRC Press, New York.

Bell Jr., G.C., Rosell, C.M. & Joslin, S.T. (1987). Rheology of Silver-Filled Glass Die Attach Adhesive for High-Speed Automatic Processing. *IEEE Transactions on Components, Hybrids and Manufacturing*, Vol. CHMT 12, No. 4, pp. 507-510.

Bender, D.K. and Ferreira, A.M. (1994) Higher Density using Diffusion Pattern Vias and Fine-Line Printing. *IEEE Trans. On Comp. ,Pack. And Manufactur Tech. Part A*, Vol. 17, No. 3, pp. 485-489.

Brookfield Engineering Labs. Inc. More Solutions to Sticky Problems, in www.BrookfieldEngineering.com.

Burnside, S., Winkel, S., Brooks, K., Shklover, V., Gratzel, M., Hinsch, A., Kinderman, R., Bradbury, C., Hagfeldt, A. & Pettersson, H. (2000). Deposition and Characterization of Screen-Printed Porous Multi-layer Thick Film Structures from Semiconducting and Conducting Nanomaterials for use in Photovoltaic Devices. *Journal of Materials Science: Materials in Electronics*, Vol. 11, pp. 355-362.

Buzby, D. & Dobie, A. (2008). Fine Line Screen Printing of Thick Film Pastes on Silicon Solar Cells. *Proceedings of the 41st International Symposium on Microelectronics (IMAPS 2008)*, Rhode Island, USA, November 2008.

Cao, K. Cheng, K and Wang, Z. (2006). Optimization of Screen Printing Process. *Proceedings of 7th International Conference on Electronics Packaging Technology*, pp. 1-4

Chen, C.-N, Huang, C.-T., Tseng, W.J., & Wei, M.-H. (2010). Dispersion and Rheology of Surfactant-Mediated Silver Nanoparticle Suspensions. *Applied Surface Science*, Vol. 257, (July 2010), pp. 650-655.

Chhabra, R.P. & Richardson, R.F. (1999). *Non-Newtonian Flow in the Process Industries: Fundamental and Engineering Applications*, Butterworth-Heinemann, Oxford

Chhabra, R.P. & Richardson, R.F. (2008). *Non-Newtonian Flow and Applied Rheology: Engineering Applications*, (2nd Ed.), Elsevier Ltd., ISBN 978-0-7506-8532-0, UK.

Chiu, K.-C. (2003). Application of Neural Network on LTCC Fine Line Screen Printing Process. *Proc. Of the International Joint Conference on Neural Network*. Vol. 2, pp. 1043-1047.

Dollen, P.V & Barnett, S. (2005). A Study of Screen Printed Yttria-Stabilized Zirconia Layers for Solid Oxide Fuel Cells. *J. Am. Ceram. Soc.* Vol. 88, No. 12, pp. 3361-3368.

Dubey, G.C. (1974). Screens for Screen Printing of Electronics Circuits. *Microelectronics and Reliability*, Vol. 13, pp. 203-207.

Dubey, G.C. (1975). The Squeegee in Printing of Electronics Circuits, *Microelectronics and Reliability*, Vol. 14, pp. 427-429.

Durairaj, R, Ekere, N.N. and Salam, B. (2004). Thixotropy Flow Behaviour of Solder and Conductive Adhesive Paste. *J. Mater. Scie: Mater Electron.*, Vol. 15, pp. 677-683.

Durairaj, R., Mallik, S., Seman, A. & Ekere, N.N. (2009a). Investigation of Wall-Slip Effect on Lead-Free Solder Paste and Isotropic Conductive Adhesives. *Sadhana*, Vol. 34, No.5, (October 2009), pp. 799-810.

Durairaj, R., Ramesh, S., Mallik, S., Seman, A. & Ekere, N.N. (2009b). Rheological Characterisation and Printing Performance of Sn/Ag/Cu Solder Paste. *Materials and Design*, Vol. 30, pp. 3812-3818.

Evans J.W. & Beddow, J.K. (1987). Characterisation of Particle Morphology and Rheological Behaviour in Solder Paste. *IEEE Trans. Compo. Hybrids Manufac. Tech.*, pp. 224-231.

Gao, Y., Kong, X., Munroe, N. and Jones, K. (2010). Evaluation of Silver Paste as a Miniature Direct Methanol Full Cell Electrode. *J. of Power Source*. Vol. 195, pp. 46-53.

Gilleo, K. (1989). Rheology and Surface Chemistry for Screen Printing. *Screen Printing*, pp. 128-132.

Harper, C.A. (2001). *Handbook of Ceramic, Glass and Diamond.* McGraw Hill, New York.

Heraeus Design Guidelines from www.heraeus.com

Hoornstra, J., Weeber, A.W., Moor, H.H.C. & Simke, W.C. (1997). The Importance of Paste Rheology in Improving Fine Line, Thick Film Screen Printing of Front Side Metallization, *Proceedings of 14th EPSEC*, Barcelona, pp. 823-826.

Hwang, S., Lee, S. & Kim, H. (2009). Sintering Behavior of Silver Conductive Thick Film with Frit in Information Display. *J. Electroceram*. Vol. 23, pp. 351–355.

Imanaka, Y. (2005). *Multilayered Low Temperature Cofired Ceramics (LTCC) Technology*, Springer, New York.

Jabbour, G.E., Radspinner, R. & Peyghambarian, N. (2001). Screen Printing for the Fabrication of Organic-light Emitting Devices. *IEEE Journals on selected topics in Quantum Electronics*, Vol. 7, No. 5. pp. 769-773.

Jackson, G.J., Durairaj, R. & Ekere, N.N. (2002). Characterisation of Lead-Free Solder Pastes for Low-Cost Flip-Chip Bumping. *Proceedings of the 27th Annual IEEE/CPMT/SEMI/ International Electronics Manufacturing Technology (IEMT) Symposium*, San Jose, CA, USA, July 2002, pp. 223-228.

Jantunen, H., Kangasvieri, T., Vahakangas, J. & Leppavuori, S. (2003). Design Aspect of Microwave Components with LTCC Techniques, *J. of Eur. Ceram. Soc.*, Vol. 23, pp. 2541-2548.

Kardashian, V.S. & Vellanki, S.J.R. (1979). A Method for the Rheological Characterization of Thick-Film Paste. *IEEE on Components, Hybrids and Manufac. Tech.* Vol. CHMT-2, No. 2, pp. 232-239.

Kim, J.W., Lee, Y.-C., Kim, J.-M., Nah, W., Lee, H.-S., Kwon, H.-C. & Jung, S.-B. (2010). Characterization of Direct Patterned Ag Circuits for RF Applications. *Microelectronic Engineering*, Vol. 87, pp. 379-382.

Lahti, M. (2008). *Gravure Offset Printing for Fabrication of Electronic Devices and Integrated Components in LTCC Modules*. PhD Thesis, University Oulu.

Lin, H.-W., Chang, C.-P., Hwu, W.-H. & Ger, M.-D. (2008). The Rheological Behaviors of Screen Printing Pastes. *Journal of Materials Processing Technology*, Vol. 197, pp. 284-291.

Lin, J.C., & Wang, C.Y. (1996). Effects of Surfactant Treatment of Silver Powder on the Rheology of its Thick-Film Paste. *Materials Chemistry and Physics*, Vol. 45, pp. 136-144.

Lin, Y.-C. and Jean, J.-H. (2004). Constrained Sintering of Silver Circuit Paste. *J. Am. Ceram. Soc.* Vol. 87, No. 2, pp. 187-191.

London, D. (2008). Temperature Dependency of Silver Paste with Regards to Print Quality and Electrical Performance of Screen-Printed Solar Cells. *Proceedings of the 23rd EU PV Conference*, Valencia, Spain, September 2008.

LTCC Production in 2003 KEKO Research Newsletter from www.keko-equipment.com

Mallik, S., Thieme, J., Bauer, R., Ekere, N.N., Seman, A., Bhatti, R. & Durairaj, R. (2009). Study of the Rheological Behaviours of Sn-Ag-Cu Solder Pastes and their Correlation with Printing Performance. *Proceedings of the 11th Electronics Packaging Technology Conference*, Singapore, December 2009, pp. 869-874.

Morissette, S.L., Lewis, J.A., Clem, P.G., Cesarano III, J. & Dimos, D.B. (2001). Direct-Write Fabrication of Pb(Nb,Zr,Ti)O3 Devices: Influence of Paste Rheology on Print Morphology and Component Properties. *Journal of the American Ceramic Society*, Vol. 84, No. 11, pp. 2462-2468.

Neidert, M., Zhang, W., Zhang, D., & Kipka, A. (2008). Screen-Printing Simulation Study on Solar Cell Front Side Ag Paste, *Proceedings of 33rd IEEE Photovoltaic Specialists Conference (PVSC '08)*, San Diego, CA, USA, May 2008.

Nguty, T.A. & Ekere, N.N. (2000). Modeling the Effects of Temperature on the Rheology of Solder Pastes and Flux System. *Journal of Materials Science: Materials in Electronics*, Vol. 11, (July 1999), pp. 39-43.

Parikh, M.R., Quilty Jr., W.F. & Gardiner, K.M. (1991). SPC and Setup Analysis for Screen Printed Thick Films. *IEEE Trans. on Components, Hybrids and Manufacturing Technology*, Vol. 14, No. 3, pp. 493-498.

Phair, J.W. & Kaiser, F.-J. (2009). Determination and Assessment of the Rheological Properties of Pastes for Screen Printing Ceramics, *Annual Transactions of the Nordic Rheology Society*, Vol. 17, Iceland, August 2009.

Pudas, M., Hagberg, J. & Leppavuori, S. (2004). Printing Parameters and Ink Components Affecting Ultra-Fine-Line Gravure-Offset Printing for Electronics Applications. *J. Europ. Ceram. Soc.*, Vol. 24, pp. 2943-2950.

Rane, S.B., Seth, T., Phatak, G.J., Amalnekar, D.P. & Das, B.K. (2003). Influence of Surfactants Treatment on Silver Powder and its Thick Film. *Materials Letters*, Vol. 57, pp. 3096-3100.

Rane, S.B., Seth, T., Phatak, G.J., Amalnerkar, D.P. & Ghatpande, M. (2004). Effect of Inorganic Binders on the Properties of Silver Thick Films. *Journal of Materials Science: Materials in Electronics*, Vol. 15, pp. 103-106.

Schwanke, D. and Polhner, J. (2008). Enhancement of Fine line Printing Resolution due to Coating of Screen Fabrics. *Proceedings of Ceram. Intercon. Microselectronic Tech.*, pp. 301-308.

Sergent, J.E. (2007). Screen Printing in *Ceramic Interconnect Technology Handbook* edited by Fred D. Barlow and Aicha Elshabini, CRC Press, pp. 199-233.

Shapee, S.M., Alias, R., Yusoff, M.Z.M., Ibrahim, A., Ambak, Z., & Saad, M.R. (2010). Screen Printing Resolution of Different Paste Rheology for Printed Multilayer LTCC Tape. *Proceedings of International Conference on Electronic Packaging 2010 (ICEP2010)*, Hokkaido, Japan, May 2010, pp. 255-258.

Shin, D.-Y., Lee. Yongshik. & Kim, C.H. (2009). Performance Characterization of Screen Printed Radio Frequency Identification Antennas with Silver Nanopaste. *Thin Film Solids*, pp. 6112-6118.

Shiyong, L., Ning, W, Wencai, X and Yong, L. (2008) Preparation and rheological behavior of lead free silver conducting paste. *Materials Chemistry and Physics 111*, pp. 20-23.

Stalnecker, Jr., S. G. (1980). Stencils Screens for Fine-Line Printing. *Electrocomponent Science and Technology*, vol. 7, pp.47-53.

Taylor, B. E., Felten, J. J., Horowitz, S. J., Larry, J.R. and Rosenberg, R. M. (1981). Advances in Low Cost Silver-Containing Thick Film Conductors. *Electrocomponent Science and Technology*, 1981, Vol. 9, pp. 67-85. Gordon and Breach Science Publishers, Inc. Printed in Great Britain

Vasudivan, S. & Zhiping, W. (2010). Fine Line Screen Printed Electrodes for Polymer Microfluidics, *Proceedings of 12th Electronic Packaging Technology Conference*, Singapore, December 2010, pp. 89-93.

Wang, G., Barlow, F. & Elshabini, A. (2005) Interconnection of Fine Lines to Micro Vias in High Density Multilayer LTCC Substrates. *Proc. of International Symposium of Microelectronics*.

Wang, S. F. & Dougherty, J.P. (1994). Silver-Palladium Thick-film conductors. *J. Am Ceram. Soc.* Vol. 77, No. 12, pp. 3051-3072.

Wang, Y., Zhang, G. and Ma, J. (2002). Research of LTCC/Cu, Ag Multilayer Substratre in Microelectronic Packaging. *Materials Science and Engineering B*, Vol. 94. pp. 48-53.

Webster, R. (1975). Fine Line Screen Printing Yields as a Function of Physical Design Parameters. *IEEE Trans. On Manufac. Tech.* Vol. MFT-4, No. 1, pp. 14-20.

Wu, S.P., Zhao, Q.Y., Zheng, L.Q & Ding, X. H. (2011). Behaviors of ZnO-doped silver thick-film and silver grain growth mechanism. *Solid States Science* , Vol. 13, pp. 548-552.

Wu, S.P., Zheng, L.Q., Zhao, Q.Y. and Ding, X. H. (2010). Preparation and characterization of high-temperature silver thick film and its application in multilayer chip inductances. *Colloids and Surface A: Physicochemical and engineering aspects*, Vol. 372, pp. 120-126.

Yin, W., Lee, D.-H., Choi, J., Park, C. & Cho, S.M. (2008). Screen Printing of Silver Nanoparticle Suspension for Metal Interconnects. *Korean Journal of Chemistry Engineering*, Vol. 25, No. 6, (April 2008), pp. 1358-1361.